Topics in Applied Physics Volume 31

Topics in Applied Physics Founded by Helmut K. V. Lotsch

1 **Dye Lasers** 2nd Edition
Editor: F. P. Schäfer

2 **Laser Spectroscopy** of Atoms
and Molecules. Editor: H. Walther

3 **Numerical and Asymptotic Techniques
in Electromagnetics** Editor: R. Mittra

4 **Interactions on Metal Surfaces**
Editor: R. Gomer

5 **Mössbauer Spectroscopy**
Editor: U. Gonser

6 **Picture Processing and Digital Filtering**
2nd Edition. Editor: T. S. Huang

7 **Integrated Optics** Editor: T. Tamir

8 **Light Scattering in Solids**
Editor: M. Cardona

9 **Laser Speckle** and Related Phenomena
Editor: J. C. Dainty

10 **Transient Electromagnetic Fields**
Editor: L. B. Felsen

11 **Digital Picture Analysis**
Editor: A. Rosenfeld

12 **Turbulence** 2nd Edition
Editor: P. Bradshaw

13 **High-Resolution Laser Spectroscopy**
Editor: K. Shimoda

14 **Laser Monitoring of the Atmosphere**
Editor: E. D. Hinkley

15 **Radiationless Processes** in Molecules
and Condensed Phases. Editor: F. K. Fong

16 **Nonlinear Infrared Generation**
Editor: Y.-R. Shen

17 **Electroluminescence** Editor: J. I. Pankove

18 **Ultrashort Light Pulses**
Picosecond Techniques and Applications
Editor: S. L. Shapiro

19 **Optical and Infrared Detectors**
Editor: R. J. Keyes

20 **Holographic Recording Materials**
Editor: H. M. Smith

21 **Solid Electrolytes** Editor: S. Geller

22 **X-Ray Optics.** Applications to Solids
Editor: H.-J. Queisser

23 **Optical Data Processing.** Applications
Editor: D. Casasent

24 **Acoustic Surface Waves**
Editor: A. A. Oliner

25 **Laser Beam Propagation in the Atmosphere**
Editor: J. W. Strohbehn

26 **Photoemission in Solids I**
General Principles
Editors: M. Cardona and L. Ley

27 **Photoemission in Solids II.** Case Studies
Editors: L. Ley and M. Cardona

28 **Hydrogen in Metals I.** Basic Properties
Editors: G. Alefeld and J. Völkl

29 **Hydrogen in Metals II**
Application-Oriented Properties
Editors: G. Alefeld and J. Völkl

30 **Excimer Lasers** Editor: Ch. K. Rhodes

31 **Solar Energy Conversion.** Solid-State
Physics Aspects. Editor: B. O. Seraphin

32 **Image Reconstruction from Projections**
Implementation and Applications
Editor: G. T. Herman

33 **Electrets** Editor: G. M. Sessler

34 **Nonlinear Methods of Spectral Analysis**
Editor: S. Haykin

35 **Uranium Enrichment**
Editor: S. Villani

36 **Amorphous Semiconductors**
Editor: M. H. Brodsky

37 **Thermally Stimulated Relaxation in Solids**
Editor: P. Bräunlich

38 **Charge-Coupled Devices**
Editor: D. F. Barbe

Solar Energy Conversion
Solid-State Physics Aspects

Edited by B. O. Seraphin

With Contributions by
J. Aranovich   J. Bougnot   A. L. Fahrenbruch
H. Fischer   H. Gerischer   K. Graff   M. Savelli
B. O. Seraphin   A. J. Sievers

With 209 Figures

Springer-Verlag Berlin Heidelberg New York 1979

Bernhard O. Seraphin, Ph. D
Optical Sciences Center, University of Arizona,
Tucson, AZ 85721, USA

TK
2960
S653

ISBN 3-540-09224-2 Springer-Verlag Berlin Heidelberg New York
ISBN 0-387-09224-2 Springer-Verlag New York Heidelberg Berlin

Library of Congress Cataloging in Publication Data. Main entry under title: Solar energy conversion. (Topics in applied physics; v. 31). Bibliography: p. Includes index. 1. Solar batteries. 2. Semiconductors. 3. Solid state physics. I. Seraphin, B. O. II. Aranovich, J., 1947–. TK 2960.S653 621.47′5 78-31978.

This work is subject to copyright. All rights are reserved, whether the whole or part of the material is concerned, specifically those of translation, reprinting, reuse of illustrations, broadcasting, reproduction by photocopying machine or similar means, and storage in data banks. Under § 54 of the German Copyright Law, where copies are made for other than private use, a fee is payable to the publisher, the amount of the fee to be determined by agreement with the publisher.

© by Springer-Verlag Berlin Heidelberg 1979
Printed in Germany

The use of registered names, trademarks, etc. in this publication does not imply, even in the absence of a specific statement, that such names are exempt from the relevant protective laws and regulations and therefore free for general use.

Monophoto typesetting, offset printing and bookbinding: Brühlsche Universitätsdruckerei, Lahn-Gießen
2153/3130-543210

Preface

This book reviews key aspects of the materials science of solar energy conversion. The objective is an identification of problem areas in which a greater engagement of the solid-state physicist will improve existing technologies.

The involvement of the materials scientist may soon be of decisive importance. Solar energy conversion will contribute to future energy needs to the extent that existing devices can be improved in performance and lowered in cost. On the engineering level, most current approaches have been tried and perfected over the past 50 years without there being much hope for further improvements. However, a novel aspect has entered during the last two to three decades with the rapid development of the physics of solids which through their optoelectronic properties play a key role in major conversion schemes. Consequently, the greater involvement of modern solid-state physics will increase the chances to move solar technology beyond the barrier of economic marginality that now hinders its progress.

In its assessment of the current technology, this volume uncovers a disturbing discrepancy between the promise, as given by theoretical estimates for the performance, and the technology realized in existing devices. The size of the gap can be reduced by a greater engagement of the materials scientist. To attract his interest, an effort must be made to identify the key problem areas in which contributions from the solid-state physicist, the electrochemist, and the metallurgist may bring about novel solutions. Without a claim to completeness or comprehensiveness, this identification has been attempted here in areas of key importance for solar technology. The goal of this volume will be fulfilled if some of the reviews serve as a stimulation to further research in the solid-state aspects of solar energy conversion.

Tucson, Ariz., December 1978 *B. O. Seraphin*

Contents

1. **Introduction.** By B. O. Seraphin 1

2. **Spectrally Selective Surfaces and Their Impact on Photothermal Solar Energy Conversion.** By B. O. Seraphin (With 39 Figures) 5
   2.1 Spectral Selectivity in Photothermal Conversion 6
      2.1.1 Energy Balance at the Converter Surface 6
      2.1.2 Spectral Profile of the Ideal Converter 8
      2.1.3 Figure of Merit for Real Surfaces 9
      2.1.4 Conditions of Operation, and the Need for Spectral Selectivity 10
         a) Effectiveness of a Simple Blackbody Absorber 10
         b) Absorptance of Merit of a Solar Converter 11
      2.1.5 Spectrally Selective Absorbers for Highly Concentrating Systems 12
   2.2 Methods of Obtaining Spectral Selectivity 14
      2.2.1 Survey of the Physical Processes Leading to Spectral Selectivity 14
      2.2.2 Spectral Selectivity Provided by a Single Material 15
      2.2.3 Metals as Absorbers and Reflectors 19
      2.2.4 Temperature Dependence of Optical Processes 21
   2.3 Realizations of Spectrally Selective Surfaces 24
      2.3.1 Absorber-Reflector Tandems 24
      2.3.2 Semiconductors as Solar Absorbers 25
         a) Spectral Profile of the Transmission 26
         b) Refractive Index of Semiconductors 27
         c) Amorphous Silicon as a Solar Absorber 29
      2.3.3 Selectivity by Wavefront Discrimination 32
   2.4 Examples of Selective Coatings 35
      2.4.1 Interference Coatings 36
      2.4.2 Tandem Stacks 39
         a) Heat Mirrors 39
         b) Dark Mirrors 41
      2.4.3 Electroplated Coatings 43
      2.4.4 Tandem Stacks Fabricated by Chemical Vapor Deposition . 47
   References 52

3. Spectral Selectivity of Composite Materials
By A. J. Sievers (With 25 Figures) 57
3.1 On the Separation of High- and Low-Frequency Excitations . . . 59
    3.1.1 Molecules 59
    3.1.2 Metals 61
    3.1.3 Insulators 65
    3.1.4 Surfaces 67
3.2 Emissivity of a Smooth Metal Substrate 69
    3.2.1 Infrared Properties of Metals 69
        a) Surface Impedance 71
        b) Classical-Skin-Effect Limit 72
        c) Surface Scattering 75
        d) Electron-Phonon Scattering 77
    3.2.2 Total Emissivities for Free-Electron Metals 78
        a) Total Normal Emissivity 78
        b) Total Hemispherical Emissivity 79
    3.2.3 Spectral Selectivity Limits 82
3.3 Composite Coatings 84
    3.3.1 The Dipole Approximation 84
    3.3.2 Composite Dielectric Function for a Drude Metal 87
        a) A Calculation with Spherical Particles 90
    3.3.3 Composite Dielectric Function for Transition Metals . . . 91
    3.3.4 Metallic Particulate Coatings on Copper Surfaces 92
3.4 Composite Metals 97
    3.4.1 Experimental Parameters for Copper 97
    3.4.2 Free-Electron Model Composite 98
    3.4.3 Infrared Emissivity 100
    3.4.4 Influence of Lattice Vibrations 102
3.5 Transforming the Solar Spectrum with Selective Surfaces 105
    3.5.1 Selective Surface Configuration 105
    3.5.2 High Temperature Selective Surfaces 107
        a) Smokes 107
        b) Cermet Films 108
    3.5.3 Outlook 109
Appendix A 110
Appendix B 111
Appendix C 111
References 112

4. Solar Photoelectrolysis with Semiconductor Electrodes
By H. Gerischer (With 35 Figures) 115
4.1 Principles of Photoelectrolysis 116
    4.1.1 The Semiconductor-Electrolyte Interface 116
    4.1.2 Electron Transfer Reactions of Semiconductor Electrodes . 122
    4.1.3 Photocurrents and Photovoltages 127
    4.1.4 The Driving Force of Photoelectrolysis 130

4.2 Photodecomposition of Semiconductor Electrodes 133
    4.2.1 Energetic and Thermodynamic Aspects 133
    4.2.2 Kinetic Aspects 137
    4.2.3 Materials for Electrochemical Solar Cells 140
4.3 Function and Efficiency of Photoelectrochemical Solar Cells . . . 147
    4.3.1 Regenerative Cells 147
        a) Principles of Operation 147
        b) System Analysis 150
    4.3.2 Storage Cells 153
        a) Principles of Operation 153
        b) Cells for Water Photoelectrolysis 159
    4.3.3 Conversion Efficiency in Relation to Materials Properties . 164
References 169

5. Carrier Lifetime in Silicon and Its Impact on Solar Cell Characteristics.
By K. Graff and H. Fischer (With 44 Figures) 173

5.1 Photovoltaic Parameters of Solar Cells as Determined by Carrier Lifetime 174
    5.1.1 Collection Efficiency 175
    5.1.2 Photogenerated Current 176
    5.1.3 Current-Voltage Characteristic 177
    5.1.4 Conversion Efficiency 179
5.2 Methods for Measuring Carrier Lifetime 180
    5.2.1 The Photoconductive Decay Method 180
    5.2.2 The Surface Photovoltage Technique 185
    5.2.3 Measurements Applying the Spectral Response of Solar Cells 187
5.3 Carrier Lifetime in as-Grown Silicon Crystals 188
    5.3.1 Comparison of Results in Czochralski-Grown and Floating-Zone p-Type Silicon 189
    5.3.2 Influence of Doping Concentration 192
    5.3.3 Local Variations of Carrier Lifetime in Silicon Crystals . . . 192
        a) Longitudinal and Radial Profiles of Carrier Lifetime in Czochralski-Grown Silicon Crystals 193
        b) Longitudinal and Radial Profiles of Carrier Lifetime in Floating-Zone Silicon Crystals 194
5.4 Carrier Lifetime in Processed Silicon Crystals 195
    5.4.1 Processing Near Room Temperature 195
        a) Effect of Sawing Silicon Samples 195
        b) Influence of Etching Silicon Crystals 197
        c) Illumination of Silicon Samples 198
        d) Electron Bombardment 200
    5.4.2 Processing at High Temperatures 200
        a) Sample Preparation and Additional Precautions 200
        b) Annealing of Czochralski-Grown p-Type Silicon 203
        c) Annealing of Floating-Zone p-Type Silicon 204

X   Contents

        d) Intentional Getter Processes 206
        e) Multiple Annealing Processes 207
5.5 Limitation of Solar Cell Parameters by Material Properties . . 208
5.6 Conclusions 210
References 210

6. Problems of the Cu_2S/CdS Cell

By M. Savelli and J. Bougnot, with the collaboration of F. Guastavino, J. Marucchi, and H. Luquet (With 30 Figures) 213

6.1 Cu_2S/CdS Heterojunction Technology 213
    6.1.1 CdS Thin-Film Technology 213
        a) Vapor Deposition 213
        b) Sputtering 214
        c) Chemical Spray Deposition 215
        d) Sintered Layers 215
    6.1.2 Formation of the Cuprous Sulfide Layer for the Cu_2S/CdS Structure 216
        a) The Dipping Process 216
        b) CuCl Evaporation 216
        c) Cu_xS Evaporation 217
        d) Electrodeposition 217
        e) The Spray Method 218
    6.1.3 Heterojunction Formation 218
    6.1.4 Fabrication of Front and Back Electrodes 218
        a) Frontwall CdS-Cu_xS Cells 219
        b) Backwall CdS-Cu_xS Cells 219
    6.1.5 Thin-Film Photovoltaic Structures 220
6.2 Properties of CdS Films 220
    6.2.1 Review of the Fundamental Properties of Bulk CdS 220
    6.2.2 Properties of Polycrystalline CdS Thin Films 224
        a) Structures 224
        b) Electrical Properties 225
        c) Optical Transmission: Photoconduction 226
6.3 Properties of Cu_2S Films 228
    6.3.1 Phase Diagram of the Cu-S System and Structural Properties of Stable Phases 228
    6.3.2 Electrical Properties of Bulk Copper Sulfides in the Range of Compositions Near the Stoichiometry Cu_2S 229
    6.3.3 Electrical Properties of Thin Copper Sulfide Layers Near the Cu_2S Composition 233
    6.3.4 Variation of the Electrical Properties of Cuprous Sulfides with Composition 235
    6.3.5 Optical Properties of Copper Sulfides 236
6.4 Photovoltaic Properties of Cu_2S-CdS Cells 238
    6.4.1 Junction Structure 238

	6.4.2 Current-Voltage (IV) Characteristics	239
	a) In Darkness	239
	b) Under Illumination	240
	6.4.3 Capacity-Voltage (CV) Characteristics	241
	6.4.4 Spectral Response	243
	6.4.5 Stability	244
6.5	Conduction Mechanisms in Cu_2S–CdS Cells	244
6.6	Conclusion	249
References		250

7. Heterojunction Phenomena and Interfacial Defects in Photovoltaic Converters. By A. L. Fahrenbruch and J. Aranovich (With 36 Figures) ... 257

7.1 The Relation of Solar Conversion Efficiency to Heterojunction Parameters ... 258
    7.1.1 The Ideal Solar Cell and Calculation of Light-Generated Current ... 258
        a) Calculation of the Light-Generated Current ... 258
    7.1.2 Critique of Assumptions ... 263
        a) $\mathscr{E}=0$ in the Quasi-Neutral Region ... 263
        b) $U=(n_p-n_{p0})/\tau_e$ ... 263
        c) The Boundary Condition at the Depletion Layer Edge ... 264
        d) Position of the Quasi-Fermi Levels Within the Depletion Layer ... 265
        e) On the Constancy of the Photogenerated Current Across the Junction ... 266
    7.1.3 Bias Voltage Dependence of Collection Efficiency ... 267
    7.1.4 The Solar Efficiency of Heterojunction Converters ... 269
7.2 Present Theories of Heterojunction Transport ... 270
    7.2.1 The Diode Parameters J_0 and A ... 271
    7.2.2 Anderson's Model for the Heterojunction ... 273
        a) The Basic Model ... 273
        b) Examination of Assumptions ... 275
        c) The Question of Abruptness ... 276
        d) Difficulties with the Anderson Model ... 276
    7.2.3 Sophistication for the Simple Heterojunction ... 277
        a) Introduction of Interface States ... 277
        b) Direct Recombination Through Interface States ... 279
        c) Combined Tunneling and Recombination ... 281
        d) Interfacial Dipoles ... 282
        e) Junction Grading ... 282
        f) The Effect of Illumination on Junction Parameters ... 283
    7.2.4 Metal-Insulator-Semiconductor Junctions ... 283
        a) Basic Schottky Barriers ... 284
        b) Consideration of the Barrier Height ... 285

XII   Contents

        c) MS and MIS Solar Cells 287
        d) Stability of the MIS Interface 291
        e) The SIS Structure 292
    7.2.5 Summary and Application to Real Heterojunctions 292
7.3 Interface Related Phenomena 295
    7.3.1 The Metal-Semiconductor Interface 296
        a) Introductary Remarks 296
        b) Ambiguities of the Parameter S 298
        c) The Origin of Surface States 298
        d) Experimental Evidence of Electrical Effects of Surface
           Quality 304
        e) Oxide Interlayers 306
    7.3.2 The Semiconductor-Semiconductor Interface 307
    7.3.3 Crystallographic Aspects of the Interface Region: Lattice
        Mismatch, Dislocations, and Electronic Behavior 311
        a) Density of Dislocations 311
        b) Dislocation Morphology 313
        c) Electronic Effects of Misfit Dislocations 318
    7.3.4 Summary 320
7.4 Conclusions 321
References 322

Several Papers on Solar Energy Physics Published in *Applied Physics* . . 327

Additional References with Titles 329

Subject Index 331

Contributors

Aranovich, Julio
  Department of Materials Science and Engineering, Stanford University, Stanford, CA 94305, USA

Bougnot, Josiane
  Centre d'Etudes d'Electronique des Solides, Université des Sciences et Techniques du Languedoc, F-34060 Montpellier Cédex, France

Fahrenbruch, Alan L.
  Department of Materials Science and Engineering, Stanford University, Stanford CA 94305, USA

Fischer, Horst
  AEG-Telefunken, Semiconductor Division,
  D-7100 Heilbronn, Fed. Rep. of Germany

Gerischer, Heinz
  Fritz-Haber-Institut der Max-Planck-Gesellschaft, Faradayweg 4–6,
  D-1000 Berlin 33, Fed. Rep. of Germany

Graff, Klaus
  AEG-Telefunken, Semiconductor Division,
  D-7100 Heilbronn, Fed. Rep. of Germany

Savelli, Michel
  Centre d'Etudes d'Electroniques des Solides, Université des Sciences et Techniques du Languedoc, F-34060 Montpellier Cédex, France

Seraphin, Bernhard O.
  Optical Sciences Center, University of Arizona, Tucson, AZ 85721, USA

Sievers, Albert J.
  Laboratory of Atomic and Solid State Physics and
  Materials Science Center, Cornell University, Ithaca, NY 14850, USA

1. Introduction

B. O. Seraphin

This volume reviews key aspects of the materials science of solar energy conversion. The objective is an identification of problem areas in which a greater engagement of the solid-state physicist will improve existing technologies.

The involvement of the materials scientist may soon be of decisive importance. Solar energy conversion will contribute to future energy needs to the extent that existing devices can be improved in performance and lowered in cost. On the engineering level, most current approaches have been tried and perfected over the past 50 years without there being much hope for further improvement. However, a novel aspect has entered during the last two to three decades with the rapid development of the physics of solids which through their optoelectronic properties play a key role in major conversion schemes. Consequently, the greater involvement of modern solid-state physics will increase the chances to move solar technology beyond the barrier of economic marginality that now hinders its progress.

The necessary effort must start with the identification of problem areas in which the contribution of the solid-state physicist could be essential. The optical properties of solids ranks high on the list. Before the solar photon can interact with the material of the converter in a technologically useful manner, it must first penetrate through the surface and be absorbed. Small reflection and large absorption over the solar emission band is therefore a common requirement for materials used in most conversion schemes.

For photothermal conversion this is not sufficient, however. As the absorber converts the intercepted solar radiation into heat, the loss by thermal reradiation must be reduced. Surfaces of the proper spectral selectivity therefore provide optimum conversion efficiency. Chapter 2 deals with such spectrally selective surfaces and their impact on the efficiency of photothermal solar energy conversion. Considering the energy balance at the surface of the converter, the key role of the spectrally selective profile of the optical properties of this surface is established. A survey of the physical processes leading to spectral selectivity places the actual realization of converter surfaces in the frame of solid-state physics. A review of the most prominent methods for making selective blacks emphasizes the empirical character of the present state of the technology, and challenges the solid-state physicist to contribute to the necessary deeper understanding.

Chapter 3, by A. J. Sievers, deals with the interesting possibility of providing spectral selectivity through the optical properties of composite materials, made up from metallic and insulating components. The optical properties of these composite materials are reviewed first, considering effects of surface texture, roughness, and surface plasmons. Considering mainly the tandem action of an absorber overlaid onto a reflective backing, the limits of such a configuration for the attainable figure of merit are investigated. The optical properties of the composite absorber are described, using the Maxwell–Garnett theory. Particulate coatings of transition metal or copper substrates appear to be promising, and a technique is suggested that makes use of surface plasmon effects. In a final section, the state of the art of high-temperature selective coatings is reviewed, supporting the conclusion that current approaches are still a long way from the physical limits established in the article.

H. Gerischer's Chap. 4, "Solar Photoelectrolysis with Semiconductor Electrodes", is interdisciplinary between electrochemistry and solid-state physics. The great promise of converting solar radiation into energy which can be stored using the semiconductor-electrolyte interface presents a strong appeal to the solid-state physicist to learn the language and concepts of the electrochemist, and vice versa. The realization of the promise in practical terms will come from materials science, and in particular from semiconductor physics. It is therefore important that the author of this chapter succeeds in presenting the electrochemical situation in the terms used by the semiconductor physicist, beginning with the principles of photoelectrolysis, the electron transfer reactions, and the driving force of photoelectrolysis. Great emphasis is placed on the susceptibility of semiconductors to decomposition, the most serious problem for their application in photochemical cells. After describing function and efficiency in regenerative cells, including cells used for water photoelectrolysis, the conversion efficiency in relation to material properties is discussed, and guidelines for the selection of optimal materials are developed in analogy to solid-state photovoltaic conversion schemes. In the tenor of nearly all authors in this volume, Gerischer concludes that present photoelectrochemical cells fall short of the theoretically estimated performance, and that a great deal more materials research will have to be performed to approach the calculated values.

With the next Chap. 5 by K. Graff and H. Fischer, the text enters an area of great concern to the solid-state physicist. The carrier lifetime and its impact on the performance of photovoltaic solar cells is a central problem, in particular to the economically attractive thin-film approach. It is in this area that solid-state physics can provide solutions most effectively. For crystalline silicon the demand for a diffusion length of the photo-generated carriers to be equal to the penetration depth of the solar radiation is particularly difficult to fulfill. The authors describe the state of our knowledge concerning the dependence of carrier lifetime upon the type of silicon crystal used, process-induced variations of this parameter, and present technological limitations.

Once solar-cell fabrication reaches the required cost level of a few dollars per watt, the price of the material and its processing will dominate. Integrated circuitly opened a new area in electronics because the density of the components could be vastly increased. Since the density of the solar flux for a given power requirement determines the area of interception, the thickness of the active material is left as the only adjustable parameter for a reduction in cost. Ideally, the thickness of the active layer should equal the penetration depth of the solar radiation and the diffusion length of the photo-pairs. To realize this in an economical thin-film method that tolerates the polycrystalline material is the central challenge.

Chapter 6 by M. Savelli and his colleagues discusses state of the art, problems and promises for the Cu_2S/CdS solar cell, one of the most attractive realizations of the thin-film approach. Conversion efficiency, stability in operation and yield of fabrication depend strongly on the parameters of preparation. Further progress will depend on studies of the electronic consequences of the interface, and the structural and compositional characterization of the material on either side of it.

A large part of the promise – and at the same time the cause of most of the problems – of the Cu_2S/CdS cell rests with its heterojunction character. Chapter 7 by A. L. Fahrenbruch and J. Aranovich reviews in depth heterojunction phenomena and interfacial defects in photovoltaic converters. The requirement for a high absorption constant and a direct band gap in materials for thin-film cells in particular suggest the heterojunction as a possible solution. This configuration provides attractive features such as the window effect, an active region far from surface recombination, and others. The authors describe the problems to be solved before the advantages can be utilized. The carrier transport is dominated by the complex phenomena near or at the metallurgical interface, involving recombination and tunneling through interfacial states. The picture is further complicated by the presence of band-profile discontinuities which may be further distorted through the presence of electrical charges in the interfacial states. This chapter gives the solid-state perspective, with strong reference to the performance of solar cells based on heterojunctions. The article culminates in the conclusion that only further coordinated research in surface, solid-state and device physics will enable the educated manipulation of the interfacial properties of heterojunction devices that can bring about substantial improvements in solar converter technology.

All the chapters in this volume end on the same note. Their assessment of the current technology uncovers a disturbing discrepancy between the promise, as given by theoretical estimates for the performance, and the technology realized in existing devices. The size of the gap can be reduced by a greater engagement of the materials scientist. To attract his interest, an effort must be made to identify the key problem areas in which contributions from the solid-state physicist, the electrochemist, the metallurgist may bring about novel solutions. Without a claim to completeness or comprehensiveness, this volume has attempted this identification in areas of key importance for solar technology. The goal of this book will be fulfilled if it serves to stimulate further research in the solid-state aspects of solar energy conversion.

2. Spectrally Selective Surfaces and Their Impact on Photothermal Solar Energy Conversion

B. O. Seraphin

With 39 Figures

Solar energy is generally recognized to be environmentally benign, and inexhaustible in quantity. As a consequence, public expectations for its successful development and early commercialization have been high. Technological feasibility taken for granted, the remaining economic difficulties are supposed to be reduced by developments on the engineering level, and, in particular, by sufficiently large mass production. Consequently, insufficient attention is paid to the necessary effort and investment in basic and applied research.

The gap between public expectation and technological reality has been of particular consequence for the potential role of material science in the development of solar energy. An effort to improve the performance of existing devices, and to lower their cost, encounters problems that involve materials and manufacturing processes. In its search for better solutions, solar energy technology must engage the various aspects of material science more effectively. This may involve parts of the field which are at present outside the mainstream of solar research, and do not use recognized solutions or approaches.

The direction of research will continue to be dictated by economic considerations, even though the basic question of the feasibility of solar energy conversion has been resolved. Most problems have more than one technologically satisfactory solution. A large number of them were, however, developed in the financially liberal context of the space programs, and are highly unsatisfactory once economic considerations are important.

Existing technologies must be reopened at a level where they touch on fundamental research. Economic gains derived from large-scale production are limited, and efforts must not be restricted simply to producing what is already available more cheaply.

While the need for increased research activity is common to all solar technologies, photothermal conversion encounters an additional handicap. In photovoltaics, the importance of solid-state physics is readily acknowledged. By comparison, photothermal conversion appears to be an "easy" technology to which the solid-state physicist can contribute little. It is the object of this article to modify this mistaken impression. Photothermal solar energy conversion is reviewed from the standpoint of the solid-state physicist, with emphasis on the optical properties of solids.

Economically attractive conversion of solar radiation into useful heat involves the optical properties of solids in a number of problem areas. Of key importance in the performance characteristic of a photothermal converter are the optical properties of the surface that intercepts the solar flux, and converts it into heat. Efforts to optimize existing devices lead in some aspects to new types of processes for which the fundamental understanding is not always sufficient. The following article aims to identify these problem areas, in which the cooperation of material scientists will be of significance.

Emphasis is placed on the optical properties of the actual converter surface, thus ignoring many other optical interactions in the system. However, research on the optical properties of the spectrally selective converter surface leaves much opportunity for exploration and improvement.

2.1 Spectral Selectivity in Photothermal Conversion

2.1.1 Energy Balance at the Converter Surface

The basic principle of photothermal solar energy conversion is illustrated in the energy-flow diagram of Fig. 2.1 [2.1]. Some optical device focuses the solar flux onto the absorber-converter, concentrating it by the ratio of the intercepting areas. At the converter surface, the concentrated flux W_1 is divided into at least three components. The part W_2 is reflected directly. A second component W_3 represents the radiation of the surface in the thermal infrared, corresponding to the temperature T_1 of the converter. The remainder W_4 of the solar input is passed on as heat to the next stage, generating Carnot work W_5 and waste heat W_6.

We shall assume that by proper engineering and choice of operating conditions, additional losses caused by convection and conduction are negligible at the converter stage. The conversion is then the more efficient, the more the useful component W_4 dominates reflection and reradiation loss. Effective suppression of the two loss components makes for high conversion efficiency. On the other hand, no power can be drawn from the converter and the efficiency is zero, if the reradiation and reflection combined equal the solar input.

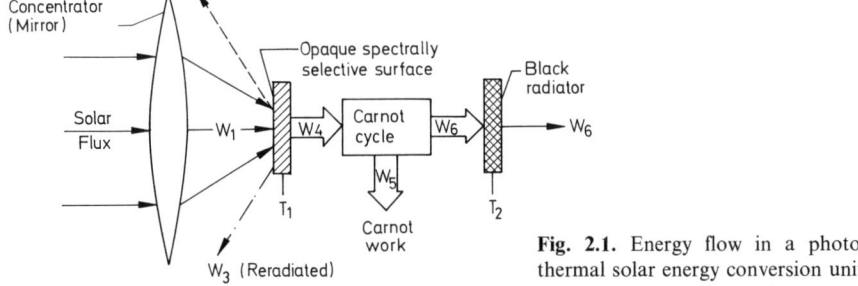

Fig. 2.1. Energy flow in a photothermal solar energy conversion unit

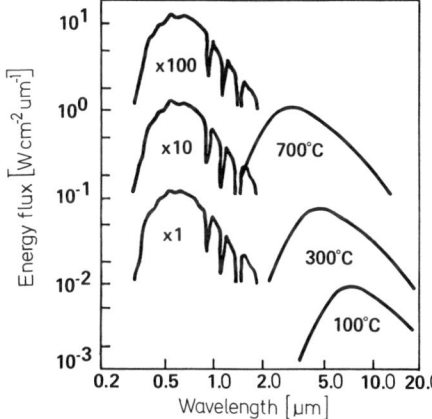

Fig. 2.2. Spectral profile of the energy flux of solar input for three concentrations, and of reradiative loss at three converter temperatures

Thus, sufficient absorption of solar radiation is only one requirement for an efficient converter. Its emittance plays an important part in determining the reradiation loss at a given temperature of operation. If reduced much below that of a blackbody, useful heat can be drawn from the converter at reasonable efficiency, and at a temperature T_1 of acceptable Carnot quality.

The seemingly contradictory requirement of having a good absorber that is simultaneously a poor emitter can be met·by spectral selectivity. The solar source of radiative temperature over 5500 °C emits 95% of its output at wavelengths smaller than 2 μm. A terrestrial receiver kept at 650 °C – an upper temperature limit for most technological applications – emits 95% of its thermal radiation at wavelengths longer than 2 μm. Thus, the spectral ranges of solar input and reradiation loss are well separated on the wavelength scale, with overlaps amounting to 5% on either side of 2 μm. This is illustrated in Fig. 2.2 [2.2].

Sufficient absorption of solar radiation can, therefore, be combined with effective suppression of the reradiative loss in converter surfaces that are spectrally selective by absorbing over the solar emission band, and emitting poorly in the thermal infrared. "Bandpass reflection filters" or "black infrared mirrors" are descriptive names sometimes used in literature for the peculiar spectral profile that improves the conversion efficiency of a surface over that of a blackbody, the improvement being trivial for some, and drastic for other applications.

We shall describe in the following section the spectral profile of an ideal surface, and discuss the gain of a spectrally selective surface over that of a blackbody as a function of concentration and operating temperature. A figure of merit will be defined for a real surface, and the physical processes that lead to spectral selectivity will be described. In keeping with the subject of this book, we shall concentrate on the optical properties that result in the desired spectral selectivity. Without further mention, conditions will be assumed that make

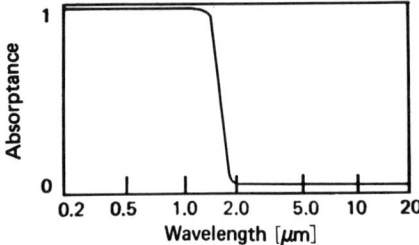

Fig. 2.3. Spectral profile of the absorptance for an ideal photothermal converter

reflection and thermal reradiation from the converter surface the dominant form of losses. Secondly, we shall keep in mind that all materials under consideration must withstand the temperature range of application. For power generation at acceptable Carnot factors, temperatures in excess of 300 °C may be considered. Since the fuel is free, we shall receive returns on the initial investment in the plant in direct proportion to the lifespan of the installation. Finally, since solar fuel is rather dilute, large intercepting areas are required to collect it, calling for low cost materials and fabrication. Spectral selectivity, long lifetime at high temperatures, and reasonable cost are, therefore, the basic requirements that the surface of a photothermal converter must meet. We shall concentrate primarily in this article on how to meet the spectral selectivity requirements.

2.1.2 Spectral Profile of the Ideal Converter

A photothermal converter operating under conditions of dominant reradiation and reflection losses performs best, if its absorptance spectrum resembles the profile shown in Fig. 2.3.

Such a surface absorbs a maximum fraction of the incident solar photons while simultaneously suppressing the emittance in the thermal infrared. Neither complete absorption nor perfect suppression is possible, whatever the absorption profile. Unavoidable overlap of the emission bands of sun and converter will cause long wavelength solar photons to be rejected, while energy will be leaked through the part of the thermal emission spectrum tailing into the absorbing region. For a nonconcentrating converter operating at 650 °C, the mutual overlap amounts to 5% of the respective emission, even if the step of the profile is perfectly sharp.

It is evident from Fig. 2.2 that the optimum location of the step on the wavelength scale is determined both by the concentration ratio and temperature of the converter. Concentrating the solar flux by X multiplies the power absorbed at each wavelength by the same factor. Varying the temperature of the converter, on the other hand, changes the reradiated power according to the Stefan–Boltzmann law, and shifts the peak emission in accordance with Wien's equation. The loss caused by spectral overlap is minimized by a profile in which the step is placed in accordance with concentration and temperature.

2.1.3 Figure of Merit for Real Surfaces

Real surfaces will, at best, approximate the ideal profile of Fig. 2.3. Sum rule considerations prove that too close an approximation must make the profile worse in some other spectral region.

Evaluation of the spectral profile of a real surface in terms of photothermal conversion leads to the solar absorptance

$$a = \frac{\int_0^\infty \alpha(\lambda) H(\lambda) d\lambda}{\int_0^\infty H(\lambda) d\lambda} \qquad (2.1)$$

and the thermal emittance

$$e(T) = \frac{\int_0^\infty \varepsilon(\lambda) \varepsilon_{BB}(\lambda, T) d\lambda}{\int_0^\infty \varepsilon_{BB}(\lambda, T) d\lambda}. \qquad (2.2)$$

The hemispherical absorptance $\alpha(\lambda)$ of a given surface is defined as the fraction absorbed from a radiative flux $H(\lambda) d\lambda$ incident hemispherically on the surface results if the solar spectrum is substituted for $H(\lambda)$. Since this varies strongly with location and time, the solar absorptance depends upon the absorption characteristic of the surface, as well as the precise nature of the solar spectrum for the situation under consideration.

We transform (2.1) into (2.2) by replacing the hemispherical absorptance by the hemispherical emittance, defined as the fraction of energy emitted by a real surface into a hemisphere as compared with that of an ideal blackbody at the same temperature T. Integration over all wavelengths gives the total hemispherical emittance, or thermal emittance $e(T)$, which in contrast to the solar absorptance, is a property of the surface alone. In both integrals, we have ignored the temperature dependence of both spectral functions $\alpha(\lambda)$ and $\varepsilon(\lambda)$ – an unreasonable assumption, as we shall see later.

The solar absorptance a is simply the percentage of radiant energy of solar origin absorbed by the surface. Raising this fraction to its maximum value, 1, is the goal – whatever the mode of operation.

The thermal emittance e designates the fraction of radiant energy emitted by the surface as compared with the radiant energy emitted by a blackbody at the same temperature. How much is gained in terms of conversion efficiency by reducing this fraction to small values depends upon the point of operation, as shown above.

The ratio a/e is often cited as a figure of merit for a selective surface. For a given concentration, this ratio determines the highest attainable temperature, and thereby the maximum Carnot efficiency. For the efficiency of the overall system, however, it is important to give a and e separately. Under nearly all conditions, a surface with $a=0.5$ and $e=0.05$ will, in overall efficiency, be inferior to a surface with $a=1.0$ and $e=0.10$, although both are characterized by the same ratio a/e.

Through their spectral profile, the optical constants of the converter surface determine the figure of merit parameters a and e. They ultimately determine the fraction of the incident energy drawn from a converter at a given temperature, and in a given configuration. Raising the spectral selectivity of the converter beyond certain threshold values of a and e will permit economic operation under conditions that would be prohibitively costly with lower selectivity. In this sense, the spectral selectivity is the key parameter in photothermal conversion. In the next section, we shall deal with its effect on the efficiency of the conversion, as a function of the two principal parameters of the operation, namely the concentration ratio X, and the absorber temperature T.

2.1.4 Conditions of Operation, and the Need for Spectral Selectivity

a) Effectiveness of a Simple Blackbody Absorber

We can ask for the range of values of concentration and temperature for which the step profile of Fig. 2.3 assures sufficient gain in efficiency over the much simpler blackbody profile. By defining the effectiveness of a blackbody in comparison to that of the ideal spectral profile as the ratio of energy absorbed and retained by the two surfaces, we arrive at the diagram plotted in Fig. 2.4. Note that for a given concentration, the need for a selective surface grows with increasing temperature. For a given temperature, a spectrally selective surface is essential unless the solar concentration is high.

Needless to say, the point of operation on Fig. 2.4 is determined by the given application and its economic, engineering and geographical context, and any other special conditions. Temperatures much in excess of 100 °C can be obtained only in concentrating configurations. The concentration ratio rises quickly with the desired temperature of operation. Such converters must be adjusted seasonally if the lowest concentration does not exceed approximately $X=10$. Concentrations from $X=10$ to 100 require tracking of the sun in one axis. Concentrations beyond that necessitate full steerability, with corresponding increase in complexity and cost. As the concentration rises above the value $X=1$ of the flat-plate converter, the amount of diffuse sky radiation collected vanishes rapidly, so that the concentrating converter is useless in cloudy weather. Apparently, we face a trade-off situation that is characteristic of solar energy conversion. However, over large areas of Fig. 2.4, the boundary conditions are such that a spectrally selective surface may render an application economically attractive whose expense would otherwise be prohibitive. In most cases, the higher cost of a spectrally selective surface is amply justified by the greater efficiency of conversion. In order to formulate the situation in

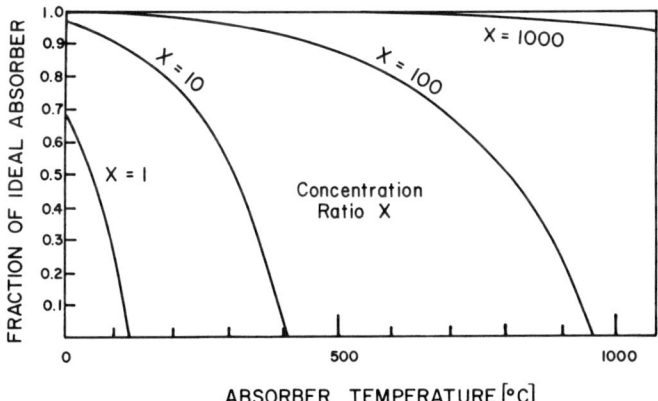

Fig. 2.4. Fraction of energy absorbed and retained by the blackbody as a function of temperature, as compared with an ideal optimized absorber, for four different solar concentrations

quantitative terms, we shall now define the "absorptance of merit", which permits us to study the effects of absorptance and emittance separately.

b) Absorptance of Merit of a Solar Converter

Some features of the performance of a photothermal converter depend on the solar absorptance and thermal emittance separately, instead of on their ratio a/e. It is meaningful, for instance, to ask for the increase in the efficiency of a converter, when its emittance is reduced at constant absorptance. To answer this, and related, questions, we define the "absorptance of merit" a_m as [2.3, 4]

$$a_m = \frac{Q}{X\Phi} = \frac{\text{Heat flux into working fluid}}{\text{Solar flux incident on converter surface}} \tag{2.3}$$

which, again ignoring losses due to convection and conduction, can be written as the difference between the actual absorptance a_s of the converter surface, and the loss due to reradiation

$$a_m = a_s - \frac{e\sigma T^4}{X\Phi}. \tag{2.4}$$

Hence, the absorptance of merit establishes an upper limit for the conversion efficiency of a converter of given solar absorptance a_s, and thermal emittance e, operating at a temperature T, where σ, X, and Φ designate

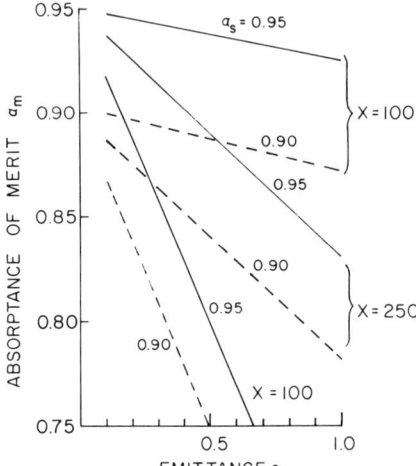

Fig. 2.5. Absorptance of merit as a function of emittance for solar absorbers of $a_s = 0.95$ and 0.90, and for various flux amplifications X, at an operating temperature of 500 °C

Boltzmann's constant, the flux amplification, and the solar flux intensity, respectively.

Figure 2.5 shows the absorptance of merit for a set of two surfaces of $a_s = 0.95$ and $a_s = 0.90$, as function of emittance for a temperature $T = 500\,°C$. For a concentration of $X = 1000$, a reduction of the emittance even to small values does not compensate for the 0.05 difference in absorptance between the two surfaces. For $X = 250$, however, a reduction of 0.4 in the emittance of the surface of inferior absorptance renders it equal to the surface with $a_s = 0.95$. For $X = 100$, a reduction of only 0.15 suffices.

A surface operating at a given temperature will heat up to a higher temperature when its emittance is being reduced. For the three concentrations in the previous figure, Fig. 2.6 demonstrates, with all other conditions equal, that a surface of $a_s = 0.95$ can operate at 500 °C instead of 400 °C, if its emittance is reduced from 1.0 to 0.6. The dependence upon concentration is much smaller in this case.

2.1.5 Spectrally Selective Absorbers for Highly Concentrating Systems

Following the preceding arguments, one could contest the need for spectrally selective absorbers for systems of large flux amplification. Concentrating a large multiple of the solar flux onto a surface that can only reradiate as one blackbody seemingly eliminates the need for emittance suppression. For receivers of the cavity design, spectral selectivity should be even less important.

Recently, this oversimplified view has tended to be replaced by an analysis that considers the value of a spectrally selective coating in terms of the cost benefits of the entire system. Based on the design parameters of the 10 MW$_e$

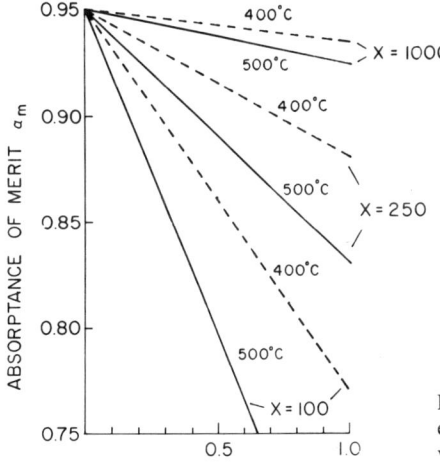

Fig. 2.6. Absorptance of merit as a function of emittance for a solar absorber of $a_s = 0.95$ at various flux amplifications X, and for $T = 400$ and $500\,°C$

central receiver plant to be completed in 1982 near Barstow, California, it has been estimated that an increase of a_s/e from 1 to 2 is equivalent to a one percent increase in a_s, and corresponds to a cost reduction of $ 7/kW [2.5]. Large solar absorptance is presently considered essential for the surfaces of the external receiver of this plant, which is expected to operate at approximately $T = 500\,°C$ and $X = 500$. Although not used in this system, it is realized that selective coatings may be able to make small but significant improvements in performance, as well as lower the cost, considering that the heliostats are a major cost item in central receiver plants. It cannot be assumed that future designs operating at even larger concentrations and higher temperatures will be of the cavity design [2.6].

In central receiver plants, coating will be exposed to flux densities in the order of $1\,MWm^{-2}$. Overheating must be tolerated in case the coolant flow fails. Defocussing the heliostat may require times of the order of 1–2 min. Ease of refurbishment, with minimum period of closure, is essential. A 30-year life with 10,000 thermal cycles, and survival in a rain and dust environment are required.

Line-focussing systems of the distributed collector type will be limited to $X < 100$. Applications in the temperature range 200–1000 °C are anticipated, making spectral selectivity indispensable. Coolant loss has resulted in overheating to 200 °C above the operating temperature before defocussing, severely restricting the use of Black Chrome in systems operating normally below 300 °C. Coating for the line absorbers must withstand long-term operation in soft vacuum of 10^{-3} Torr or less, and tolerate the bending and flexing resulting from thermal expansion. Normal incidence cannot be assumed in a line focus absorber, so that the coating must perform under a wide acceptance angle [2.7].

2.2 Methods of Obtaining Spectral Selectivity

2.2.1 Survey of the Physical Processes Leading to Spectral Selectivity

Reference to spectral selectivity was implicitly made in most early papers on the subject of photothermal conversion. However, the connection with the performance characteristic was clearly established, and related work brought to practical fruition, only after the presentations by *Tabor* [2.8], and by *Gier* and *Dunkle* [2.9], at the Tucson Conference in 1955. In the years since, a variety of physical processes have been employed in the development of spectrally selective surfaces. None of the various mechanisms can individually generate a good approximation to the desired spectral profile. It takes the tandem action of two or more processes to produce sufficient selectivity. While this is usually understood, little effort has been invested in a systematic investigation of the fractional contributions of the various processes present in the optical action. Most studies were product-oriented, and carry an empirical note. As a result, the designer of solar energy converters has a number of surfaces to chose from – fabricated by various methods, and acceptable for various applications. Few, if any, of the methods are understood sufficiently to make attempts at optimization fruitful. In this respect, the technology of making spectrally selective surfaces resembles that of the oxide cathode, and many other products that work well, but are not well understood. This is acceptable, as long as the recipe provides products which perform the applications for which they were developed. However, the need for improvement or adaptation to a new set of conditions renders most recipes useless, and makes systematic studies necessary.

In our survey of the various mechanisms, we will first investigate the possibility of providing spectral selectivity by one single material. We find several candidates in nature sufficiently close to the desired profile in their optical spectrum. The question now arises as to which direction the development of synthesized materials must move in order to improve on these candidates.

Metals provide, in their high infrared reflectance, the required suppression of the thermal emittance. However, their solar absorptance is insufficient. To a first approximation, the optical properties of metals can be understood as the tandem action of free and bound electrons, each operating in its own wavelength range. There is no inherent reason for the two phenomena to be coupled in the sense the experimental data seem to indicate. The band structure mechanisms involving free and bound electrons, roughly correlated to thermal emittance and solar absorptance, respectively, can be invoked independent of each other. The few optical data on transition-metal and rare-earth compounds indicate that the variety of properties is indeed promising, and much wider than the simple metals seem to indicate.

Existing metals require a boost of their solar absorptance by overcoating them with a layer of sufficient solar absorptance, and good infrared transparency. Semiconductors are well suited, if their fundamental absorption edge is located in the proper spectral region. The intrinsic absorption caused by interband transitions provides the solar absorptance, while the metal reflector underneath shows through in the thermal infrared. We will see, however, that the spectral location of the absorption edge in the semiconductors prescribes a limitation with respect to its refractive index. Such a tandem must be made non-reflecting by interference or by the texture of the surface.

Coating the metal substrate with several carefully-tailored layers provides the necessary solar absorptance by means of interference. Coatings of exceptionally high spectral selectivity have been made in this manner. However, the "tuned-cavity" character of an interference filter renders it sensitive to "detuning" by very slight changes in any of the cavity parameters. Long-term degradation in operation at high temperatures, sensitivity of performance to the angle of incidence, as well as cost and complexity of fabrication, will restrict the large-scale use of interference filters.

Processes based on texture of a surface can greatly assist absorption and interference. *Tabor* [2.10] summarizes these effects under the name "spectral selectivity by wavefront discrimination". We shall treat them in two categories, roughly depending on the dimension of the texture with respect to the wavelength of light. We shall speak of "reflective scattering", if grooves or pores simply increase the absorbing area through multiple reflections. Selected patterns can distinguish between different wavelengths, absorbing the solar input, but appearing smooth and reflective to the thermal infrared. The reduction of the refractive index in a porous material can be placed in the same category. A sufficient density of voids can reduce the effective refractive index by a factor of three, greatly enhancing the absorption.

Textural effects of a different kind are based on "resonant scattering". Deposition of very small metal particles on a highly reflective substrate, or dilute dispersion of particles in a host matrix, are characteristic examples of this approach. Depending upon the size and the optical constants of the particles and matrix, a spectral profile of the reflectance results that is close to the one desired.

2.2.2 Spectral Selectivity Provided by a Single Material

Taking the simplest approach, we want the desired spectral profile to be provided by the intrinsic optical excitation spectrum of a single material. Sum rules place restrictions on the approximation preventing it being too close to the required step function. Compensating structure, however, could be located sufficiently far from the two emission bands to make the profile acceptable.

Unfortunately, to the best of our knowledge, there is no such material in nature. If there had been one on the surface of the earth a long time ago, it would have run consistently at a higher temperature than its surroundings under solar exposure and decomposed more quickly. Natural surfaces balance the solar input against the reradiative loss at a temperature well below the decomposition temperature of proteins. One can safely touch all surfaces – even at high noon on a summer day in the desert – although a metal surface coated with zinc makes us painfully aware of the existing differences in spectral selectivity.

However, we do find some materials which, in their optical properties, approximate the desired characteristic sufficiently close to raise the question about whether synthetization could effect an improvement. Can material science synthesize a material for optical applications as successfully as it has synthesized thousands of materials for electronic applications?

As recently as ten years ago, this question could not have been asked with any reasonable hope for an answer. Our understanding of the optical properties of solids, on the basis of their electronic structure, is just emerging. We can readily interpret the observed spectra, but we are still far from predicting or designing deliberate changes in optical performance. A challenge to future research will be to improve on existing optical spectra by proper chemical and crystal engineering. The theoretical foundation for such an approach is found in recent developments that successfully correlate the physical chemistry of solids with their electronic structure [2.11]. The subsequent insights into the theory of bonds and bands, as well as into the electronic structure of disordered phases, make the concept of "optical engineering on the level of solid-state physics" a promising one.

The question of an ideal spectrally selective metal was extensively discussed at a symposium in Tucson in 1975. *Ehrenreich* [2.12] speculated on the electronic structure of a material, "Elysium", as consisting of a partly-filled conduction band having a width of about 0.1 eV, and a plasma frequency of about 1 eV, the former being separated from the unfilled band immediately above by a gap of about 10 eV. These requirements are unrealistic, since the conduction bandwidth is sufficiently small so that the electrons would probably be localized rather than itinerant. However, even in this case, the overall efficiency (including Carnot) is only about 23%, not much greater than that achieved by presently available but more complicated selective coatings. *Kohn* [2.13] pointed out that in semiconductors, a carrier density greater than 10^{20} cm^{-3}, and reasonable effective masses, are required in order to achieve the required reflectivity profile. Because of the Kramers–Kronig relations and sum rules, the introduction of ultraviolet interband processes affects the optical properties in the intraband region. However, there are no theoretical considerations, based on causality alone, that suggest that there is an intrinsic obstacle to producing such a material. The efficiency is also limited by scattering processes (due to temperature, or imperfections), and the fact that the Drude free-electron reflectivity edge has a finite width even in the absence of damping. Silicon, for example, would appear to be a good candidate material because of its abundance and stability at high temperatures. On the other hand, it would be very difficult, if not impossible, to achieve the high level of doping required to bring the plasma edge near 2 μm.

This restriction probably holds for semiconductors in general. Although the plasma edge offers a suitable spectral profile, its spectral placement near 1–2 μm

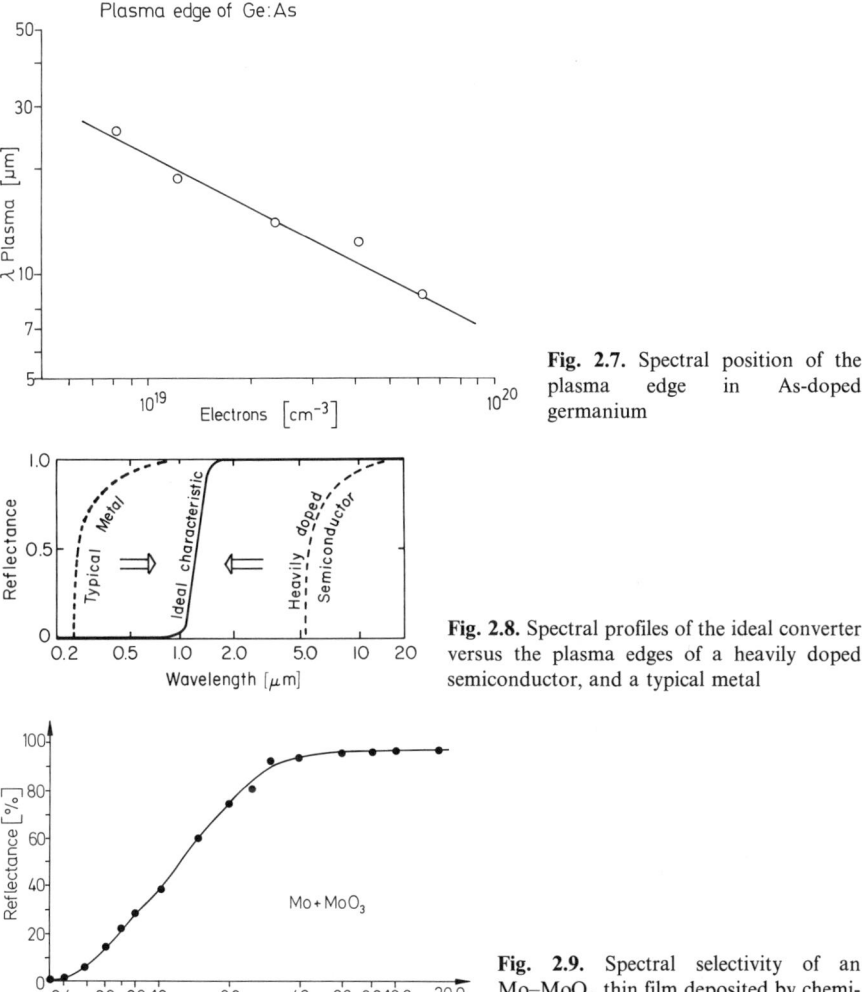

Fig. 2.7. Spectral position of the plasma edge in As-doped germanium

Fig. 2.8. Spectral profiles of the ideal converter versus the plasma edges of a heavily doped semiconductor, and a typical metal

Fig. 2.9. Spectral selectivity of an Mo–MoO$_3$ thin film deposited by chemical vapor decomposition

requires an excessively high level of doping, as demonstrated for the case of As-doped germanium in Fig. 2.7.

The ideal characteristic can be approached from the opposite spectral direction, as schematically shown in Fig. 2.8. If the optical mobility of the free electrons of a typical metal could be "frustrated", the plasma edge would move closer to the desired characteristic.

Such "frustration" can be accomplished by scattering, as evidenced in the reflectance profile of a molybdenum film containing inclusions of MoO$_3$, as shown in Fig. 2.9. Note that the reflectance in the visible is sufficiently lowered to make this film a promising candidate for a single-material converter with a solar absorptance of 0.74 [2.14–16].

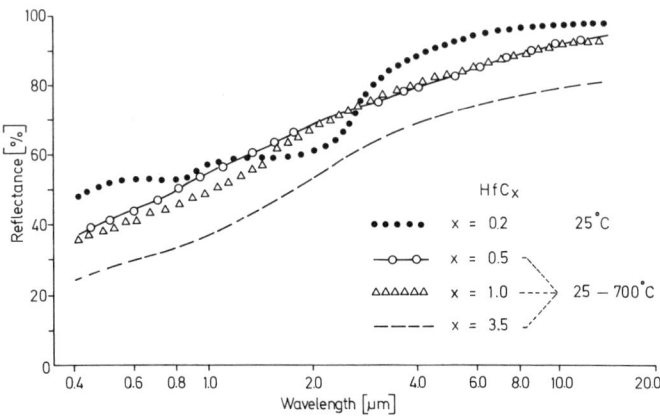

Fig. 2.10. Near-normal-incidence reflectance of HfC$_x$ as a function of temperature and composition

An alternate approach hybridizes the d electrons of a transition metal in a bond with the sp electrons of carbon or nitrogen. Although the refractory transition-metal carbides and nitrides have recently attracted interest because of their extreme hardness and high melting point [2.17], their optical properties have scarcely been investigated. Certain relations with their other physical properties and underlying mechanisms, however, can be established. Scattering of the outer electrons is common to electrical and optical properties. We are dealing with a group of materials that possess a very high density of scattering centers because carbon sites are left unoccupied. Scattering on carbon vacancies is known to be responsible for variations in most physical properties of the MeC$_x$ group over the range $0.5 < x < 1$, where Me signifies a transition metal. Therefore, although no experimental data are presently available, we would anticipate a pronounced effect of scattering upon the optical properties as well. In addition to the scattering by carbon vacancies, the distribution of the electron charge in the bond between neighbors affects the optical response. The existing band structure calculations disagree with respect to the exact nature of the electron transfer upon the formation of this bond. They agree, however, that the d electrons of the transition-metal atom are hybridized during bond formation, and that the electron density near the carbon atom increases [2.18]. This hybridization and charge transfer causes the valence electrons of the metal to be reduced in their optical activity by lowering their effective density. This, in turn, should lead to a shift of the reflectance drop, corresponding to the plasma edge in the metal, towards longer wavelengths, as is actually observed. This could explain the location of the shallow reflectance step in the infrared that is unlike the spectrum of the parent transition metal, and that qualifies the carbides as potential solar converters. Thus, variations in bond type and scattering mechanism will permit us to modify the spectral profile in this region of the spectrum. Preliminary results on HfC$_x$ films deposited by reactive sputtering, as shown in Fig. 2.10, confirm the speculation that candidates for single-material converters will probably be found among the refractory transition metal compounds [2.19]. The peculiar electron dynamics in these compounds produce a number of unique characteristics, most of them of importance for optical properties. Their variation offers significant freedom in the deliberate design of optical properties. The carbides, for which the name "naturally degenerate semiconductor" has been used, present an intermediate case between metals and semiconductors. The tolerance of the metal sublattice to a large and adjustable fraction of carbon vacancies introduces a large density of scattering centers. The interaction of the d electrons with the sp electrons of the carbon permits a variety of bond types, resulting in changes of the optical activity of the electrons participating in the bond. The fractions of ionic, metallic and covalent contributions will depend upon the metal–metal or metal–metalloid interaction [2.20]. Itinerant versus localized character of the electrons will correspondingly affect their optical response.

2.2.3 Metals as Absorbers and Reflectors

Metals attract interest for solar applications because their high reflectance in the thermal infrared effectively suppresses the emittance. The good conductors and reflectors such as Au, Ag, Cu, and Al show thermal emittance values e in the range 0.02–0.04, up to temperatures of 300 °C. Metals of higher resistance reflect the thermal infrared less effectively, and have a higher thermal emittance. This is to be expected, since the Drude–Zener theory interprets both phenomena in terms of the same set of free electrons. Deviations are numerous, however, and are particularly apparent in the temperature dependence of the optical properties.

It is more difficult to understand why poor infrared reflectance appears to be correlated to higher solar absorptance. The noble metals absorb between 0.10 and 0.14 of the solar emission, in contrast to Ni, Fe, Sn, and Ti, which absorb up to 0.50 at a thermal emittance of 0.10–0.15. Alloys suggest that this correlation is characteristic for elements only. Kovar absorbs 0.50 at an emittance of 0.14, surpassed by 98% Fe:2% Mn which has $a=0.58$ and $e=0.08$ [2.21]. A wide range of values appears to be accessible, depending on composition.

This is not surprising, since the combined action of free and bound electrons [2.22], responsible, to a first approximation, for the optical properties of metals, does not postulate a spectral relationship between the two. Proceeding on the spectral scale from small photon energies upward, the metal is first characterized by the Drude–Zener action of the free electrons, more or less influenced by the action of the interband transitions following at higher energies. At photon energies generally too far into the middle of the solar emission band, collective electron effects cause the reflectance to drop drastically in the plasma edge. However, through hybrid resonances between the conduction band and lower-lying bands, the spectral location of the plasma edge can be shifted considerably. Toward higher photon energies, and often overlapping with the plasma region, interband transitions determine the optical properties. Ultimately, sum rules dictate the spectral correlation of the various optical processes, fortunately within wide limits. Width, relative position, and the occupation of the higher energy bands will to a greater extent determine the spectral profile of the optical properties of a metal.

The interband absorption may be improved by proper choice of crystal symmetry leading to several sets of Bragg planes with substantial band gaps [2.23]. These interband transitions can be shifted by alloying, which can introduce new Brillouin zone boundaries, and hence, additional absorption edges, as well as changes in electron concentration, and the electron scattering time τ. The Mg–Li system has been extensively studied and exhibits many of these features. Another example is the case of ordered β-brass. The absorption peak at about 2.5 eV can be shifted up to 0.1 eV with only a small change in the Cu/Zn ratio, without appreciably affecting the scattering rate, and hence the infrared emittance. Even though these materials may not be significant for photothermal solar applications in the form of single coatings, they may be useful as alloy constituents, as dispersed particles in composite layers, and as substrates in multicomponent systems.

The optical properties of a wide variety of transition metals and their systematics are not well understood [2.24]. Because the Fermi energy intersects the complex d-band structure, interband transitions set in at very low photon energies. In fact, no true Drude region exists below about 0.1 eV. Thus, there is no microscopic significance to fits using the Drude model over more extended wavelength ranges, despite the fact that the data may fit reasonably well. Strong interband absorption associated with the d electrons does not occur until relatively high photon energies (6–10 eV, depending on the crystal structure), which is well outside the region of interest for solar applications. At higher energies, there are again strong transitions from the d electrons to relatively flat bands arising from f states. The optical properties depend significantly on the crystal structure. The magnetic properties have no major influence on the spectra. The details of the observed spectral profile can be qualitatively understood on the basis of existing band calculations, but effects associated with optical matrix elements are also important. Interband critical points play a role, but do not give rise to prominent structure by themselves. The optical properties of various disordered substitutional alloys exhibit no dramatic changes in structure with respect to those of the constituent elements, despite the fact that in cases such as ScCo, calculations would predict that such changes should occur.

The transition-metal oxides demonstrate how wide this range of physical properties can be, and how sensitively they react to relatively small changes in the energy-band structure. The incomplete d shell of the metal ion is the common feature of this group, and it is the localization of these electrons in transition-metal oxygen complexes, or their itinerant character in energy bands, that is responsible for this wide range of properties, optical as well as many others.

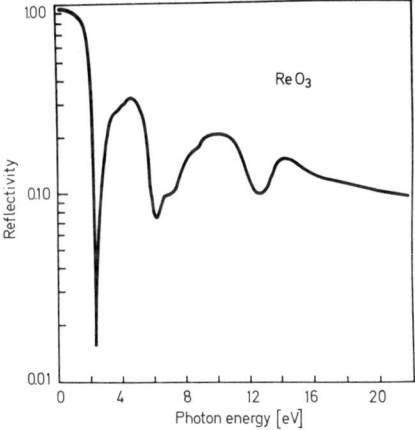

Fig. 2.11. Near-normal-incidence reflectance of ReO_3

A case in point is the position of rhenium trioxide, ReO_3, vis-à-vis the sodium tungsten bronzes Na_xWO_3. Figure 2.11 shows the reflectance of a single crystal, ReO_3 [2.25]. Note the large infrared reflectance leveling off at 94% for small photon energies, as well as the sharp plasma edge at 2.3 eV responsible for the lustrous red color of the oxide. A sequence of interband transitions follows the steep drop down to 1% reflectance.

The character and occupation of the partially filled conduction band responsible for the metallic nature of ReO_3 are derived from a comparison with the similar oxide of tungsten WO_3. With one less electron on the metal, tungsten trioxide is an insulator. As sodium is added in going to the sodium tungsten bronze, Na_xWO_3, the metallic character increases according to the fraction, x, of the added sodium electron. Analogously, it can be expected that for $x = 1$, an electron configuration similar to ReO_3 is reached, compatible with the model of a single electron in a wide d conduction band.

We have cited this analogy extensively because it demonstrates the variability of the optical properties in the transition-metal oxides because of the interaction and occupancy of the uppermost energy bands. Adjusting the optical characteristic anywhere between that of an insulator and that of a metal by proper choice of the composition, provides a degree of freedom useful in meeting a specific requirement.

2.2.4 Temperature Dependence of Optical Processes

The temperature dependence of the optical properties of a material designed to be used over a large range of temperatures is of great importance. A scan of the literature reveals the scarcity of optical data above room temperature mentioned in the introduction. In addition, the data scatter widely when more than one measurement is reported for the same material. This reflects the great difficulty of making accurate reflectance or emittance measurements over an

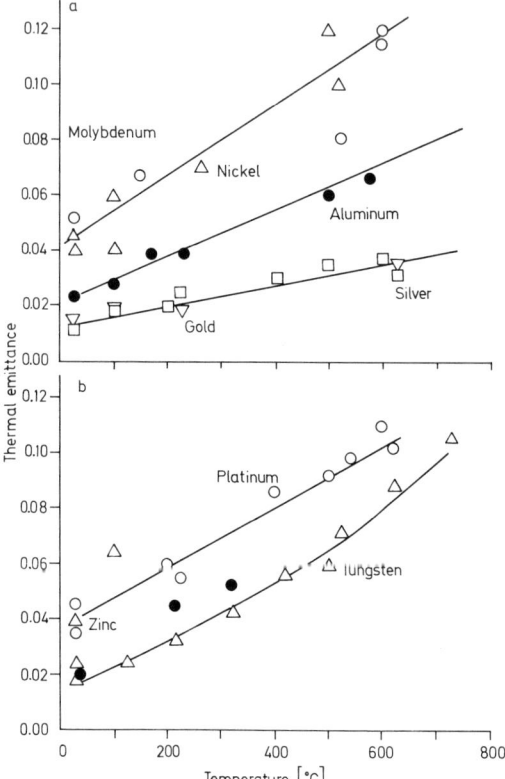

Fig. 2.12a and b. Thermal emittance as functions of temperature for some metals. Data sources: Metals Handbook (1949), AIP Handbook (1957), Smithsonian Physical Tables (1934), Engineering Heat Transfer, Welty (1974), and NASA Conf. Rpt. SP-31 (1963)

extended range of temperatures. Due to thermal expansion, systematic errors are difficult to minimize, since the surface condition of a sample may change irreversibly.

With these reservations in mind, we show the emittances for a number of metal films over a range of temperatures of interest for solar energy applications in Figs. 2.12a, b. Gold and silver have the lowest values, and copper is very close, but hard to measure since it oxidizes readily at high temperatures. Aluminum has twice the emittance of gold and silver, while nickel and molybdenum lie about an equal distance farther away from gold. Tungsten also has a low emittance when used in bulk form, but it is very difficult to evaporate with an electron beam or sputterer. It is difficult to prepare thin films of an emittance less than twice the value of the bulk material. Platinum has the same emittance as Mo and Ni, while zinc is close to aluminium.

The increase in emittance with rising temperature is very important for the behavior of selective surfaces. The net result is that the selectivity a/e measured at room temperature decreases as the temperature increases. It is therefore necessary to measure a selective coating at the temperature at which it will be used.

Besides the dependence upon temperature, the optical effects associated with defects in metals, such as vacancies, interstitials, dislocations and grain boundaries, should be determined. Such measurements would require a combination of metallography and optics.

The paucity of high-temperature optical measurements is accompanied by an insufficiently developed theoretical picture. Preoccupied with the study of elementary excitations, the solid state theorist has neglected the temperature factors entering into the dielectric function [2.26]. Essentially, two major effects of temperature on the optical constants need to be considered. They arise, respectively, from the scattering of electrons from thermally induced disorder (optical absorption with phonon creation and absorption), and the changes in the electronic energy levels due to disorder. In semiconductors, the change of population in the conduction and valence bands with temperature must also be considered. The phonon contribution to the optical properties has a temperature dependence due to multiphonon processes, and frequently extends into the relevant wavelength range at high temperatures. This may be important in connection with composite materials or in multilayer films in which absorbing layers are placed on infrared reflectors. Similar problems have already received considerable attention in connection with the study of laser windows.

Some temperature dependent effects in semiconductors remain unexplained. Among these is the fact that in many families of semiconductors, the temperature dependence of fundamental gaps is nearly the same, and independent of the particular minimum studies. By contrast, the pressure coefficient of the gap is highly dependent on the minimum in question. In amorphous semiconductors, the temperature dependence of the gap differs little from that of the crystal [2.27].

The changes in the Drude relaxation time in metals due to phonon scattering are qualitatively similar to those arising in alloys. The considerable body of experience with alloys should be useful in disentangling similar effects involving temperature dependence in metals.

While temperature-dependent band calculations have achieved some degree of success, these calculations are complicated both conceptually and technically [2.28]. A more general understanding of the temperature dependence might require the adoption of some kind of chemical viewpoint instead of the conventional band structure approaches. The Frank–Condon effect, and the fact that optics measures free energy differences between levels, must also be considered.

The increase in emittance with rising temperature is very important for the behavior of selective surfaces. The decrease in performance due to a rise in temperature is necessary for a Drude-type reflector. The optical properties are determined by the plasma frequency, the optical frequency, and the relaxation time of the electrons, which are in turn determined by the electron-phonon collision frequency, suitably averaged over the electron distribution in k space. In the framework of the Drude–Zener theory, the temperature coefficient of the emittance is necessarily positive, as long as the temperature coefficient of the electrical resistance is positive [2.29].

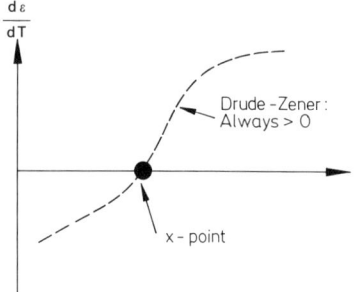

Fig. 2.13. Schematic illustration of X point

Experiments contradict this prediction by observing sign changes of the temperature coefficient of the emittance on the wavelength scale [2.30]. For the majority of metals, the temperature coefficient is indeed positive for wavelengths larger than the so-called X point, while being negative on its short-wavelength side [2.31] (Fig. 2.13). Evidently, the X point reflects the effect of interband transitions on the optical properties to the extent that the temperature coefficient changes sign. In the case of transition metals, in which there are several partially filled bands, and, accordingly, several types of conduction electrons with different values of the plasma and relaxation frequencies, the reason for the appearance of an X point may also be the difference in the relaxation frequencies of electrons belonging to different bands. Consequently, we may speculate on the possibility of a deliberate shift of the X point in an alloy which is more attractive when the sign inversion is located near the crossover of solar and thermal emission bands, covering the range 0.8–1.3 μm in standard refractory metals [2.32]. The temperature dependence of the performance of a photothermal converter may be designed in a particular way according to the spectral position of the X point.

We have emphasized the X point as an example of one of the many aspects of the temperature dependence of optical processes that are insufficiently understood. The challenge to the optical solid-state physicist is particularly urgent.

2.3 Realizations of Spectrally Selective Surfaces

2.3.1 Absorber-Reflector Tandems

The search for a single material with the optical properties of an ideal photothermal converter has not produced a satisfactory candidate. We have found various materials of high reflectance in the thermal infrared, but we are always confronted with the need to boost the solar absorptance.

Going to the next level of increasing complication, we can use two materials in tandem to provide the two basic functions of the converter, absorption and

Absorber-reflector tandem

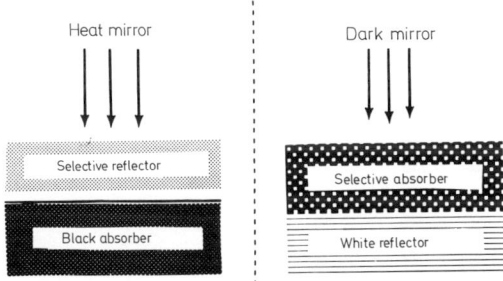

Fig. 2.14. The two basic configurations of the absorber-reflector tandem

reflection. Two basic configurations are possible, shown schematically in Fig. 2.14.

If the reflector intercepts the sunlight first, we call the configuration a "heat mirror". If the absorber intercepts the sunlight first, we have the "dark mirror", which is the configuration used in the majority of photothermal converters.

Heat Mirrors. An overlying reflector material that is highly reflective in the thermal infrared but transparent over the solar emission range can greatly reduce the requirements on the underlying absorber. Since converters operating at high temperatures must be enclosed in vacuum anyway, the heat mirror can be combined with the housing. Some highly doped semiconductors such as indium oxide, tin oxide, or cadmium stannate serve well. Systematic studies are underway [2.33] to improve the transparency of such materials, as insufficient transparency is presently the chief limitation to solar absorptance of heat mirror systems.

Dark Mirrors. When the absorber layer covers the reflector base, the former is required to be spectrally selective. At the crossover of the two emission bands of the sun and the hot converter, the absorber must become transparent to longer wavelengths so that the reflector can "see through" and suppress the emittance in the thermal infrared [2.34] (Fig. 2.15).

The surface of the absorber facing the sun must be of low reflectance to let the solar photons penetrate. We shall see below that this poses a fundamental problem. One cannot combine in a single material good absorptance, proper spectral cutoff, and low reflectance. Much of the presently existing technology is characterized by a plethora of ill-understood contraptions designed to circumvent this contradiction. We shall return, in Sect. 2.4.2, to the technological realizations of the dark mirror.

2.3.2 Semiconductors as Solar Absorbers

High absorption in the range of solar emission and low absorption in the range of thermal emission must be accomplished in a material of as low a refractive index as possible in order to minimize the front-surface reflectance. The desired

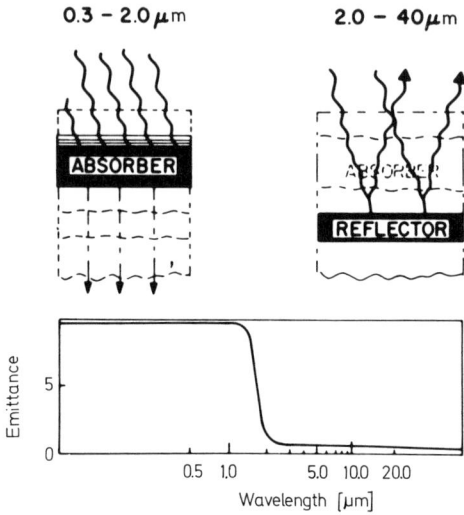

Fig. 2.15. Configuration of a dark mirror and the resulting absorptance characteristic

absorption profile is most easily provided by a semiconductor, for which a fundamental absorption edge located in the proper spectral range gives the necessary spectral profile of the absorption constant. Therefore, we shall now discuss the optical processes in semiconductors – ending up with a contradiction between proper spectral location of the edge and small refractive index. Most approaches taken to date resolve this contradiction by adding a third optical mechanism to the twofold action of the absorber-reflector tandem.

a) Spectral Profile of the Transmission

If the absorber-reflector tandem is to be effective, the absorber layer must have a step-function absorption profile and low refractive index. Semiconductors generally meet these requirements.

A generalized absorption profile for a semiconductor is shown in Fig. 2.16. Starting from short wavelengths, we first encounter a region of strong absorption (Region A), caused by interband transitions of the electrons. The oscillator strength of these transitions is large, leading to absorption coefficients of more than 10^3cm^{-1}. The intrinsic nature of this absorption renders it insensitive to doping within wide limits – an advantage with respect to poisoning of the performance by interfacial diffusion of impurities.

The absorption drops in an edge located roughly at the photon energy required to cross the forbidden gap. Its profile depends upon a number of parameters. Direct versus indirect character of the fundamental transition, as well as the $E(k)$ profile of valence and conduction bands, is important. Final-state interactions, such as excitons, determine the line shape of the edge. Purity and structural perfection of the material are influential. A steep absorption edge is to be preferred, placing a premium on a direct-gap material of high purity.

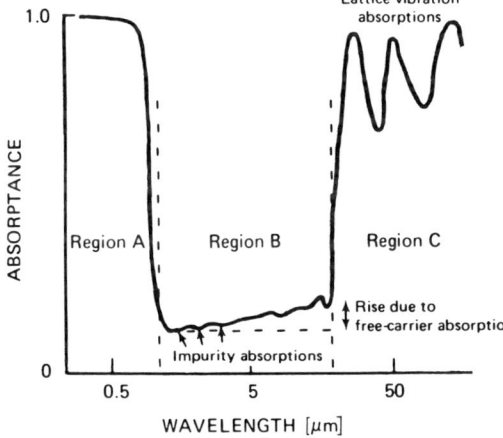

Fig. 2.16. Absorption profile characteristic for a semiconductor

Most amorphous semiconductors, although otherwise advantageous, have a shallow edge profile; fortunately, however, there are exceptions.

The region of infrared transparency (Region B) is important, as it permits the metal base to "see through" the absorber and suppress the emittance by its high reflectance. Free carrier absorption can decrease the transparency, especially at high temperature. Semiconductors deposited by chemical vapor deposition (CVD) are pure enough to minimize this increased absorption. The infrared absorption of CVD-deposited silicon, for example, is only 5–10% higher at 500 °C than at room temperature [2.34]. This is in contrast to most materials deposited as thin films in vacuum [2.35].

Even if the free-carrier absorption is minimized, lattice vibrations ultimately limit the transmission range towards the longer wavelengths. The mass of the lattice constituents, and the forces between neighbors, determine the onset of the absorption. The higher the atomic weight, the further out extends the region of transparency. The stiffer the bond (as determined by its nature and length), the shorter the wavelength of the transmission cutoff. Unfortunately, the otherwise promising oxides cutoff in the near infrared, covering the range from 4.5 μm (SiO_2) to 16 μm (CeO_2). Sulfides transmit up to 12 μm, selenides to 15, and tellurides to 20 μm [2.36].

b) Refractive Index of Semiconductors

A low refractive index, possibly of a value 2 or less, constitutes the second requirement for the absorber in the tandem approach. We shall first review the factors that determine the refractive index, and then establish a contradiction between the absorption profile of a semiconductor, and the request for a small index of refraction.

At optical frequencies, the refractive index, as given by the polarizability of the valence electrons, is determined by the atomic weight of the lattice constituents, and the type of bond acting between them. If the atomic number is high, the valence electrons are well shielded from the nuclear charge, leading, in element lattices, to a high polarizability and a large refractive index. Going down a column of the periodic table, the refractive index increases. Comparing neighboring columns, however, it is evident that the net charge on the ion core, rather than the atomic number by itself, must be considered. The arrangement of the core electrons is important, and the insertion of subshells is of influence.

In compounds, the bond type greatly influences the polarizability. A covalent bond is more easily polarized than a comparable bond with ionic contributions. An approximate measure is given by the difference in electronegativity. Large differences between neighbors lead to a polar bond and a small refractive index, and vice versa for the covalent bond. If one of the constituents is of multivalent character, the higher valence state normally leads to the greater polarizability, since more electrons are involved.

In this simplified manner, we have reduced the value of the refractive index to a question of atomic weight and ionicity of the bond. Again, oversimplifying a situation that is studded with exceptions, we can now reduce the band gap to an interpretation that leads to an unfortunate contradiction between proper band gap and small refractive index. By and large, a low atomic number and strong ionicity of the bond makes for a wide-gap material. The relationship can be reduced to the equation [2.37]

$$n^4 E_g = 77, \tag{2.5}$$

where the gap is given in electron volts. On the level of dispersion relations, a similar interrelation between the dispersive part of the dielectric function ε_1 at small frequencies ω, and the spectral location of the first sizable contribution to the absorptive part ε_2 (as represented by the band-gap transition), can be expressed by the sum rule

$$\varepsilon_1(0) = 1 + \frac{2}{\pi} \int_0^\infty \frac{\varepsilon_2(\omega') d\omega'}{\omega'}. \tag{2.6}$$

The experimental basis for this interrelationship can be visualized with the help of Fig. 2.17. It is apparent that most materials with a refractive index of $n=2$ or less are transparent in the visible. To place the absorption edge in the vicinity of 2 μm requires a material of a refractive index in excess of three. It is this contradiction that forces all approaches based on the absorber-reflector tandem to resort to an additional mechanism that either antireflects the absorber, or lowers its refractive index by structural or textural effects. This leads to complex structures that are costly, cumbersome, and subject to deterioration at high operating temperatures. In keeping with the leitmotif of this article, we must ask to what extent the present limitations can be overcome by deliberate variation of the bond type in a compound material that combines the proper band gap with as small a refractive index as the relationship (2.5) permits.

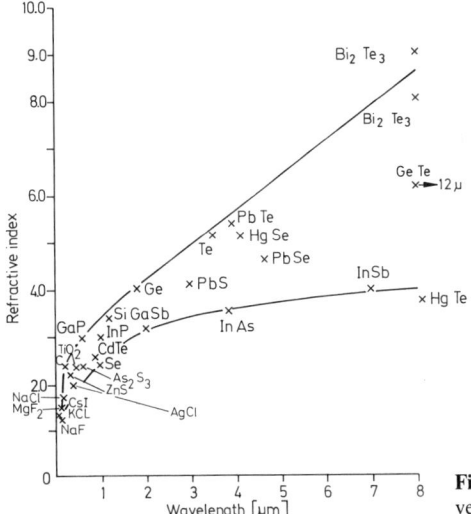

Fig. 2.17. Refractive index of various materials versus wavelength of the absorption cutoff

c) Amorphous Silicon as a Solar Absorber

Many of the applications envisioned for amorphous materials during the past ten to fifteen years have not been realized because of the lack of fundamental understanding of the amorphous state, and of its electrical and optical properties. The greatest difference between the amorphous and the crystalline states is the existence of electron states in the gap between valence and conduction bands in an amorphous substance. Their character, density on the energy scale and their communication with the band states largely determine the electrical and optical properties of an amorphous material [2.38].

In the case of amorphous silicon (a-Si), wide differences in the properties of samples prepared under different conditions or by different techniques were characteristic of early work in the field. In particular, a-Si prepared by rf-assisted decomposition of silane was markedly different from that deposited by either vacuum evaporation or sputtering. Within the last two groups, samples deposited at different rates, substrate temperature, or deposition pressure differed somewhat in their electrical and optical properties. Post-deposition anneal was of considerable influence [2.39].

Many of the differences have been interpreted as variations of the density-of-states profile resulting from the different methods of preparation. More recently, it was realized that the states can be associated either with randomness of the network and its lack of long-range order, or with structural defects. States intrinsic to the amorphous lattice fill the energy regions close to the band edges, while structural defects create states deeper in the gap. Post-deposition anneal and/or differences in preparation affect mainly the structural defects, voids, and dangling bonds. They are therefore of consequence for the states

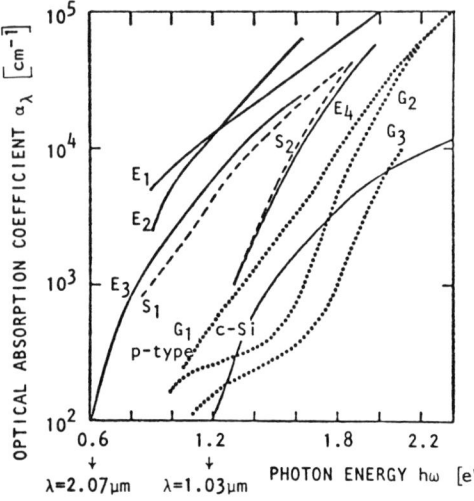

Fig. 2.18. Absorption edge of a-Si films prepared by rf glow discharge (G), by sputtering (S), or by electron-beam evaporation (E), as well as that of crystalline silicon

deep in the gap and in between the intrinsic states [2.40], and thereby substantially affect the absorption profile.

The role of introducing hydrogen, oxygen, or other atoms from the ambient atmosphere into the amorphous material during growth or anneal became increasingly evident [2.41]. In particular, the incorporation of hydrogen into the silicon network during rf-assisted decomposition of silane assumed a key role [2.42]. Saturating the dangling bonds that seemed to be inherent to a-Si produced by other methods reduced the density of deep-gap states to an extent where doping of the material became feasible [2.43]. The hydrogenated a-Si could be doped n- and p-type, and the way was open to the fabrication of a photovoltaic cell made from a-Si [2.44]. The present state of the art is a solar cell of 6% conversion efficiency, which it is believed possible to double [2.45].

The recent developments had essentially two consequences for the use of a-Si as photothermal absorbers. First, they resulted in the preparation of a material of superior optical performance. While the absorption edge in crystalline silicon is too close to the visible to trap more than 80% of the incident solar spectrum, an analysis of the situation predicts a trapping of more than 90% by properly prepared a-Si [2.46]. The survey of the absorption profiles of a-Si prepared by different methods is taken from this study (Fig. 2.18). The large variation of the spectral position of the absorption edge with conditions of preparation permits tailoring a material best matched to the solar spectrum [2.47]. In addition, the temperature shift of these edges into the infrared is about twice that of the crystalline material, giving even more efficient trapping of the solar photons as the converter heats up.

While it had long been realized that the optical properties of a-Si are superior to those of crystalline Si (c-Si) for use in photothermal converters [2.48, 49], the insufficient temperature stability of the amorphous phase of

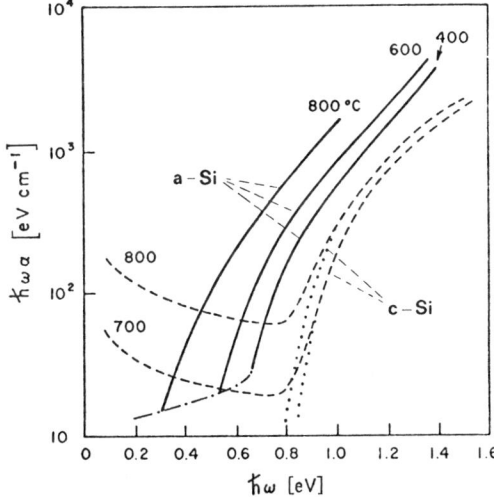

Fig. 2.19. Absorption coefficient of a-Si (——) and c-Si (---) at different temperatures. Tail (···) and residual absorption (–·–·–) are marked separately

evaporated or sputtered material prevented its use, however. With the systematic study of the incorporation of hydrogen into the a-Si lattice it became apparent that hydrogenated a-Si crystallizes at a much higher temperature. Hydrogen dispersed in an unclustered, single Si–H bond manner escapes at temperatures in the neighborhood of 650–680 °C, with subsequent crystallization [2.50]. The transition from amorphous to crystalline is observed in the same temperature range for films deposited by CVD [2.51]. The stabilizing role is not confined to hydrogen, however. Films deposited by rf sputtering in high purity argon show a stable absorption edge insensitive to anneal to temperatures as high as 800 °C. There was also no evidence of crystallization up to 800 °C. In an interpretation of their remarkable results, Tauç and co-workers credit this extraordinary temperature stability to the incorporation of Ar and O impurities into the network of their a-Si films [2.52]. We show as Fig. 2.19 their results, which demonstrate the superiority of a-Si over c-Si in three aspects. First, the absorption edge is shifted further into the infrared, permitting a larger fraction of the solar photons to be absorbed. Second, the subsequent rise of the absorption in c-Si is absent in a-Si, due to the "trapping" of the free carriers in localized states. And third, the already beneficial shift of the absorption edge into the infrared is further enhanced by a faster temperature shift of the edge in a-Si. It is indeed remarkable that all three advantages can be utilized up to temperatures of 800 °C, thereby eliminating the previous key problem of an early crystallization. The remaining absorption in the order of $\alpha = 50\,\mathrm{cm}^{-1}$ at the bottom of the edge is of little concern in films 1 μm thick or less, and is in any case smaller than the free carrier absorption in even the purest c-Si at comparable temperatures. It can also be expected that systematic control of the type and density of the stabilizing impurity will lower this residual absorption even further.

The promise of *a*-Si as a solar absorber must also be seen in connection with the ease of economic large-scale fabrication. Common to all the deposition methods described in this section is the feature that all operate in vacuum and require the fabrication to proceed under corresponding restrictions. This is of great consequence to the dimensions of the product and the continuity of the fabrication process. It has been stated clearly that the high cost of vacuum equipment does not suggest a future for the industrial manufacture of large-area coatings fabricated by vacuum-based methods [2.53]. It is, therefore, significant that the preparation of *a*-Si has been recently accomplished by CVD [2.54], a method for which the capital cost is predicted to be considerably below those of sputtering, vacuum, or rf discharge systems. We shall describe the details in Sect. 2.4.4.

2.3.3 Selectivity by Wavefront Discrimination

Selectivity by wavefront discrimination is an important mechanism for the generation of a spectrally selective profile and is treated in detail in the contribution of *Sievers* to this book. A conceptual description will therefore suffice in the context of this chapter. *Tabor* [2.10] was the first to realize that the textural profile of a surface can discriminate between the pencil-beam character of solar input and the hemispherical wavefront of the thermal reradiation. If of proper dimensions and configuration, such a surface profile can result in spectral selectivity. Many solar absorbers owe their absorptance to the structure of a surface that is neither smooth nor homogeneous in depth.

Structural surface effects are difficult to categorize. The underlying mechanism may be the size and separation of the structural elements as compared with the wavelength of light, or the properties and distribution of inhomogeneities.

At the extreme of large dimensions of surface texture are the corrugated specular surfaces that accomplish spectral discrimination by directional selectivity [2.55, 56]. Such corrugated surfaces utilize the fact that the solar input is directional in character whereas the radiation loss is emitted hemispherically. The dimensions of the corrugation are large compared to the wavelength. This effect is most helpful in cases where the increase in solar absorptance is more important than the concomitant increase in emittance.

When the geometrical parameter of the texture is of the order of the wavelength of light, we have surfaces that are absorbing for visible light but reflecting for longer wavelengths. In this region, we are approaching the transition from "reflective scattering", in which selectivity is obtained simply by multiple reflections within the geometry of the surface, to "resonant scattering", in which the geometry and the physical properties of the scatterer combine to produce the selectivity.

All effects caused by the surface roughness profile are presumably borderline cases between the two categories. The scattering properties of a surface as

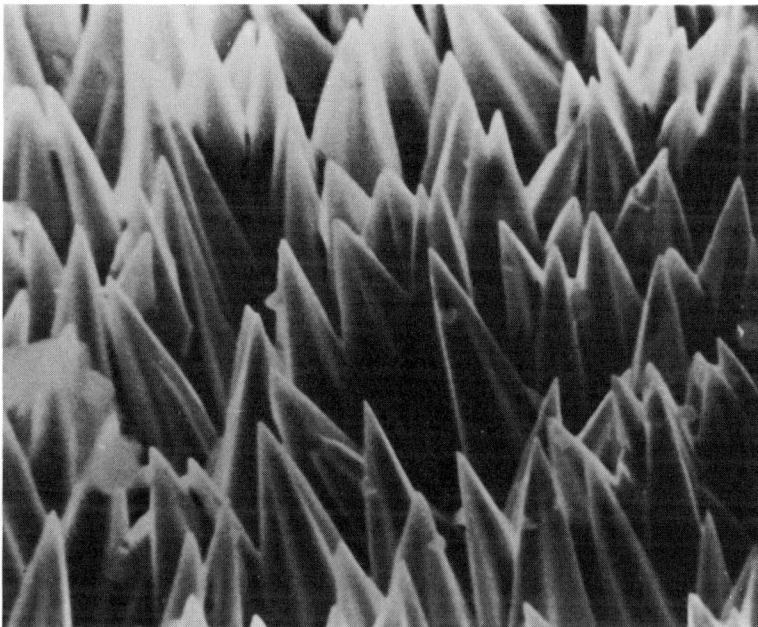

Fig. 2.20. Scanning electron micrograph of a forest of rhenium whiskers deposited by CVD. Magnification 5650 × (subsequently reduced to 25% of original size). Photograph courtesy of K. P. Murphy, University of Arizona

expressed by its albedo (ratio of scatter coefficient to total extinction coefficient) affect the hemispherical emittance and can be used to establish selectivity [2.57, 58]. The surface roughness can be simplified in a forest of metal whiskers deposited by chemical vapor deposition. Figure 2.20 shows such a reflectively scattering surface made of rhenium whiskers [2.59].

The line between reflective and resonant scattering separates the *gold-black* deposits from the *gold-smoke* filters [2.60, 61]. These are similar in their physical structure, particle size, and degree of aggregation. Both absorb strongly in the visible and near infrared. Gold-black deposits, consisting of submicron particles of pure gold deposited on insulating substrates, retain the metallic properties of the bulk material. Gold-smoke deposits are evaporated under conditions that add tungsten oxide to the otherwise similar aggregate, rendering it an insulator, Consequently, gold-black absorbs strongly in the infrared through the action of the free metal electrons, whereas gold-smoke is transparent to infrared radiation. The small fraction of tungsten oxide in the gold-smoke particles suggests that structural and alloy effects can drastically change the optical properties of a material that is dispersed in a colloidal phase.

Resonant scattering in colloidal dispersion is also likely to be the cause of the spectrally selective properties of doped gold films. These films are deposited from an organic solution containing organometallic compounds resulting, after firing, in the addition of a total of 10% of the oxides of rhenium, bismuth, chromium, silicon, and barium [2.62]. The doped gold film retains the high reflectance of a pure gold film of similar thickness but drops in reflectance near the 1.5 μm wavelength.

Dispersion of conductive particles in host matrices, either dielectric or conductive, provides spectral selectivity through resonant scattering, characterized by a dependence not only upon the geometry, but on the material parameters as well. A dispersion of vanadium, calcium, or niobium in copper produces a broad resonance in the visible region of the spectrum, but retains high reflectance in the thermal infrared. The lowest-frequency near-resonance of small (100 Å) particles of transition metals in dielectric composite structures can be adjusted to occur in the center of the solar spectrum by controlling optical properties and density of the particles and the medium in which they are embedded [2.63]. Similar effects can be obtained in a dispersion of titanium particles of variable density in MgO or Al_2O_3 matrices.

Most of the effects of resonant scattering are difficult to interpret. Part of the difficulty lies in the insufficient characterization of the system with respect to degree of dispersion, shape and uniformity of the particles, and their physical properties as dependent upon size. However, the theoretical frame must often be stretched as well, and still does not reproduce the data well. Maxwell-Garnett theory fails for the inhomogeneous metal dispersions in dielectric hosts, even if adapted for nonspherical shapes and variable relaxation times.

Although complicated in their theoretical interpretation, and difficult to reproduce reliably on the experimental level, textural effects in spectrally selective surfaces are of the utmost technological importance. The most successful "selective blacks" are of the semiconductor-on-metal type, which would be inferior if textural effects did not reduce the refractive index of the semiconductor. The Maxwell-Garnett theory of porous materials applies here. Thin films are unavoidably of a porous structure, and a porous coating may have substantially better optical properties for spectrally selective applications than the corresponding solid materials. An example is the PbS-on-metal configuration, which should reflect strongly owing to the high refractive index (4.1) of PbS. If the PbS is applied as solid material, such a tandem should have a solar absorptance of only 0.36. If, however, it is deposited as a porous structure with a sufficiently large void density, the effective refractive index is lowered to 1.8, resulting in solar absorptances greater than 0.90 [2.64]. Many films fabricated by electrodeposition are, in their solar absorptance, dependent upon a very complex surface texture that represents a composition of reflective and resonant scattering, assisted by the lowering of the refractive index typical of a porous structure. Figure 2.21 shows the surface texture of a "selective chromium black".

The complex surface structure responsible for the high solar absorptance of many selective blacks makes it difficult to fabricate such surfaces without strict control of the critical parameters. Such surfaces will also be extremely sensitive to abrasion, such as simply being touched. Little research effort has been

Fig. 2.21. Surface texture of selective chromium black on dull nickel (Harshaw Company) as recorded under the SEM microscope at 3330 × magnification (photograph subsequently reduced to 25% of original size). Photograph courtesy of K.P. Murphy, University of Arizona

applied to determining critical parameters or to the fractional admixture of the various processes responsible for the absorptive action. Such an understanding will be the necessary first step toward optimizing the performance.

2.4 Examples of Selective Coatings

We described in the preceding sections a number of features that are characteristic of spectrally selective surfaces suitable for photothermal conversion. No single material provides the desired profile in sufficient approximation. It is necessary to employ a tandem of an absorber overlaid onto a good reflector. The absorber must absorb sufficiently over the solar emission band, but be transparent in the thermal infrared. The simultaneous requirement of a small refractive index severely restricts the use of semiconductors. The absorber-reflector tandem must be assisted by a third mechanism that boosts the solar absorptance, antireflects the semiconductor absorber, or lowers its refractive index. Interference or textural surface effects are widely used for this purpose.

We shall briefly describe some successful attempts to fabricate spectrally selective surfaces, using the various combinations of mechanisms required to generate the desired profile. The examples are selected as illustrations of typical approaches. Comprehensive coverage is not intended, and omission does not represent value judgement. None of the examples fits squarely into one category only. A mixture of several mechanisms makes for a good deal of overlap, although this is never clearly determined. In essence, this is the point we wanted to make.

2.4.1 Interference Coatings

In the cases of intrinsic materials or absorber-reflector tandems, solar absorptance is caused in the former, by a single passage through the optically active medium, or, in the latter, by the return passage after reflection by the underlying mirror surface. In the case of the interference stack, the desired effect is the net result of a multiplicity of passages through the dielectric portion of the stack lying between the two reflective surfaces, the upper one of which is partially transparent. Careful tuning of the layer thicknesses is necessary to get a good broad-band selective surface. The basic idea for a selective interference stack starts with a sufficiently polished metal substrate, or a substrate covered by an opaque reflective metal layer, for which the reflectivity is high in the thermal infrared and gradually drops in the visible. When the first layer of dielectric is added it tends to reduce the reflection at the metal interface. The "selective" effect is not as strong as the desired one since, in general, the reflectivity of the dielectric is weak at its air interface, so the interference effect it produces is weak. Some dielectrics can be selected, such as chromium oxide which does provide good interface reflection, plus some internal absorption, to make a reasonable "two-layer" interference stack. In general, one needs to add a second reflective layer, partially transparent as in the traditional Fabry-Perot interferometer, to strengthen the reflected wave so as to maximize the internal interference in the dielectric of the second layer. Layer three is generally thin – about 50 Å – so that its actual optical properties are quite different from the bulk metal. This property of the third layer is deliberately enhanced to increase its effect, and considerable speculation exists as to whether this thin metal layer is really pure metal or some mixture of metal atoms and impurities from the vacuum chamber. Vacuum "art" plays a role in getting good third-layer performance [2.65, 66]. The final layer of a typical four-layer interference stack is a dielectric, and its role is to broaden the wavelength region of high absorption. Multiples of the basic stack can be used, but the gains are not cost-effective.

An example of this four-layer design is given by the Al_2O_3–Mo–Al_2O_3 (AMA) coatings, successfully developed for high-temperature space applications by *Schmidt* and co-workers at Honeywell [2.67, 68]. The coating, for which the reflectance profile after successive anneals is shown in Fig. 2.22, consists of a semitransparent molybdenum layer sandwiched between layers of aluminium oxide. The high cost of the massive molybdenum substrate prohibits the use of the otherwise so successful AMA coating ($a_s = 0.85$, $e = 0.11$ at 500 °C) for terrestrial applications.

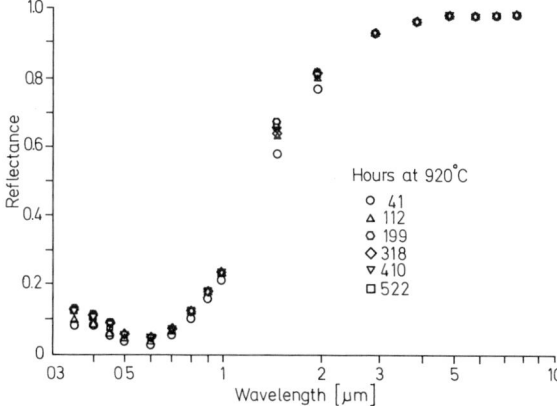

Fig. 2.22. Spectral reflectance of AMA (Al$_2$O$_3$–Mo–Al$_2$O$_3$) coating on Mo after heating at 920 °C for varying periods of time

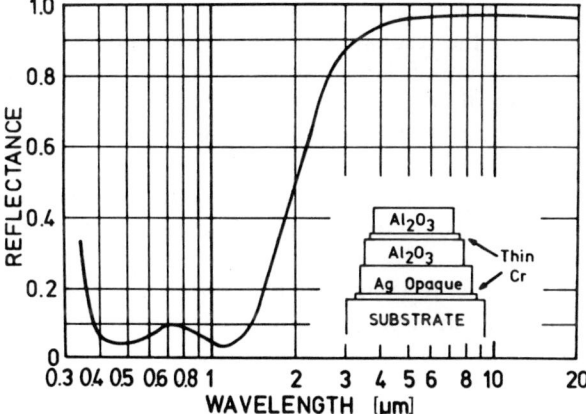

Fig. 2.23. Spectral reflectance of an Al$_2$O$_3$/Cr/Al$_2$O$_3$/Ag coating

Efforts to replace the costly Mo substrate by stainless steel overcoated with thin molybdenum films reduced the performance of the reflector due to the notorious difficulties of evaporating or sputtering thin films of the refractory metals. Out-diffusion of the component metals of the stainless steel substrate, as revealed by Auger studies [2.69], reduced the lifetime of these stacks at high temperature.

On the basis of the AMA design, *Meinel* et al. [2.70] studied not only the reflective film at the substrate surface, but also the role of the semitransparent metal layer at the center of the Al$_2$O$_3$ sandwich on top. A wide range of metals including Cu, Ag, Au, Al, Cr, and Mo were used for the reflector and semitransparent layers, while a larger group of potential materials including MgF$_2$, SiO, Al$_2$O$_3$, CeO$_2$, etc., were used for the dielectric layers. A four-layer coating using Ag/Al$_2$O$_3$/Cr/Al$_2$O$_3$ was found to give excellent spectral performance as shown in Fig. 2.23. This coating exhibits the usual double reflectance minimum in the solar region with a sharp transition leading to high infrared reflectance, and the cutoff is easily shifted as desired within the 1.2–2.5 µm wavelength interval. The measured solar absorptance values ranged between 0.90 and 0.95 with emittance values of 0.02–0.04. A significant feature was the excellent repeatability of the spectral properties for successive depositions. These surfaces have shown high stability after 1000 h at 150 °C in air with probable stability as high as 300 °C. Tests in vacuum have shown no spectral deterioration after brief periods at temperatures as high as 600 °C.

Evaporation and sputtering have been used for the realization of selective interference stacks on the basis of Ni, Fe, Sn, Ti on Al, or Cu on Al [2.71], or metal carbides on Cu [2.72]. Solar absorptances in excess of 0.90, with emittances between 0.1 and 0.2, are typical performance values for these stacks.

Evaluating the prospects of a large-scale application of interference coatings to high-temperature photothermal conversion, we must keep the following limitations in mind:

I) The effectiveness of the interference filter rests with its character as a tuned optical cavity. The sharper the resonance, the better the performance. Detuning and subsequent degradation of the performance can be caused by a variety of effects, especially in operation at high temperatures. The temperature variations of the optical constants, as well as the dimensions of the stack, can detune the cavity. Corrosion at interfaces spoils the balance of the partial reflectances. Diffusion across these interfaces can poison the optical properties of any of the layers. The list can be continued.

II) The sensitivity to degradation of the performance leads to a stringent selection of the materials with respect to their refractive properties, expansion coefficients, chemical compatibility, vapor pressure, etc. The selection can result in rare and costly materials as the only choice.

III) Vacuum deposition is recommended as the technology backed up by the largest experience. This leads to restrictions with respect to large-scale fabrication. Chemical vapor deposition of optical multilayer elements has been tried successfully [2.34] and may eliminate this restriction by fabricating interference filters in open-tube, atmospheric pressure processes of a flow-through character.

IV) The interference and reinforcement effects of multilayer stacks are sensitive to the angle of incidence of radiation. If used in the concentrating collectors that are necessary for reaching high temperatures, the interference filter reduces its effectiveness.

Observations on actual interference coatings confirm this concern. If the total radiative loss into a hemisphere is extrapolated from the measurement of the emittance at normal incidence, losses too low by a factor of 1.5 to 2 are predicted [2.70, 73, 74]. Thin surface films and/or the morphology of the radiating surface are probably responsible for this descrepancy. After all, this extrapolation only works well for a smooth mathematical interface – a configuration hardly ever obtained in a real thin film system.

Most of these arguments refer to the sensitively tuned interference filter. If used in coarser configurations of one or two layers, interference will be

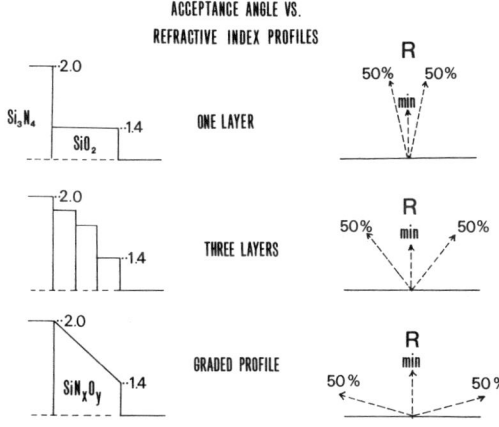

Fig. 2.24. Schematic representation of the correlation between the acceptance angle of an antireflection coating, and the spatial profile of its refractive index

indispensable in providing the additional optical mechanism that the absorber-reflector tandem calls for. Nearly all spectrally selective surfaces use some form of interference to boost the solar absorptance. The reverse arrangement – depositing an interference filter of high infrared reflectance and solar transparence on top of a black substrate – is in principle possible, but has rarely been used. Metals are too convenient for the suppression of the thermal infrared.

Interference works even better in surfaces that do not consist of a sequence of discrete layers, but generate the optical inhomogeneity in a layer for which the refractive index varies throughout the thickness, as shown schematically in Fig. 2.24. Such a profile resembles a stack of many layers, but has, in addition, the advantage of a wider optical-acceptance angle [2.75]. In most methods for the electrodeposition of selective surfaces, the parameters are changed as the layer grows, generating this useful variation of the optical constants. The effectiveness of the resulting product critically depends on this programming of the deposition.

Chemical vapor deposition of multilayer stacks provides a particularly elegant variation. By changing the composition of the carrier gas as the film grows, the nature of the deposit can be altered in a continuous manner, resulting in a preprogrammed profile of the refractive index [2.34, 76]. For instance, by gradually changing the fractional composition of the gas stream, the deposition can go through the sequence silicon-silicon nitride-silicon oxynitride-silicon dioxide. In response to the ratio of H_2O to NH_3, the refractive index of the deposit changes through the layer in a graded profile, as shown in Fig. 2.25. Such a graded-index profile gives antireflective action over a large angular field of view.

2.4.2 Tandem Stacks

a) Heat Mirrors

In the heat mirror, the two functions of the absorber-reflector tandem are generated by an infrared reflector of high solar transparency overlaid on a non-

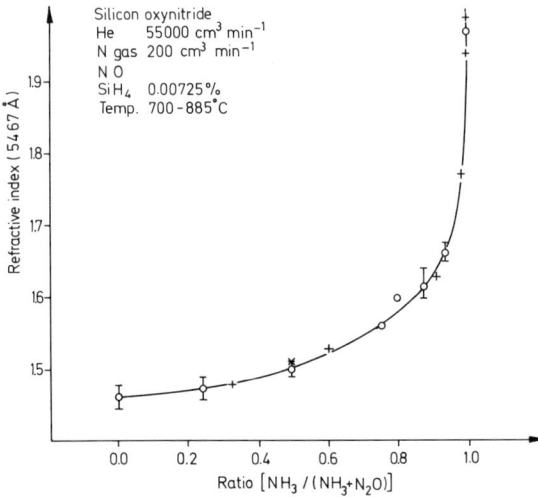

Fig. 2.25. Refractive index of silicone oxynitride deposited from a vapor phase of variable composition

selective black absorber (Fig. 2.14). This configuration offers advantages in the design of a converter. Flatplate collectors with non-selective, inexpensive absorbers can be upgraded by mounting them behind a cover glass or enclosure of high infrared reflectance [2.77]. The reflective shield of visible transparency can be placed in direct contact with the absorber as well, however.

We have already discussed the difficulties of placing the otherwise suitable plasma edge into the proper spectral position, for both metals and highly doped semiconductors [2.78]. The best approximation to a solution of the problem is represented by semiconductors of a band gap greater than 3 eV for sufficient solar transmittance, and highly doped for the generation of the infrared reflectance by free carriers. The necessary coupling of the basic optical processes limits the heat mirror in its performance. Nevertheless, the need for transparent conductive coatings has given the related technology a boost [2.79, 80].

Films of tin oxide, indium oxide, or mixtures of the two have served so well that technology has coined the acronym ITO (Indium-Tin-Oxide) [2.81–86]. Doping plays a large role in most applications, and post-deposition anneal, sometimes in reactive gases, often improves the performance. Figure 2.26 shows a typical reflectance and transmittance profile of such an ITO film. The high visible reflectance – a necessary consequence of the presence of free carriers – usually limits the solar absorptance, unless an antireflectance layer is provided.

The ITO film can be sprayed directly onto a silicon absorber, serving as an antireflection layer for the silicon in the solar emission region, and as an emittance suppressor in the infrared [2.85]. Figure 2.27 compares the reflectance of such an ITO-coated silicon with that of the uncoated material.

Indium-Tin-Oxide films are stable up to temperatures in excess of 300 °C. They adhere well to glass, resist chemical and environmental attack, and are applied in an economically attractive method. These advantages should

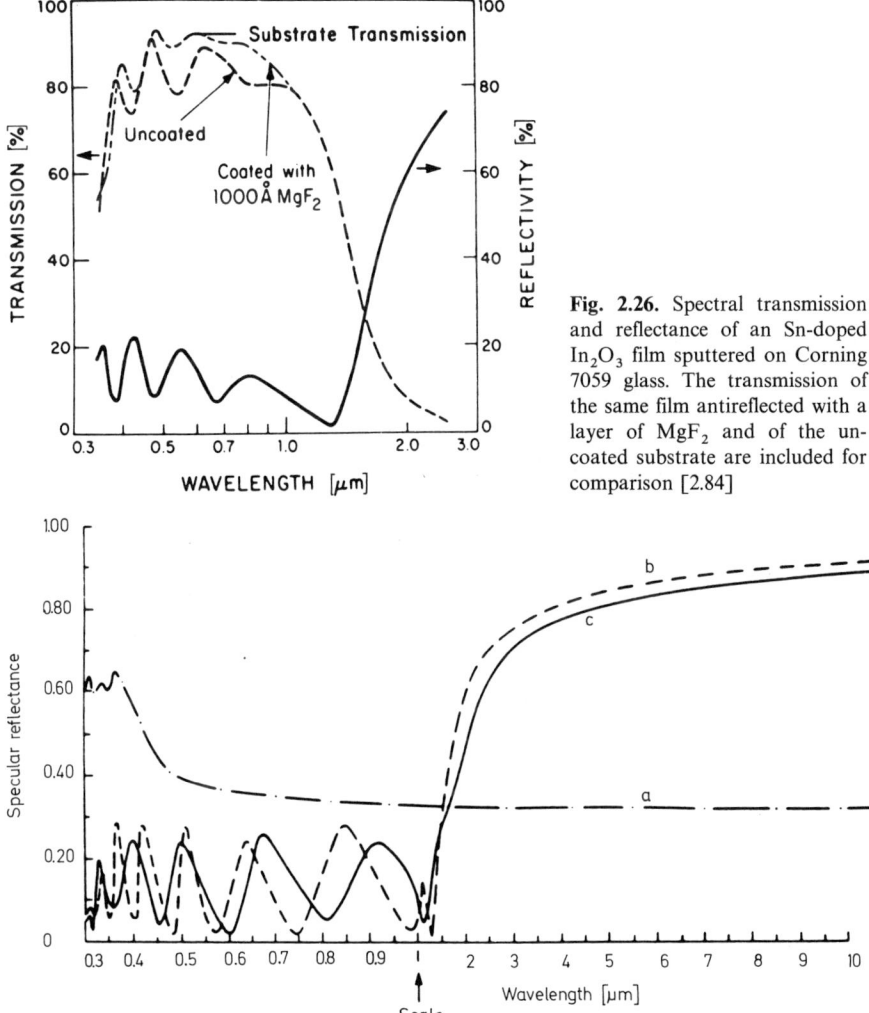

Fig. 2.26. Spectral transmission and reflectance of an Sn-doped In_2O_3 film sputtered on Corning 7059 glass. The transmission of the same film antireflected with a layer of MgF_2 and of the uncoated substrate are included for comparison [2.84]

Fig. 2.27. Specular spectral reflectance of (a) uncoated silicon, (b) and (c) silicon of different thickness coated with 0.79 µm of spraydeposited indium tin oxide [2.85]

guarantee continuation of the present development efforts, the more so as most of the doping and post-deposition effects are not sufficiently well understood to predict limitations on the eventual performance of heat mirrors in general.

b) Dark Mirrors

Reversing the sequence of absorber and reflector leads to the configuration of the dark mirror, shown in Fig. 2.15. Selectivity is now required of the absorber

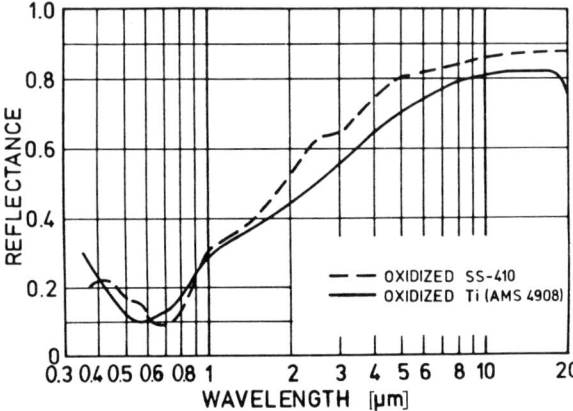

Fig. 2.28. Spectral reflectance of oxidized SS-410 and Ti (AMS 4908) [2.95]

facing the sun, permitting the white reflector underneath to "see through" the absorber, and reducing the emittance in the infrared. Most of the commercially available selective blacks are of the dark-mirror type. Selectivity is easier to realize, since most solids absorb at high photon energies, turning transparent at small photon energies. Requiring the opposite spectral dependence, the heat mirror is more difficult to adapt to the electronic band structure of solids.

In its simplest form, the dark mirror can be realized by changing the chemical or morphological habitat of a metal surface such that the initial infrared reflectance of the metal is preserved, while chemical composition or texture of the metal surface provides the necessary solar absorptance. The semiconductor nature of most metal oxides makes them suitable candidates for selective absorbers. The oxidation can be accomplished by heating, dipping in baths, or other methods of chemical conversion, most of them readily expanded to economical large-scale fabrication. The optical action of the thin absorber layer results from a collaboration of effects due to intrinsic semiconductor absorption, interference, graded profile of the refractive index, and surface texture.

Oxidation has been used to produce copper oxide by dipping and heating [2.87, 88], thermal decomposition of sprayed nitrate solutions [2.89, 90], and direct oxidation [2.91]. Additional oxides formed on various metal substrates include cobalt oxide [2.89, 91–93], nickel oxide [2.94], manganese oxide [2.89], antimony oxide [2.89], titanium oxide [2.95], zirconium oxide [2.94], and an aluminum/aluminum oxide mixture [2.96]. Some representative results by *Edwards* et al. [2.95] for oxidized titanium and stainless steel are shown in Fig. 2.28. Measurements reveal that most of these oxide coatings have moderate selectivity ($a_s \cong 0.60$–0.85 and $e \cong 0.10$–0.40) at low operating temperatures. However, the very shallow spectral transition between the region of high absorptance and low emittance leads to very low thermal collection efficiencies at high temperature [2.95]. Layer thicknesses, variations of film deposition parameters [2.90], material phase changes [2.93], diffusion effects and the

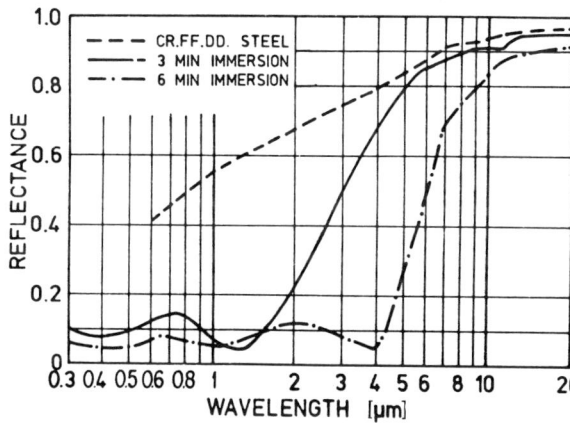

Fig. 2.29. Spectral reflectance of a selective black on steel for different treatment times [2.88]

general suitability of transition metal oxides [2.91] and substrate impurities [2.96] influence the spectral and physical properties of these films. Absorbers can also be formed by chemical conversion of a metal surface. The various Ebanol[1] treatments have been used to treat copper [2.95, 97, 98] and steel [2.88, 95, 98] to produce coatings of CuO/Cu_2O and Fe_3O_4, respectively. A typical spectral reflectance curve for a treated steel surface is shown in Fig. 2.29. These surfaces usually have high solar absorptance ($a_s \cong 0.80$–0.90) with moderate emittance values (typically 0.15–0.20), although values as low as 0.07 at approximately 100 °C have been reported [2.88]. The mechanical and optical properties have also been examined [2.98]. Copper treated with Ebanol-C produces a surface with a dendritic structure which increases the solar absorptance. Heating in air (245 °C for 244 h) caused a slight increase in solar reflectivity ($\sim 3\%$) with no detectable change in chemical composition. However, there was an additional oxidation at the film-substrate interface which caused a loss of film adhesion. By contrast, there were no changes in film reflectivity when heated in vacuum, and the films were stable under uv radiation. Carbon and stainless steels treated with Ebanol-S and Ebanol-SS, respectively, exhibited no change in surface morphology and were stable under heating in air (230 °C) and uv treatment [2.99].

Two types of dark mirrors, the electroplated coatings and tandem stacks fabricated by CVD, have acquired such prominence that we shall deal with them separately in the following two sections.

2.4.3 Electroplated Coatings

Black chromium, a promising coating for the low and intermediate temperature range, is perhaps the best known among the electroplated selective blacks [2.100]. It consists of a variable composition of metallic chromium particles

---

[1] Trade name of Enthone, Inc., New Haven, Connecticut.

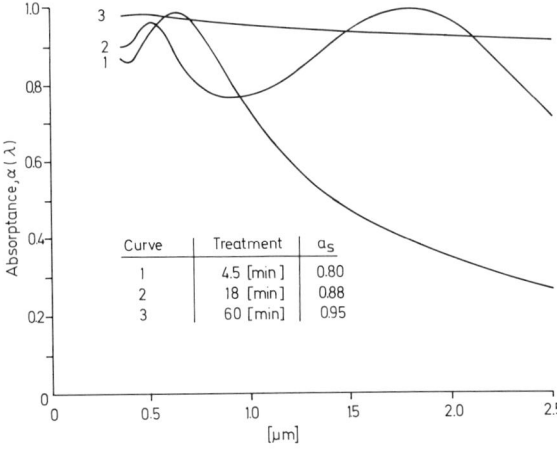

Fig. 2.30. Spectral absorptance as a function of plating time for an electroplated black nickel coating deposited in bright nickel [2.113]

embedded into Cr_2O_3 electrolytically deposited onto a nickel-coated steel substrate [2.101]. The method was first used for selective blacks by *Tabor* [2.102], who deposited nickel-zinc sulfide systems on nickel reflectors or on galvanized iron [2.103], and already speculated on the usefulness of black chromium. Later work deposited copper oxide on copper and aluminum substrates [2.9, 104] or cobalt oxide [2.105], and chromium-nickel-vanadium alloys [2.106].

During the plating process, the conditions are varied in a systematic, but largely empiric, fashion. The current density is programmed, the pH value varied, or the temperature of the bath changed (Fig. 2.30). The programmed variation results in one or several of three effects, all designed to increase the solar absorptance. First, the composition of the growing deposit can be varied, exploiting interference effects in a multilayer stack. Second, the refractive index can be programmed to vary throughout the thickness, resulting in a profile resembling a stack of numerous, finely divided layers. Third, the structure of the surface can be textured to antireflect the absorber by pores, voids, multiple reflections, or resonant scattering. In all practical cases, a simultaneous action of all three effects is characteristic for the selective black, giving the art a strong empiric note. Optical control is characteristic for most techniques, as exercised by placing a reflectance minimum caused by absorption and interference acting in tandem in the optimum position of the solar emission.

The basic features of the art were developed 10–15 years ago. Modern work simply varies the earlier accomplishments. The NiS–ZnS coating now gives $a=0.96$ and $e=0.07$ at 100 °C [2.68] (Fig. 2.31). Chromium oxides on a nickel base reach similar values [2.107]. Also in this range are nickel-blacks electroplated on an aluminium base [2.108]. Alternating layers of nickel and silicon dioxide deposited on copper or aluminium bases give suitable black mirrors [2.71].

For most selective blacks, it is difficult to estimate quantitatively the influence of the surface morphology. The situation is clearer for absorber films

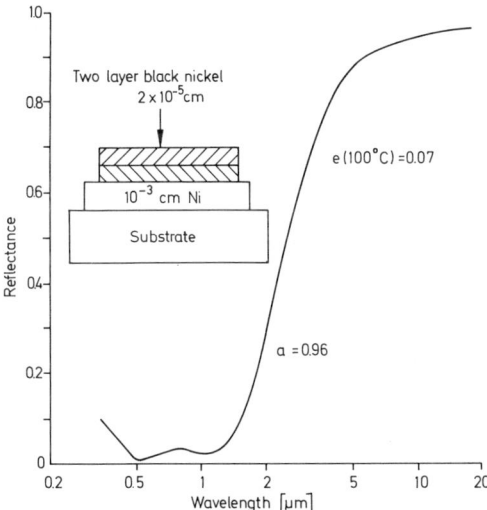

Fig. 2.31. Spectral reflectance of an electroplated two-layer black nickel (NiS–ZnS) coating on a nickel-plated substrate [2.68]

made by depositing PbS. In its solid phase, PbS has a refractive index of 4.1, resulting in a reflectance of 37%, with both values averaged over the solar emission range. Used in a non-homogeneous phase, however, the effective refractive index can be lowered to $n=2$ or less in void-rich deposits [2.109] or particulate suspensions in organic binders [2.107]. Recent work gives values of $a/e=40$ in films of 1000 Å thickness, emphasizing the strong importance of the surface morphology [2.110].

The low cost of fabrication suggests selective blacks for flat-plate, low-temperature applications characterized by large intercepting areas and low thermal efficiencies. As shown in Fig. 2.32, the solar absorptance does not vary over a large range of the angle of incidence, probably due to textural effects.

The advantages of satisfactory performance at low temperature and the low-cost fabrication both suggest widespread use of the electroplated coatings, in particular black chromium. Their durability at high temperatures must consequently be of great concern. Unfortunately, the rapid deterioration of most electroplated coatings, including black chromium, reduces the benefits of good performance and economic fabrication. Black nickel, although of acceptable optical performance, is recognized to be unstable in air at temperatures as low as 300 °C, and at even lower temperatures if moisture is present as well [2.111]. Thermal aging effects in black chromium largely depend upon the exact conditions of fabrication. Some coatings withstand well extended operation at 350 °C, others are severely degraded at temperatures of 300 °C or below [2.112]. Studies of a large number of coatings from different manufacturers do not reveal significant differences in either composition or surface morphology between stable and unstable coatings [2.113]. It is evident that presently known process control limits which are adequate for decorative black chrome are not sufficient for solar coatings [2.114]. Greater efforts should be made to specify

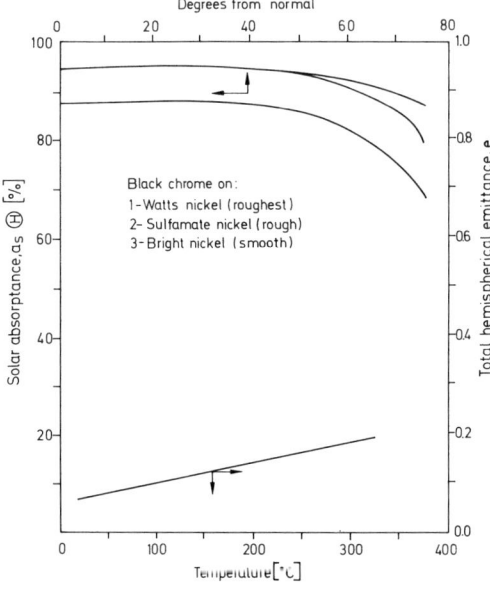

Fig. 2.32. Measured solar absorptance (air mass = 2) as a function of incidence angle for a black chrome coating on substrates of various indicated roughnesses [2.116]

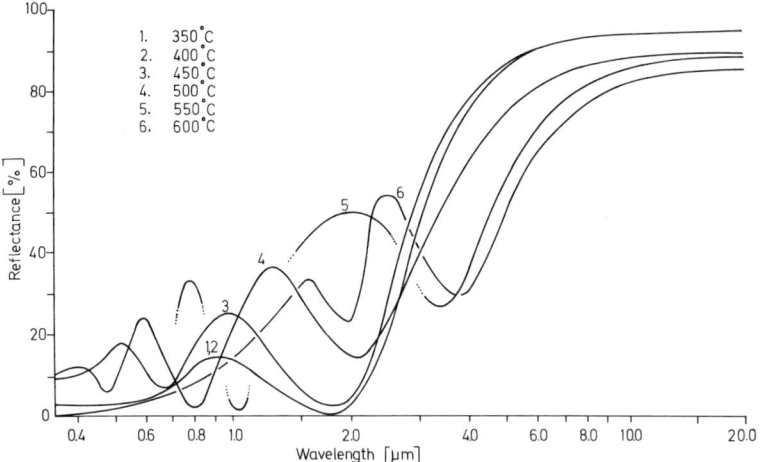

Fig. 2.33. Spectral reflectance of an electroplated black chrome coating on bright nickel after heating at a pressure of about 10^{-2} Torr for successive 15 h intervals at temperatures between 350 and 600 °C [2.115]

the plating process more accurately in order to obtain a reliable fabrication of stable coatings, particularly in view of the operating temperature of 300 °C for the distributed collector concept, and the necessity of protecting the coatings from temperature transients in case of coolant loss. Beyond 400 °C, even good coatings of black chromium fail, as evidenced by the successive anneals shown in Fig. 2.33. In the final section of this chapter, we will describe a tandem coating that withstands extensive exposure to temperatures in excess of 500 °C.

2.4.4 Tandem Stacks Fabricated by Chemical Vapor Deposition

In this final section, the author reports on work of his own group, in which a novel technology for the deposition of optical multilayer stacks for high-temperature photothermal conversion was developed. The work demonstrated the feasibility of fabricating converters of intermediate selectivity ($a/e = 12$ at 500 °C, with $a = 0.80$), long-term survival at elevated temperatures (several thousand temperature cycles to 500 °C, several hours at 700 °C), and in a manner that can readily be expanded to economically attractive largescale manufacture [2.34].

The characteristic feature of the approach is the use of chemical vapor deposition (CVD) techniques to fabricate the absorber-reflector tandem. The stainless steel substrate to be coated with silicon is placed into the hot zone (>700 °C) of a furnace and exposed to a silane/carrier gas mixture. If all parameters are properly chosen, the silane breaks up through the transfer of thermal energy at the surface of the host substrate, leaving behind a thin layer of the silicon absorber. If repeated through different reaction zones under different conditions, a sequence of successive layers of different materials and functions can be deposited in a similar fashion, and a multilayer stack results at the end of the line.

CVD techniques are well under control for specific applications in semiconductor device technology, corrosion resistant coatings, and specific coatings for use in the atomic energy field. In all of these areas, little or no attention has been paid to the optical properties of the resulting layers and their dependence on the CVD process parameters. The optical technology had thus to be developed from a fundamental level.

The resulting transfer of CVD methods to the fabrication of optical multilayer elements signals the availability of a novel technology for large-area optical applications operating at high temperatures, such as photothermal solar energy conversion.

The major advantages of CVD fabrication can be summarized as follows:
I) Sequential flow-through fabrication.
II) Open-tube process at atmospheric pressures.
III) Fabrication at temperatures >650 °C.
IV) Economic availability of starting materials.

The following disadvantages of the method have largely been overcome by our work of the last few years:
I) Chemical vapor deposition of optical elements requires adaptation of existing technology.
II) Uniform deposition from turbulent gas phase requires close control of process parameters.
III) High deposition temperature poses problems of thermal expansion mismatch, agglomeration and interaction of component layers, etc.

The state of the art is characterized by the performance of the stack, shown in Fig. 2.34. It soon became obvious that the agglomeration of the thin silver reflector at the temperature of the CVD process presented a serious problem

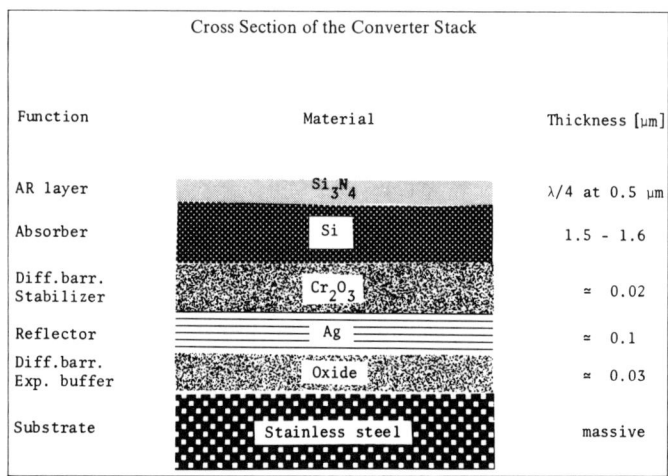

Fig. 2.34. Cross section of a multilayer tandem stack predominantly fabricated by CVD

[2.117, 118]. A thin silver film annealed to as low a temperature as 250 °C agglomerates quickly, as shown in the scanning electron micrograph of Fig. 2.35.

We have investigated a number of materials, such as Cr_2O_3, SiO_2, Al_2O_3, and Si_3N_4, that stabilize thin metal films against high-temperature agglomeration, as evidenced by the photomicrograph of an interface between stabilized and unstabilized films, as shown in Fig. 2.36.

The stabilizers act, at the same time, as diffusion barriers, as evidenced by the Auger profile of a stack that has been annealed for 150 h at 540 °C, shown in Fig. 2.37. The steep flanks of the composition profile are characteristic of the resolution of the Auger spectrograph, and indicate the absence of a diffusion of layers into each other, even after longtime high-temperature anneal [2.69].

As a basis for the evaluation of the optical performance of the stacks, their near-normal reflectance is routinely measured at room temperature and at 500 °C in our high-temperature reflectometer. Traces such as the one shown in Fig. 2.38 are planimetrically evaluated with respect to a and e.

Proving the feasibility of the process on the laboratory level, the development program has resulted in the following situation as of the end of 1977.

I) The spectral profile of the reflectance, if integrated numerically in the usual manner, gives a spectral selectivity $a/e = 15$ at room temperature.

II) Reflectance measurements actually performed at 500 °C result in $a/e = 12$–14. This slight reduction in a/e at 500 °C, compared to the room temperature, is typical for a semiconductor absorber, and sets it aside from other converter types.

III) Stacks cycled several thousand times to 500 °C did not degrade in their optical performance. Sample sets annealed at 700 °C for several hours resulted in changes of their infrared reflectance of less than 2.5%.

Spectrally Selective Surfaces 49

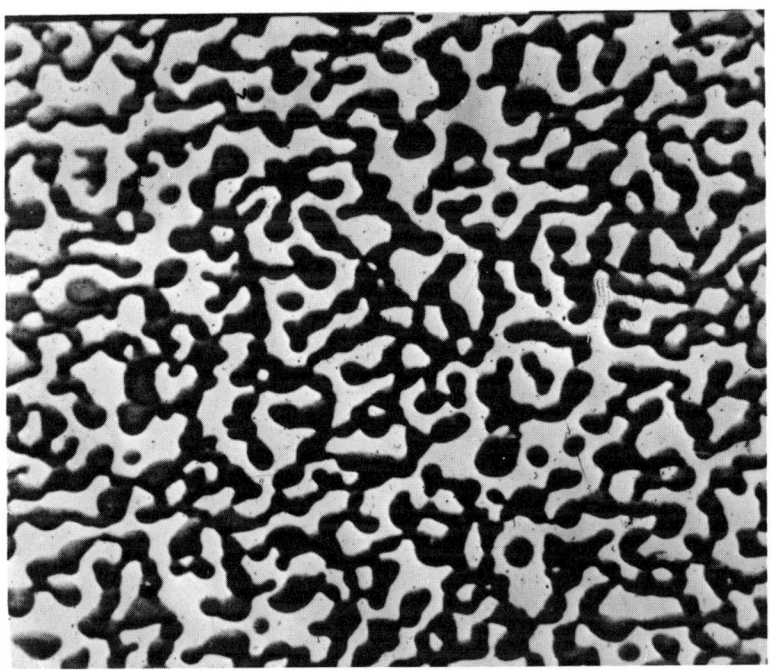

Fig. 2.35. Scanning electron micrograph of a silver film annealed briefly at 250 °C

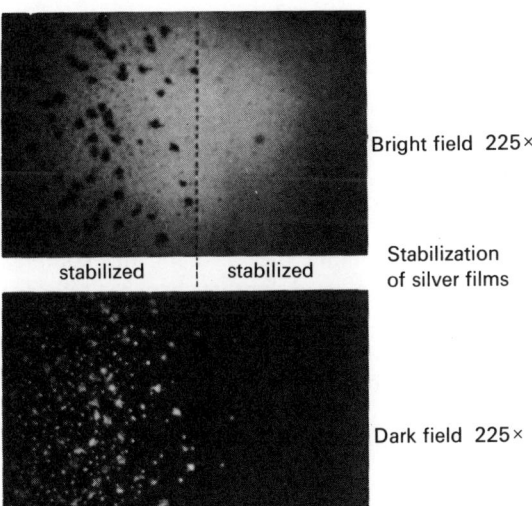

Fig. 2.36. Optical micrographs of the interface between stabilized and nonstabilized thin-film silver reflector after anneal at 700 °C

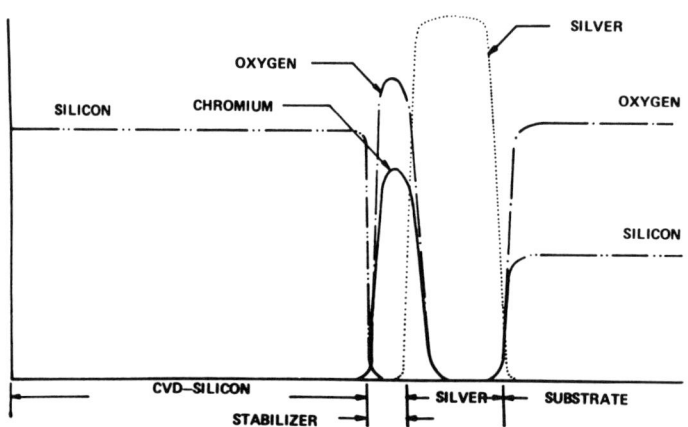

Fig. 2.37. Auger profile of the stack of Fig. 2.34 after 150 h anneal at 540 °C

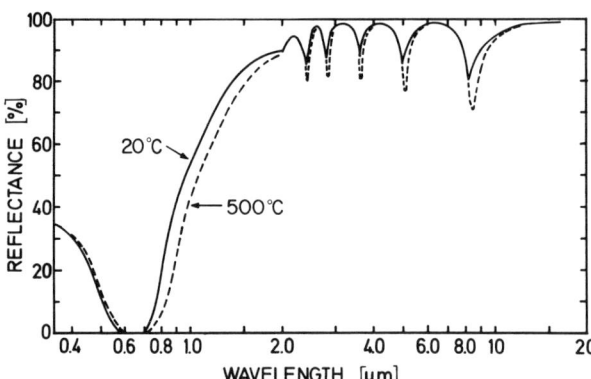

Fig. 2.38. Near-normal-incidence reflectance of a CVD stack at 20 °C and at 500 °C

Work in progress attempts first to eliminate the need for the complex stabilization of the silver reflector presently used in the absorber-reflector tandem. Refractory metals are more stable than the noble ones, and do not need a stabilizer. Unfortunately, the refractory metals, such as molybdenum and tungsten, are prohibitively expensive in bulk form and must be disqualified for large-scale applications. If evaporated or sputtered as thin layers, however, their strong gettering action and/or structural modifications give the films a considerably lower reflectance [2.121]. If, in contrast, the metal is deposited in the highly-reducing atmosphere of the chemical vapor deposition process, a film results with a reflectance close to that of the bulk metal [2.122]. This

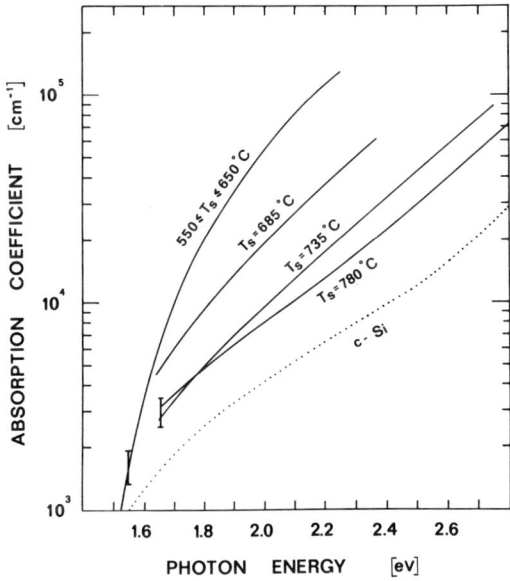

Fig. 2.39. Absorption coefficient of amorphous silicon films deposited at different temperatures $T_s = 550$, 590, 635, 685, 735, 780 °C, and of different thicknesses $d = 0.58$, 1.79, 1.67, 1.66, 0.85, 0.80 μm, respectively [2.54]

development introduces the refractory metals to thin-film technology. High temperature applications are possible, since molybdenum films can be passivated against oxidation. Without degradation of their infrared reflectance, such Mo films have been exposed to open air at 500 °C for several hours.

In a second development underway, the moderate solar absorptance of the stacks is improved by replacing the crystalline silicon absorber with a layer of amorphous silicon. Although its potential for increased absorptance was recognized early by our group, amorphous silicon was known to crystallize above 500 °C, rendering it unsuitable for high-temperature application. However, fabrication by CVD produces a material that is more stable and still has the favorable infrared absorption of amorphous silicon deposited at low temperature and under the restrictions of vacuum-based methods (Fig. 2.39).

Besides the attraction of potentially continuous fabrication at atmospheric pressure, chemical vapor deposition of multilayer stacks provides a particularly elegant feature. By changing the composition of the carrier gas as the film grows, the nature of the deposit can be altered in a continuous manner, resulting in a preprogrammed profile of the refractive index [2.123–125]. We have shown, for instance, that by gradually changing the fractional composition of the gas stream, the deposition can go through the sequence: silicon-silicon, nitride-silicon, oxynitride-silicon dioxide. Such graded-index profiles antireflect over a large angular field of view, giving the resulting stack superior performance in concentrating collectors of large optical acceptance angles [2.75].

The successful transfer of CVD technology to the fabrication of spectrally selective surfaces represents a promising addition to solar conversion technology. The two features under development – greater solar absorptance through the use of CVD amorphous silicon, and simplified fabrication through incorporation of CVD refractory-metal reflectors – will increase this potential. If finally developed into the continuous, flow-through mode of operation, the method will provide an attractive technique for economic large-scale deposition of photothermal converters of high-temperature durability.

References

2.1   C.H. Liebert, R.R. Hibbard: Sol. Energy **6**, 84 (1964)
2.2   R.E. Petersen, J.W. Ramsey: J. Vac. Sci. Technol. **12**, 471 (1975)
2.3   H.S. Gurev, R.E. Hahn, K.D. Masterson: Int. J. Hydrogen Energy **2**, 259 (1977)
2.4   C.M. McCulloch, R.E. Treadwell: SANDIA Tech. Rpt., SAND 74–0124, 14 (Albuquerque, N. M. 1974)
2.5   R.D. Tobin: Proc. Thermal Power Systems Workshop on Selective Absorber Coatings (Solar Energy Res. Inst., Golden, Colo. 1977) p. 73
2.6   A.C. Skinrood: Proc. Thermal Power Systems Workshop on Selective Absorber Coatings (Solar Energy Res. Inst., Golden, Colo. 1977) p. 50
2.7   J.A. Leonard: Proc. Thermal Power Systems Workshop on Selective Absorber Coatings (Solar Energy Res. Inst., Golden, Colo. 1977) p. 90
2.8   H. Tabor: Trans. Conf. Use Solar Energy **2**, 1 A, 32 (1955)
2.9   J.T. Gier, R.V. Dunkle: Trans. Conf. Solar Energy **2**, 1 A, 41 (1955)
2.10 H. Tabor: Bull. Res. Counc. Isr. A **5**, 119 (1956)
2.11 J.C. Phillips: Bonds and Bands in Semiconductors (Academic Press, New York 1973)
2.12 H. Ehrenreich: Symp. Fund. Opt. Prop. Relevant to Solar Energy Conversion, ed. by B.O. Seraphin, H. Ehrenreich, Tucson, Ariz. (1975); NSF Tech. Rpt. DMR 74–18134 (1976)
2.13 W. Kohn: Symp. Fund. Opt. Prop. Solids Relevant to Solar Energy Conversion, ed. by B.O. Seraphin, H. Ehrenreich, Tucson, Ariz. (1975); NSF Tech. Rpt. DMR 75–18134 (1976)
2.14 H.S. Gurev, G.E. Carver, B.O. Seraphin: Solar Energy (Electrochemical Society, Princeton 1976)
2.15 H.S. Gurev: Optics in Solar Energy Utilization, Vol. 85 (Society of Photo-Optical Instrumentation Engineers 1976)
2.16 B.O. Seraphin: In *Physics and Contemporary Needs* Vol. 1, ed. by Riazuddin (Plenum Press, New York, London 1977) p. 187
2.17 W.S. Williams: Prog. Solid State. Chem. **6**, 11 (1957)
2.18 J.F. Allward, C.Y. Fong, M.L. Batanouny, F. Wooten: Phys. Rev. B **12**, 1105 (1975)
2.19 B.O. Seraphin, J. Spitz: Tech. Rpt. NSF-INT 76–02664 (1977)
2.20 E. Dempsey: Philos. Mag. **8**, 285 (1963)
2.21 M.D. Kudryashova: Geliotekhnika **5**, 47 (1969)
2.22 H. Ehrenreich, H.R. Philipp: Phys. Rev. **128**, 1622 (1962)
2.23 J.A. Ashcroft: Symp. Fund. Opt. Prop. Solids Relevant to Solar Energy Conversion, ed. by B.O. Seraphin, H. Ehrenreich, Tucson, Ariz. (1975); NSF Tech. Rpt. DMR 75-18134 (1976)
2.24 D.E. Lynch: Symp. Fund. Opt. Prop. Solids Relevant to Solar Energy Conversion. ed. by B.O. Seraphin, H. Ehrenreich, Tucson, Ariz. (1975); NSF Tech. Rpt. DMR 75-18134 (1976)
2.25 J. Feinleib, W.J. Scouler, A. Ferretti: Phys. Rev. **165**, 765 (1968)
2.26 J.J. Hopfield: Symp. Fund. Opt. Prop. Solids Relevant to Solar Energy Conversion, ed. by B.O. Seraphin, H. Ehrenreich, Tucson, Ariz. (1975); NSF Tech. Rpt. DMR 75-18134 (1976)
2.27 W. Paul: Symp. Fund. Opt. Prop. Solids Relevant to Solar Energy Conversion, ed. by O.B. Seraphin, H. Ehrenreich, Tucson, Ariz. (1975); NSF Tech. Rpt. DMR 75-18134 (1976)
2.28 M.L. Cohen: Symp. Fund. Opt. Prop. Solids Relevant to Solar Energy Conversion, ed. by B.O. Seraphin, H. Ehrenreich, Tucson, Ariz. (1975); NSF Tech. Rpt. DMR 75-18134 (1976)
2.29 G.A. Boloshin: Opt. Spectrosc. **18**, 423 (1965)

2.30 K. Ugihara: J. Appl. Phys. **43**, 2376 (1972)
2.31 D.J. Price: Proc. Phys. Soc. London **59**, 131 (1947)
2.32 B.T. Barnes: J. Opt. Soc. Am. **56**, 1546 (1966)
2.33 J.C.C. Fan: NSF Workshop on Surface Coatings, Minneapolis, Minn., Aerospace Rpt. ATR-75 (7523-02)-1, 13 (1975)
2.34 V.A. Wells, B.O. Seraphin, L.S. Raymond: Proc. 4th. Conf. CVD (Electrochemical Society, New York 1973) p. 512
2.35 B.O. Seraphin: Thin Solid Films **39**, 87 (1976)
2.36 P.W. Black, J. Wales: Infrared Phys. **8**, 209 (1968)
2.37 T.S. Moss: *Optical Properties of Semiconductors* (Butterworth, London 1959)
2.38 M.H. Cohen, H. Fritzsche, S.R. Ovshinsky: Phys. Rev. Lett. **22**, 1065 (1969)
2.39 M.H. Brodsky, D.M. Kaplan, J.F. Ziegler: Appl. Phys. Lett. **21**, 305 (1972)
2.40 P.G. LeComber, A. Madan, W.E. Spear: J. Non-Cryst. Solids **11**, 219 (1972)
2.41 M. Abkowitz, P.G. LeComber, W.E. Spear: Commun. Phys. **1**, 175 (1976)
2.42 J.J. Hanser: Solid State Commun. **19**, 1049 (1976)
2.43 W.E. Spear, P.G. LeComber: Solid State Commun. **17**, 1193 (1975)
2.44 W.E. Spear, P.G. LeComber, S. Kinmond, M.G. Brodsky: Appl. Phys. Lett. **28**, 105 (1976)
2.45 D.E. Carlson, C.R. Wronski: Appl. Phys. Lett. **28**, 671 (1976)
2.46 R.W. Griffith: Sharing the Sun – Solar Technology in the Seventies, Winnipeg, Canada (1975), Intern. Solar Energy Soc., Vol. 6 (1976), p. 205
2.47 B.O. Seraphin, H. Ehrenreich (eds.): Symp. Fund. Opt. Prop. Solids Relevant to Solar Energy Conversion, Tucson, Ariz. (1975); NSF Grant DMR-75-18134 (1975)
2.48 B.O. Seraphin (ed.): Proc. Symp. on Material Sciences Aspects of Thin Film Systems in Solar Energy Conversion, Tucson, Ariz. (1974); NSF-RANN Grant GI-43-795, 7 (1974)
2.49 B.O. Seraphin: J. Jpn. Soc. Appl. Phys. **44**, 11 (1975)
2.50 M.H. Brodsky, M.A. Frisch, J.F. Ziegler, W.A. Lanford: Appl. Phys. Lett. **30**, 561 (1977)
2.51 N. Nagasima, N. Kubota: Jpn. J. Appl. Phys. **14**, 1105 (1975)
2.52 D.E. Ackley, A.P. DeFonzo, J. Tauç: Proc. 13th Int. Conf. on the Physics of Semiconductors, Rome (1976), p. 993
2.53 P. Baumeister: NSF Workshop on Surface Coatings, Minneapolis, Minn., Aerospace Rpt. ATR-75(7523-02)-1, 49 (1975)
2.54 M. Janai, D.D. Allred, D.C. Booth, B.O. Seraphin: Solar Energy Materials **1** (to be published)
2.55 E.R.G. Eckert, E.M. Sparrow: Int. J. Heat Mass Transfer **3**, 42 (1961)
2.56 K.G.T. Hollands: Sol. Energy **7**, 108 (1963)
2.57 H.C. Hottel, A.F. Sarofine, E.J. Fahimian: Sol. Energy **11**, 2 (1967)
2.58 H.E. Bennett: In Proc. Symp. on Material Science Aspects of Solar Energy Conversion, ed. by B.O. Seraphin, Tucson, Ariz. (1974); NSF-RANN Grant GI-43795, 145 (1975)
2.59 B.O. Seraphin: In Proc. Symp. on Material Science Aspects of Solar Energy Conversion, ed. by B.O. Seraphin, Tucson, Ariz. (1974); NSF-RANN Grant GI-43795, 18 (1975)
2.60 L. Harris, J.K. Beasley: J. Opt. Soc. Am. **42**, 134 (1952)
2.61 L. Harris, R.T. McGinnies, B.M. Siegel: J. Opt. Soc. Am. **38**, 582 (1948)
2.62 R.C. Langley: In Proc. Symp. on Material Science Aspects of Solar Energy Conversion, ed. by B.O. Seraphin, Tucson, Ariz. (1974); NSF-RANN Grant GI-43795, 321 (1975)
2.63 A.J. Sievers: Presented to Material Research Council Summer Conference, La Jolla, Calif. (1974)
2.64 D.A. Williams, T.A. Lappin, J.A. Duffin: J. Eng. Power **5**, 213 (1963)
2.65 L.F. Drummeter, G. Hass: Phys. Thin Films **2**, 305 (1964)
2.66 G. Hass, H.H. Schroeder, A.F. Turner: J. Opt. Soc. Am. **46**, 31 (1956)
2.67 R.N. Schmidt, K.C. Park, J.E. Janssen: Tech. Rpt., Wright-Patterson Air Force Base, ML-TDR-64-250 (1964)
2.68 R.E. Peterson, J.R. Ramsey: J. Vac. Sci. Technol. **12**, 471 (1975)
2.69 G.K. Wehner: NSF Workshop on Surface Coatings, Minneapolis, Minn., Aerospace Rpt. ATR-75(7523-02)-1, 47 (1975)
2.70 A.B. Meinel, D.B. McKenney, W.T. Beauchamp: Tech. Rpt., NSF-RANN/SE/GE-41895 (1975)

2.71 M.M. Koltun: Zh. Prikl. Spektrosk. **12**, 350 (1970)
2.72 G.L. Harding: J. Vac. Sci. Technol. **13**, 1070 (1976)
2.73 D.M. Mattox: Opt. News **2**, 12 (1976)
2.74 R.E. Hahn, B.O. Seraphin: In Phys. Thin Films, Vol. 10, ed. by G. Haas (Academic Press, New York 1978)
2.75 A. Donnadieu, B.O. Seraphin: J. Opt. Soc. Am. **68**, 292 (1978)
2.76 M.J. Rand, J.F. Roberts: J. Electrochem. Soc. **120**, 446 (1973)
2.77 R.M. Winegarner: Proc. SPIE Conf. Opt. Sol. Energy Util., San Diego, Calif. **68**, 154 (1975)
2.78 J.C.C. Fan, T.B. Reed, J.B. Goodenough: Proc. Intersoc. Energy Conversion Eng. Conf., 9th, San Francisco, Calif. (1974), p. 341
2.79 Z.M. Jarezebski, J.P. Marton: J. Electrochem. Soc. **123**, 199C, 299C, 333C (1976)
2.80 J.L. Vossen: Phys. Thin Films **9**, 16 (1977)
2.81 W.W. Molzen: J. Vac. Sci. Technol. **12**, 182 (1975)
2.82 A.V. Sheklein: Geliotekhnika **4**, 42 (1968)
2.83 N.B. Rekont, A.V. Sheklein: NTIS(AD-755250) (translated from Russian)
2.84 J.C.C. Fan, F.J. Bachner: Appl. Opt. **15**, 1012 (1976)
2.85 R.B. Goldner, H.M. Haskal: Appl. Opt. **14**, 2328 (1975)
2.86 G. Redaelli: Appl. Opt. **15**, 1122 (1976)
2.87 H. Tabor, J. Harris, H. Weinberger, B. Doron: U.N. Conf. New Sources Energy, Rome, Pap. No. E/35/S/46 (1961)
2.88 E.A. Christie: Int. Sol. Energy Soc. Conf., Melbourne (1970), Pap. 7/81, Aust. N.Z. Sect. ISES, Parkville, Victoria (1970)
2.89 H. Tabor: Research on Optics of Selective Surfaces, Final Rep. under U.S.A.F. Contract AF 61(052)-659 (1964)
2.90 H.C. Hottel, T.A. Unger: Sol. Energy **3**, 10 (1959)
2.91 P. Kokoropoulos, E. Salam, F. Daniels: Sol. Energy **3**, 19 (1959)
2.92 R.B. Gillette: Sol. Energy **4**, 24 (1960)
2.93 P. Kokoropoulos, M.V. Evans: Sol. Energy **8**, 69 (1964)
2.94 R.L. Lincoln, D.K. Deardorff, R. Blickensderfer: Proc. SPIE Conf. Opt. Sol. Energy Util., San Diego, Calif. **68**, 161 (1975)
2.95 D.K. Edwards, J.T. Gier, K.E. Nelson, R.D. Roddick: Sol. Energy **6**, 1 (1962)
2.96 J.H. Powers, A.G. Craig, Jr., W. King: Proc. Joint Conf. Am. Sect. Int. Sol. Energy Soc. Sol. Energy Soc. Can., Vol. 6, Cape Canaveral, Fla. (1976), p. 166
2.97 A.B. Meinel, M.P. Meinel: *Applied Solar Energy* (Addison-Wesley, Reading, Mass. 1976)
2.98 D.M. Mattox, R.R. Sowell: J. Vac. Sci. Technol. **11**, 793 (1974)
2.99 L. Melamed, G.M. Kaplan: J. Energy **1**, 100 (1977)
2.100 G.E. McDonald: Sol. Energy **17**, 119 (1975)
2.101 A. Ignatiev: Proc. Thermal Power Systems Workshop on Selective Absorber Coatings, Solar Energy Res. Inst., Golden, Color. (1977), p. 189
2.102 H. Tabor: Proc. Acad. Science (USA) **47**, 127 (1961)
2.103 H. Tabor, J. Harris, H. Weinberger, B. Doron: United Nations Conf. New Sources of Energy, Rome, Paper E/35/S/46 (1961)
2.104 H.C. Hottel, T.A. Unger: Sol. Energy **3**, 10 (1959)
2.105 P. Kokoropoulos, E. Salam, F. Daniels: Sol. Energy **3**, 19 (1959)
2.106 R.B. Gillette: Sol. Energy **4**, 24 (1960)
2.107 D.M. Mattox, R.R. Sowell: J. Vac. Sci. Technol. **11**, 793 (1974)
2.108 G.E. McDonald: NASA Technical Memorandum TMX-71730 (1974)
2.109 D.A. Williams, T.A. Lappin, J.A. Duffin: J. Eng. Power **5**, 213 (1963)
2.110 T.J. McMahon, S.N. Jasperson: Appl. Opt. **13**, 2750 (1974)
2.111 E.M. Sparrow, J.W. Ramsey, G.K. Wehner: NSF/RANN Tech. Rpt. SE/GI-34871 (1974)
2.112 G.E. McDonald, J.O. Curtis: NASA Tech. Memo. TMX-71731 (1975)
2.113 R.B. Pettit, R.R. Sowell: J. Vac. Sci. Technol. **13**, 596 (1976)
2.114 R.R. Sowell, R.B. Pettit: Proc. Thermal Power Systems Workshop on Selective Absorber Coatings, Solar Energy Res. Inst., Golden, Colo. (1971), p. 175
2.115 K.D. Masterson, B.O. Seraphin: NSF/RANN Tech. Rpt., SE/GI-36731X (1975)

2.116 D.M. Mattox: J. Vac. Sci. Technol. **13**, 127 (1976)
2.117 P.H. Smith, H.S. Gurev: Int. Conf. Metallurgical Coatings, San Francisco, Calif. (1977) (Elsevier, Lausanne 1977), p. 159
2.118 H.S. Gurev: Int. Solar Energy Soc., Orlando, Fla. (1977)
2.119 S.O. Sari, P.H. Smith, H.S. Gurev: Phys. Rev. B**15**, 4817 (1977)
2.120 K.D. Masterson: SPIE **68**, 147 (1975)
2.121 H.E. Bennett: Semi-Annual Rpt. No. 6, ARPA Order 2175, 9 (1975)
2.122 G.E. Carver, H.S. Gurev, B.O. Seraphin: J. Electrochem. Soc. **125**, 1138 (1978)
2.123 H.S. Gurev, B.O. Seraphin: Proc. 5th Int. Conf. CVD (Electrochemical Society 1975), p. 667
2.124 B.O. Seraphin, A.B. Meinel: In *Solar Energy Conversion and the Optical Properties of Solids; Optical Properties of Solids – New Developments*, ed. by B.O. Seraphin (North-Holland, Amsterdam 1976), p. 927
2.125 R.E. Hahn, B.O. Seraphin: J. Vac. Sci. Technol. **12**, 905 (1975)

3. Spectral Selectivity of Composite Materials

A. J. Sievers

With 25 Figures

The use of spectrally selective surfaces for collectors of solar energy in power generation systems has been proposed by *Tabor* [3.1], *Shaffer* [3.2], *Gier* and *Dunkle* [3.3], and others [3.4].

Maximum conversion of solar radiation to useful heat requires surface properties with a maximum total absorptivity a for solar wavelengths (0.3–2.0 µm) and a minimum emissivity ε for longer wavelengths appropriate to thermal reradiation (2–30 µm).

Tabor [3.5] had divided the kinds of selective surfaces into two classes. The first class consists of a smooth, low-emissivity, metal base covered by a thin surface layer which is visibly dark but substantially transparent in the infrared. The second class pertains to those systems which are entirely metallic so that the infrared emissivity is naturally low, but, because of color, or a finely divided structure, the absorptivity in the visible spectrum is larger.

Selective surfaces of the first class usually consist of a polished metal surface covered with a very thin layer of a black semiconductor, having a thickness around 10^{-4}–10^{-5} cm. This black outer surface is about the thickness of one wavelength of visible light and absorbs the solar radiation. In the infrared, the coating thickness is small compared to a wavelength, so it is transparent, and the emissivity is determined by the nature of the bright metallic surface under the coating. The experimental work on these coatings has been reviewed by *Daniels* [3.6]. Typically a/ε ratios between 5 and 10 have been obtained using this technique.

A somewhat different approach first used by *Edwards* et al. [3.7] is to coat the surface with a thick (100 µm) film semiconductor (Si) coating. The absorption coefficient of a semiconductor increases greatly with increasing photon energy for energies near the band gap. The film is then absorbing in the short-wavelength (visible) region and is transparent in the long wavelength (infrared) region. Recently, *Seraphin* [3.4] has obtained a/ε ratios of 10–12 using this technique.

Another contribution to this class is based on the principle of the interference filter, in particular, interference effects in metal-dielectric multilayer stacks. A particularly promising interference stack has been developed by *Schmidt* and *Park* [3.8]. Again, an a/ε ratio between 10 and 15 is obtained.

As the temperature is increased for each of the coatings described above, the a/ε ratio decreases. At 300 °C, the a/ε ratio typically decreases by a factor of two from the room temperature value. Moreover, the multilayered coatings have the additional problem that diffusion between layers occurs, eutectics form, and mechanical changes, such as roughening or flaking, have been observed.

These degradation factors will be kept to a minimum if a one-component metallic selective surface of the second class is used. Little work has been done on such systems. *Tabor* [3.5] noted that polished zinc absorbs about 55% of the solar spectrum but has a low emissivity in the infrared. *Irvine* et al. [3.9] measured the selective radiation characteristics of fine metal mesh. They found an a/ε ratio of 2.2 which was temperature independent up to 260 °C.

A selective coating of the second class which has intriguing possibilities consists of metal particles on a shiny metal substrate. It has been known for some time that small metallic particles have very different optical properties [3.10] than those of the bulk material from which the particles are made. In particular, absorption bands in the visible region of the spectrum have been identified with metallic colloids dispersed in various optically transparent media [3.11–15].

Because the Fermi temperature of the electrons in the metallic particles is much larger than the temperatures of interest and the mean free path of the electrons is determined by the particle size, the optical properties of the metallic particles remain temperature independent. A solar collector consisting of metallic particles on a metallic surface could have an a which is essentially temperature independent. In addition, if the metallic particles consist of the same metallic element as the substrate, the problems associated with chemical and mechanical change would be minimized.

Small metallic particles can be routinely produced [3.16–19] by evaporation of the bulk metal in argon gas at low pressures. The size of the particles is controlled to a large extent by the argon pressure. These factors have provided the motivation for this paper on the optical properties of composite media with particular emphasis on the spectral selectivity of small metal particle coatings on a smooth metal substrate.

Composite coatings are usually made up from metallic and insulating components. We begin this chapter by reviewing the optical properties of these substances. We show that the electromagnetic response is quite straightforward in both materials. We next review the concept of surface plasmons on metal surfaces. Although these modes are not normally electromagnetically active, surface texture or roughness changes the picture completely.

In Sect. 3.2, we describe the emissivity of a smooth metal substrate since this is the low emissivity backing material for the composite coatings. We find that the hemispherical emissivity is 4/3 times the normal emissivity for good conductors. We also show that the metal backing and the step function absorptivity behavior of the selective coatings place stringent limits on the physically accessible figure of merit for composite coatings.

The optical properties of composite coatings are derived in Sect. 3.4 using the Maxwell–Garnett theory. We show that transition-metal particulate coatings on copper substrates should produce spectrally selective surfaces.

In Sect. 3.4, we describe a little-used technique which utilizes the surface-plasmon effect. For composite surfaces in which ultrafine dielectric spheres are embedded in a copper surface, we show that the infrared spectral emissivity is less than would be calculated from the measured dc resistivity of the material.

Finally, in Sect. 3.5, we show how ideal selective surfaces could be used to transform the solar spectrum for photovoltaic applications. The state of the art of high-temperature selective surfaces is reviewed, and we conclude that current materials are still a long way from the physical limits presented in the text.

3.1 On the Separation of High- and Low-Frequency Excitations

Composites are intrinsically a many-body problem. As such, an exact treatment of the dynamics is practically impossible, even though the basic theory of solids seems well under control. Fortunately, it is possible to view much of solid-state physics in terms of certain elementary excitations that interact only weakly with one another. This view has proven to be very useful not only theoretically, but also experimentally. One starts by assuming that the crystalline state at very low temperatures is the ground state. Now, even though the crystalline solid consists of strongly interacting particles, its behavior at finite temperature and its response to small external perturbations may be described in terms of weakly interacting quasiparticles which are the first few excited states of the many-body system relative to the ground state.

In beginning our study of excitations in composite media, we note that both electrons and nuclei must contribute to the energy of a solid. Born and Oppenheimer first showed in 1927 that the great disparity in electronic and nuclear masses allows the problem of nuclear and electronic motion to be largely separated.

3.1.1 Molecules

To illustrate this disparity, let us examine the excitations associated with a diatomic molecule, albeit a simple one. Consider one electron bound to two protons which are separated by a distance d. We approximate this system by an electron in a one-dimensional box of length d. The lowest momentum state allowed for the electron is

$$p = \frac{h}{2d}, \tag{3.1}$$

so the energy associated with the electron motion is

$$E_e = \frac{p^2}{2m} = \frac{h^2}{8md^2}. \tag{3.2}$$

For d of the order of a few angstroms, the transitions between energy levels are in the uv part of the electromagnetic spectrum.

What about the vibrational energy associated with the motion of the nuclei in this molecule? If we treat the molecule as a simple harmonic oscillator, then the vibrational energy is

$$E_v = \hbar \left(\frac{K}{M}\right)^{1/2}, \tag{3.3}$$

where M is the effective mass of the molecule and K is the phenomenological spring constant. Now, to relate the vibrational energy to the electronic energy, we must estimate the magnitude of K.

For displacement along the normal coordinate direction from d to $2d$, the electronic energy change is

$$\Delta E_e = \frac{h^2}{8md^2}\left(1 - \frac{1}{4}\right) = \frac{h^2}{8md^2}\left(\frac{3}{4}\right) = \frac{3}{4}E_e.$$

This change must just be equal to the vibrational potential energy, so

$$Kd^2/2 = \frac{3}{4}E_e$$

hence

$$K = \frac{3}{2}\frac{E_e}{d^2}.$$

From (3.2), we know that

$$d^2 = \frac{h^2}{8mE_e}$$

so

$$K = \frac{12mE_e^2}{h^2}. \tag{3.4}$$

Substituting (3.4) into (3.3) we find

$$E_v = \frac{\sqrt{3}}{\pi} E_e \left(\frac{m}{M}\right)^{1/2}. \tag{3.5}$$

The vibrational energy is typically a factor 60 smaller than the electronic energy.

In a treatment of a more general system, we can use this notion to treat the Hamiltonian associated with nuclear motion as a small perturbation on the electron Hamiltonian. This separation of the problem is physically reasonable. The light electron orbits around so rapidly compared with the motion of the heavy nuclei that we can approximately solve for the electron states considering the nuclei as fixed. We then include the energy of these electron states in the effective potential for the quasi-static nuclear motion. To summarize, we must first solve the electronic problem for a given internuclear distance, next find the equilibrium internuclear distance (so we have the minimum ground state energy), and finally look at excitations above this ground state.

3.1.2 Metals

Since we are interested in the solid state, let us not focus our attention on the total energy, but, instead, only on the amount the energy of the solid differs from the energy of a system of free atoms. This difference is usually called the cohesive energy. For a stable solid,

$$E(\text{solid atom}) < E(\text{free atom})$$

at the same temperature. If one can be content with considering the difference between solids and free atoms, the problem can be recast in a simple form which takes into account the fact that some electrons have practically the same energy per atom in the solid as in the free atom.

Consider sodium as an example. The ground-state configuration of a normal sodium atom is

$$1s^2\, 2s^2\, 2p^6\, 3s,$$

that is, closed K and L shells and a $3s$ electron left over. The radial charge densities in the various shells for the free atom are given in Fig. 3.1a. Also given in the figure is r_s, the radius whose associated volume is the mean volume per sodium atom in the solid. The experimental evidence indicates that the K and L shells are confined to a radius less than 1 Å. Since core electrons have energies much greater than the vaporization energy, it is reasonable to suppose that when a cloud of free atoms condense to form a solid, the K and L shells are not altered significantly. We can better categorize the electronic states of the metal by picturing the electronic states associated with N atoms, each separated by a large distance from its neighbor. These degenerate electron levels spread out into bands as the distance between atoms decreases, as shown in Fig. 3.1b. The outermost $3s$ electron only half fills the $3s$ band conduction band. Cohesion, then, must be due principally to a redistribution of conduction electrons. We should no longer regard conduction electrons as being associated with a particular atom.

The main difficulty in solving the next problem is due to the electron-electron repulsion. Experimentally, it has been observed that the one-electron

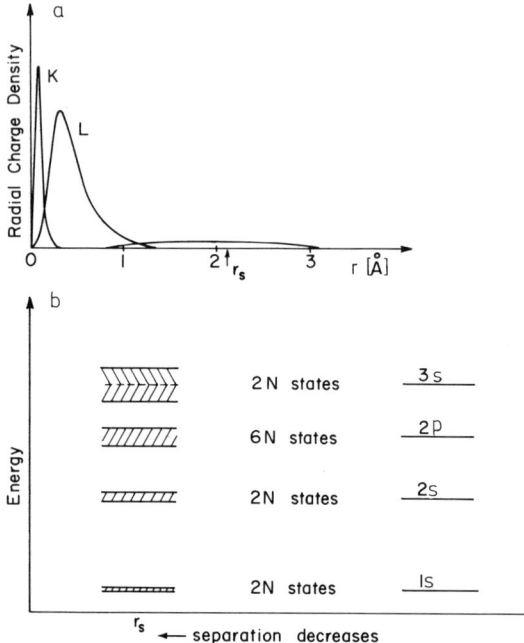

Fig. 3.1a. Radial charge densities of the K and L shells and of the $3s$ electron in a free sodium atom. **b** Schematic representation of the appearance of energy bands as N atoms are brought close to each other. For sodium, the $3s$ band is only half full

approximation is a very successful first approximation for the conduction band; in other words, each electron can be represented by its own wave function as it is a function only of the coordinates of that electron.

How can one take proper account of the Coulomb correlation between electrons when, experimentally, it is better to ignore all correlations of the conduction electrons than to treat some of them? The principal difficulty, of course, arises because the Coulomb interaction of the electrons is so strong that its effect cannot be calculated by perturbation theory.

During the past 20 years, an explanation of the interaction of electrons in metals has been developed which offers a simple justification of the independent electron approximation. This progress stems from the work of *Bohm* and *Pines* [3.20] on the theory of plasma oscillations.

The fundamental idea is to note that plasma oscillations can occur in a metal just as in an ionized gas. The oscillation frequency can be derived simply. We simulate a block of sodium by n electrons per cm^3, together with a positive charge background of infinite mass, but equal charge density. In the long wavelength limit, a shift of n electrons by a vector r with respect to the position background produces a polarization P where

$$P = ner. \tag{3.6}$$

Now any vector field can be separated into its solenoidal, P_t, and irrotational, P_l, parts, so that

$$P = P_t + P_l, \tag{3.7}$$

where

$$\mathbf{\nabla} \cdot \mathbf{P}_t = 0 \quad \text{and} \quad \mathbf{\nabla} \times \mathbf{P}_l = 0. \tag{3.8}$$

The fields and polarizations are related by Maxwell's equations for solenoidal fields in a neutral material with $\mu = 1$:

$$\mathbf{\nabla} \cdot \mathbf{E}_t = 0$$
$$\mathbf{\nabla} \cdot \mathbf{B}_t = 0$$
$$\mathbf{\nabla} \times \mathbf{E}_t + \frac{1}{c}\frac{\partial \mathbf{B}_t}{\partial t} = 0,$$

and

$$\mathbf{\nabla} \times \mathbf{B}_t - \frac{1}{c}\frac{\partial \mathbf{E}_t}{\partial t} - \frac{4\pi}{c}\frac{\partial \mathbf{P}_t}{\partial t} = 0 \tag{3.9}$$

while, for irrotational fields,

$$\mathbf{\nabla} \cdot \mathbf{E}_l = 4\pi \mathbf{\nabla} \cdot \mathbf{P}_l$$
$$\mathbf{\nabla} \cdot \mathbf{B}_l = 0$$
$$\frac{\partial \mathbf{B}_l}{\partial t} = 0$$
$$\frac{\partial \mathbf{E}_l}{\partial t} = -4\pi \frac{\partial \mathbf{P}_l}{\partial t}. \tag{3.10}$$

Since the Coulomb field is irrotational, we examine this second set of equations more closely. The last relation gives us a connection between the longitudinal field and the longitudinal polarization. If we neglect damping, then by Newton's second law,

$$Nm\ddot{\mathbf{r}}_l = Ne\mathbf{E}_l$$

so

$$N\dot{\mathbf{v}}_l = \frac{Ne}{m}\mathbf{E}_l. \tag{3.11}$$

With $\mathbf{E}_l, \mathbf{v}_l \sim \exp(i\omega t)$, then

$$eN\mathbf{v}_l = (Ne^2/i\omega m)\mathbf{E}_l = \frac{\partial \mathbf{P}_l}{\partial t}. \tag{3.12}$$

Substituting this expression into (3.10),

$$i\omega E_l = -4\pi \frac{Ne^2}{i\omega m} E_l$$

$$\omega_l^2 = \omega_p^2 = \frac{4\pi Ne^2}{m}. \tag{3.13}$$

Note that, for a metal, $\hbar\omega_p \sim$ many $eV \gg kT \sim (1/40)\,eV$. The *Bohm* and *Pines* result is that the long wavelength Fourier components of the electron density do oscillate at the plasma frequency. Since the excitation energy of a plasma is very much larger than the thermal energy which any electron is able to give up at normal temperature, the plasmons remain in their ground state, and this degree of freedom is not normally excited. In other words, plasmons take no active part in thermal excitations. Now the long range part of the Coulomb interaction is responsible for plasmons. If one can ignore plasmons when treating many transport processes, then presumably one can also ignore the long range part of the Coulomb interaction in the corresponding electron-electron interaction.

Bohm and *Pines* showed that the remaining electron-electron interaction has an effective range of about 1 Å, which is so small that it can be neglected as well. The resultant simple picture of a metal is then of an electron gas of non-interacting particles which obey Fermi statistics.

Electromagnetic waves interact with this medium in a very characteristic manner. From (3.9) the wave equation is

$$-\nabla^2 \mathbf{E}_t + \frac{1}{c^2}\frac{\partial^2 \mathbf{E}_t}{\partial t^2} - \frac{4\pi}{c^2}\frac{\partial^2 \mathbf{P}_t}{\partial t^2} = 0 \tag{3.14}$$

with

$$Nm\ddot{\mathbf{r}}_t = Ne\mathbf{E}_t$$

so

$$\nabla^2 \mathbf{E}_t + \frac{\omega^2}{c^2}\left(1 - \frac{\omega_p^2}{\omega^2}\right)\mathbf{E}_t = 0. \tag{3.15}$$

Solutions of the form $\mathbf{E}_t \sim \exp[i(\omega t - \mathbf{k}\cdot\mathbf{r})]$ gives the dispersion relation

$$k^2 = \frac{\omega^2}{c^2}\varepsilon_m, \tag{3.16}$$

where

$$\varepsilon_m = 1 - \omega_p^2/\omega^2. \tag{3.17}$$

For $\omega < \omega_p$, the EM wave does not propagate through this medium. It is this behavior which we shall use to produce a low emissivity substrate.

3.1.3 Insulators

What about the low-frequency properties of insulators? We can categorize the electronic states of an insulator in the same way as for sodium metal. Again, we rely on Fig. 3.1b; however, in contrast to the sodium case, we now have enough electrons available so that the most loosely bound band is completely full and electronic conduction cannot occur because the next empty band is several electronvolts away (many kT). Cohesion is now due to a redistribution of valence electrons. The electronic degrees of freedom drop out of the analysis of the low-frequency properties of insulators. We now turn to consider one of the remaining degrees of freedom which does influence the low-frequency properties of solids, namely, the ion cores.

The interaction of electromagnetic radiation with lattice vibrations can be cast in the same form as the plasma oscillation problem treated earlier. If we assume a diatomic lattice of positive and negative ions such as NaCl, then the polarization produced by the relative displacement of unlike charges in the long wavelength limit is

$$P = Ner. \tag{3.18}$$

Again, the irrotational polarization and corresponding electric field are related by (3.10) but now the positive and negative charges have a finite mass and there is a natural restoring force, $M\omega_0^2 r$, where M is the reduced mass of the two ions. Newton's second law now gives

$$NM\ddot{r}_l + NM\omega_0^2 r_l = NeE_l; \tag{3.19}$$

hence

$$P_l = \frac{(Ne^2/M)E_l}{\omega_0^2 - \omega^2} \tag{3.20}$$

and

$$\omega_l^2 = \omega_0^2 + \omega_p^2. \tag{3.21}$$

These excitations, of course, will be thermally excited at normal temperatures, hence the long range part of the Coulomb interaction cannot be ignored as we were able to do in metals.

Equation (3.14) again describes the interaction of electromagnetic waves with this medium. The transverse polarization now has the form of (3.20), so the

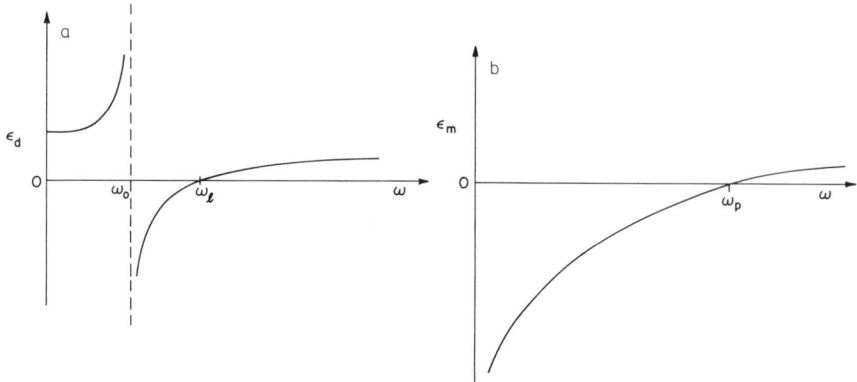

Fig. 3.2a. The real part of the dielectric function ε_d for a diatomic ionic lattice. Since damping has been ignored, the imaginary part would consist of a Dirac delta function at ω_0. (**b**) The real part of the dielectric function, ε_m, for a simple metal. Again, without damping, the imaginary part would consist of a delta function at zero frequency

dispersion curve is again given by (3.16), but with a different dielectric function, namely,

$$\varepsilon = 1 + \omega_p^2/(\omega_0^2 - \omega^2) \tag{3.22}$$

or

$$\varepsilon = (\omega_l^2 - \omega^2)/(\omega_0^2 - \omega^2). \tag{3.23}$$

For $\omega_0 < \omega < \omega_l$, the EM wave does not propagate through the medium, a result which is similar to what we found for a metal. The dielectric function of a simple metal and a diatomic insulator are sketched as a function of frequency in Fig. 3.2a, b. Aside from the scale, the main difference in the response of the two systems to EM waves stems from the absence of a transverse restoring force for the electrons in a metal. We set the scale for the two kinds of systems in Fig. 3.3. Here energies are given in meV, in μm, and in wavenumbers. The top curve illustrates the low temperature absorption coefficient of KCl versus frequency. The main loss mechanism at low frequencies is by means of the long-wavelength transverse optic lattice mode at about 20 meV. Electronic transitions begin at about 10 eV. The lower curve shows the absorptivity of a free electronlike metal on the same frequency interval. At low frequencies, the electrons effectively screen the EM wave from the interior of the medium because there is no transverse restoring force associated with their motion. Interband transitions associated with electronic excited states begin at about 2 eV. The difference between the insulator and the metal is most pronounced at low frequencies.

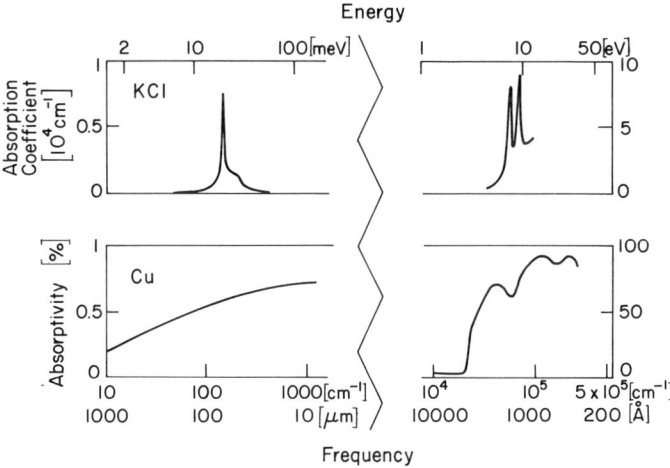

Fig. 3.3. Absorption spectrum for insulators and metals. Top curve: Absorption coefficient versus frequency for KCl. The intrinsic lattice absorption occurs at roughly 20 meV, while the intrinsic electron absorption begins at about 10 eV. Lower curve: Absorptivity versus frequency for Cu. At low frequencies, the infrared radiation is almost completely reflected by the free electronlike metal. Strong absorptions associated with interband transitions begin at a few eV

3.1.4 Surfaces

We have found that the electromagnetic response of a metallic or dielectric medium can be described by the simple dielectric function, ε, given by (3.23), namely,

$$\varepsilon = (\omega_l^2 - \omega^2)/(\omega_0^2 - \omega^2).$$

For a free electron metal, $\omega_0 = 0$; while for the ionic lattice, ω_0 is finite.

If a boundary is now introduced, (3.13, 16, 21, 23) no longer describe all of the possible solutions to Maxwell's equations. To see this most easily, we combine the equation of motion for the charged particles, (3.20), where

$$\boldsymbol{P} = \frac{(Ne^2/m)\boldsymbol{E}}{\omega_0^2 - \omega^2}$$

together with Maxwell's equations in the long-wavelength limit, so

$$\varepsilon \boldsymbol{\nabla} \cdot \boldsymbol{E} = 0$$
$$\boldsymbol{\nabla} \times \boldsymbol{E} = 0.$$

From these four equations, we obtain

$$(\omega_l^2 - \omega^2)\boldsymbol{\nabla} \cdot \boldsymbol{P} = 0 \tag{3.24}$$

and

$$(\omega_0^2 - \omega^2)\nabla \times \mathbf{P} = 0. \tag{3.25}$$

Setting $\nabla \times \mathbf{P} = 0$ and $\omega^2 = \omega_l^2$ describes the longitudinal or irrotational motion, while setting $\nabla \cdot \mathbf{P} = 0$ and $\omega^2 = \omega_0^2$ gives us again the transverse or solenoidal motion. There is another solution when $\nabla \cdot \mathbf{P} = 0$ and $\nabla \times \mathbf{P} = 0$. For infinite media, this gives the trivial result that $\mathbf{P} = 0$. However, for finite media $\mathbf{P} \neq 0$ and the polarization is associated with surface modes. The surface mode frequencies can be determined by matching the fields at the boundaries. As with bulk waves, one requires that longitudinal $\mathbf{E}$ and normal $\mathbf{D}$ be continuous at the boundaries.

As an example, consider a plane-bounded electron gas where ε_d is the dielectric constant of the dielectric and ε is the dielectric constant of the electron gas.

$$\nabla \cdot \mathbf{D} = 0$$

implies that at the boundary

$$\varepsilon_d \mathbf{E}_+ \cdot \mathbf{n} = \varepsilon \mathbf{E}_- \cdot \mathbf{n},$$

where $\mathbf{n}$ is the unit normal to the surface and we have assumed that the surface does not contain any external surface charge. Since the electric field arises only from polarization charges on the boundary, then by symmetry [3.21]

$$\mathbf{E}_+ \cdot \mathbf{n} = -\mathbf{E}_- \cdot \mathbf{n}$$

and

$$\varepsilon = -\varepsilon_d. \tag{3.26}$$

The interface has naturally occurring modes if ε has sufficient dispersion, so that this equation is satisfied at some frequency. If we neglect dampling, then the surface mode frequency is

$$\omega_s = \sqrt{\frac{\omega_l^2 + \varepsilon_d \omega_0^2}{(\varepsilon_d + 1)}}. \tag{3.27}$$

For a metal surface, this expression simplifies to

$$\omega_0 = 0$$

and

$$\omega_s = \frac{\omega_p}{\sqrt{\varepsilon_d + 1}}. \tag{3.28}$$

Ruppin [3.22] has shown in general for any finite solid that there exists a series of surface modes whose frequencies are given by equations of the form:

$$\varepsilon(\omega_n) = -\alpha_n \quad n = 1, 2, 3, \ldots, \tag{3.29}$$

where the α_n are positive and depend on the shape of the specimen. These surface mode frequencies always occur between ω_l and ω_0 for lattice modes and between ω_p and 0 for the electronic modes.

To derive the general response function valid in the region where retardation effects are important, one must use inhomogeneous waves [3.23] for the fields, so that

$$\boldsymbol{E} = \boldsymbol{E}_0 \exp(-\boldsymbol{a} \cdot \boldsymbol{r}) \exp[\mathrm{i}(\omega t - \boldsymbol{k} \cdot \boldsymbol{r})]$$

in each of the media, and $\boldsymbol{a}$ is no longer parallel to $\boldsymbol{k}$. All of Maxwell's equations are now needed as well as the usual boundary conditions on $\boldsymbol{E}$ and $\boldsymbol{H}$. The resonance condition is

$$\frac{\omega_s^2}{c^2 k^2} = \frac{\varepsilon + 1}{\varepsilon}. \tag{3.30}$$

This expression is plotted in Fig. 3.4 for a plane-bounded electron gas [3.24]. ω_s is tangent to the line $\omega_s = ck$ as $k \to 0$ and increases monotonically to the asymptotic value $\omega_s = \omega_p/\sqrt{2}$. Thus the phase velocity ω_s/k is always less than the speed of light in vacuum. A plane homogeneous electromagnetic wave incident on the surface will not couple to this surface plasmon because both surface momentum and energy cannot be conserved simultaneously. This restriction is removed by surface structure which breaks the translational invariance of the $\boldsymbol{k}$ vector. Surface roughness is an important mechanism for coupling surface plasmons onto and off surfaces, and consequently is an important quantity in determining the optical properties of metals [3.25].

3.2 Emissivity of a Smooth Metal Substrate

3.2.1 Infrared Properties of Metals

To calculate the thermal emissivity of metals, we need to know the absorptivity of metals in the blackbody spectral region. By Kirchhoff's law, this spectral absorptivity at a given temperature is equal to the spectral emissivity at the same temperature [3.26]. This spectral emissivity is folded together with the blackbody spectrum to find the total emissivity.

The radiant spectral power emitted by a surface element of a blackbody in a given direction per unit frequency interval per unit solid angle is given by Planck's law

$$L_\omega = (1/\pi)(1/2\pi c)^2 \hbar \omega^3 / \{\exp(\hbar\omega/kT) - 1\}, \tag{3.31}$$

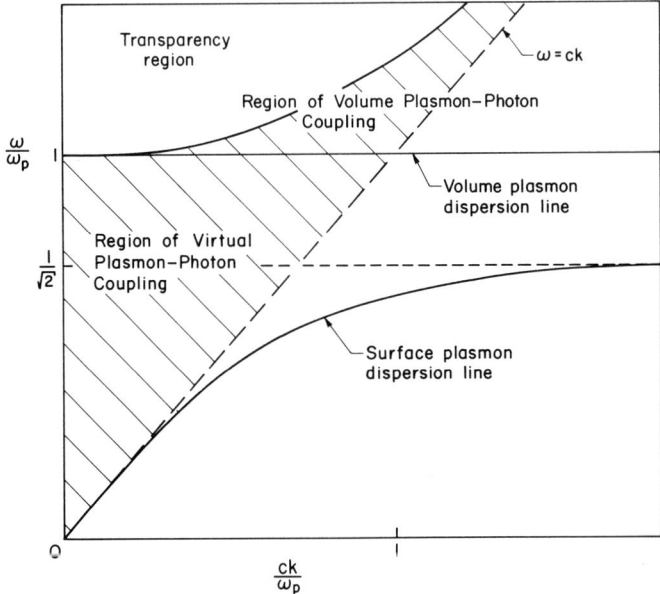

Fig. 3.4. Frequency-wavenumber space appropriate to a semi-infinite plane-bounded electron gas. The wavevector k is parallel to the bounding surface. The line labeled "volume-plasmon dispersion line" is sketched for the special case of a volume plasmon with zero momentum perpendicular to the surface. There is actually a continuum of possible $\omega-k$ values lying above this line corresponding to various volume-plasmon momenta perpendicular to the surface; this continuum is indicated by the legend "region of volume-plasmon-photon coupling" [3.24]

where ω is the angular frequency, c is the velocity of light in vacuum, $\hbar$ is Planck's constant divided by 2π, and k is Boltzmann's constant.

A characteristic frequency of this blackbody spectrum which is useful in analyzing the emissivity problem is the centroid frequency $\langle \omega \rangle$ given by

$$\langle \omega \rangle = \frac{\int_0^\infty \omega L_\omega d\omega}{\int_0^\infty L_\omega d\omega}. \tag{3.32}$$

It is convenient to normalize frequency with respect to temperature, so we define new variables

$$x = \frac{\tilde{v}}{T}, \quad m = \frac{\langle \tilde{v} \rangle}{T} \, [1/\text{cm K}], \tag{3.33}$$

where $\omega = 2\pi c \, \tilde{v}$, and $\tilde{v}$ is the frequency measured in cm^{-1}. We find using (3.32, 33)

$$\langle \tilde{v} \rangle = 2.665 \, T \quad [\text{cm}^{-1}]. \tag{3.34}$$

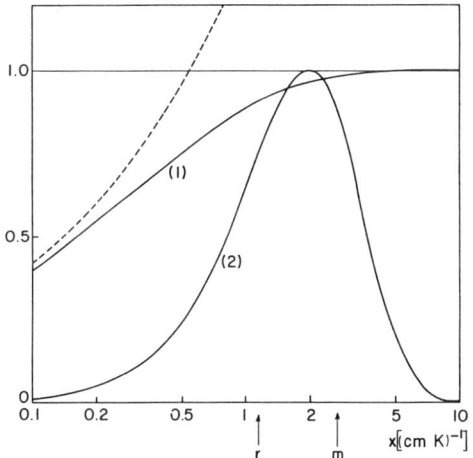

Fig. 3.5. Normalized black body and metal absorptivity functions used in the text. Curve 2 gives the normalized spectral distribution of the blackbody where m is the centroid value. The normalized spectral distribution of $g(x,r)$ for a particular value of $r(1.14)$ is given by curve *1*. The dashed line gives the Hagen–Rubens result for the same relaxation frequency

The centroid of a room temperature blackbody spectrum is at 800 cm^{-1}. The normalized spectral distribution of the blackbody spectrum is given by curve 2 in Fig. 3.5. Most of the spectral weight associated with curve 2 comes from a decade of x values centered around m. The scaling of all frequencies with temperature dictates that the infrared properties of metals, not the far infrared or optical properties, will play the dominant role in determining the high temperature emissivities of metals.

a) Surface Impedance

The concept of surface impedance is often used to describe the interaction of electromagnetic radiation with a metal when the frequency of the radiation is much less than the plasma frequency [3.27]. The usual definition of surface impedance ignores the displacement current and uses only the conduction current. At normal incidence, $Z(\omega)$ is defined as

$$Z(\omega) = R(\omega) + iX(\omega) = E_0 / \int_0^\infty j(z)dz, \quad (3.35)$$

where E_0 is the parallel component of the electric field at the metal surface (which is perpendicular to the z axis) and $j(z)$ is the current density at a depth z beneath the surface.

The electromagnetic fields within a metal are described by Maxwell's equations together with an equation relating the electric field and the current density. For cases in which the electromagnetic field (of skin depth δ) does not vary appreciably over the range of an electron mean free path ($l = v_F \tau$), or the distance an electron travels during one radian of the radiation (v_F/ω), a local conductivity $\sigma(\omega)$ can be defined where

$$j(z, \omega) = \sigma(\omega) E(z, \omega)$$

and $j(z, \omega)$ and $E(z, \omega)$ are the current density and electric field at the position z and the frequency ω. We often have for pure metals that $\delta \ll v_F/\omega$ (anomalous skin effect), or that $v_F/\omega \ll \delta \ll l$ (surface-scattering effect), and a nonlocal relationship between the current density and electric field must be used. Very complicated wave forms occur for the fields inside the metal. The surface-impedance concept permits a reformulation of the problem in terms of a boundary condition on the fields at the metal surface.

The boundary condition is easily obtained from Maxwell's equations and the definition of surface impedance. Because of the rapid attenuation of the fields within the metal, it is necessary to retain only the drivatives normal to the surface in Maxwell's equations. With this restriction, the fields are tangential to the metal surface. Integrating the induction equation, we find

$$E_t(0) = \frac{Z}{Z_0} H_t(0) \times n, \tag{3.36}$$

where the subscript t signifies tangential fields, Z_0 is the impedance of free space, and n is the normal to the metal. The coefficient Z/Z_0 in (3.36) is the only quantity characterizing the metal which must be known in order to find the external electromagnetic field. This boundary condition is valid even though the field inside the metal may not be described by the usual macroscopic Maxwell equations [3.28]. This surface impedance concept should be contrasted with the corresponding description of the absorptivity of a metal in terms of a complex index of refraction. This latter description is only reasonable when the fields are attenuated exponentially within the metal [3.29].

The reflectivity ϱ and absorptivity α of electromagnetic radiation normally incident on a metal surface with surface impedance Z are given by

$$\varrho = \left|\frac{1-Z/Z_0}{1+Z/Z_0}\right|^2 \tag{3.37}$$

and

$$\alpha = 1 - \varrho = \frac{4R/Z_0}{(1+R/Z_0)^2 + (X/Z_0)^2}. \tag{3.38}$$

For most metals, $R/Z_0, X/Z_0 \ll 1$, so that the absorptivity reduces to

$$\alpha \cong 4R/Z_0 \tag{3.39}$$

which by Kirchhoff's law, is the spectral emissivity (3.26).

b) Classical-Skin-Effect Limit

In this domain the electron mean free path is much less than the skin depth, so the conductivity can be treated as local. The Drude model conductivity is

$$\sigma(\omega) = \sigma_1 + i\sigma_2 = \frac{\sigma_0}{1+i\omega\tau}, \tag{3.40}$$

where $\sigma_0 = \omega_p^2 \tau/4\pi$ and τ is the relaxation time. The corresponding surface impedance is

$$\frac{Z(\omega)}{Z_0} = (1+i)\left(\frac{\omega}{2\omega_p} \cdot \frac{1}{\omega_p \tau}\right)^{1/2} (1+i\omega\tau)^{1/2} \tag{3.41}$$

so that

$$\frac{R}{Z_0} = \frac{1}{2\omega_p \tau}(2\omega\tau)^{1/2}\left[(1+\omega^2\tau^2)^{1/2} - \omega\tau\right]^{1/2} \tag{3.42}$$

and

$$\frac{X}{Z_0} = \frac{1}{2\omega_p \tau}(2\omega\tau)^{1/2}\left[(1+\omega^2\tau^2)^{1/2} + \omega\tau\right]^{1/2}. \tag{3.43}$$

By (3.39) the absorptivity is directly related to the real part of the surface impedance, hence

$$\alpha = \left(\frac{2}{\omega_p \tau}\right)^{1/2} \left[(1+\omega^2\tau^2)^{1/2} - \omega\tau\right]^{1/2}. \tag{3.44}$$

In the nonrelaxation region where $\omega\tau \ll 1$, the absorptivity is

$$\alpha = \left(\frac{2}{\omega_p \tau}\right)(2\omega\tau)^{1/2} \tag{3.45}$$

which is the Hagen–Rubens result. In the extreme relaxation region where $\omega\tau \gg 1$, the absorptivity is

$$\alpha = \frac{2}{\omega_p \tau} \tag{3.46}$$

which is the Mott–Zener result.

For copper at room temperature, $\langle\omega\rangle\tau = 3.7$ and the infrared spectral absorptivity and emissivity are characterized to a first approximation by (3.46). At high temperatures, the resistivity of pure metals is dominated by electron-phonon scattering [3.30]. For temperatures large compared to the characteristic Debye temperature Θ, of the solid, the number of phonons scattered is proportional to T, so

$$1/\tau \sim T \quad \text{for} \quad T > (2/3)\Theta.$$

Since $\langle\omega\rangle$ is proportional to T, and τ is inversely proportional to T, then, to a good approximation, $\langle\omega\rangle\tau = 3.7$ for copper at all temperatures.

To take advantage of this temperature-invariant behavior, we normalize the relaxation frequency by temperature in the same manner as was done for the blackbody frequencies [see (3.33)].

The normalized relaxation frequency, r, of the metal is

$$r = 1/2\pi c \tau T \quad [(\text{cm K})^{-1}]. \tag{3.47}$$

Combining (3.33) and (3.47), we find that $m/r = 3.7$ for copper.

For metals which are not characterized by the extreme relaxation region, the general expression given by (3.44) must be used to calculate the spectral emissivity. Substituting (3.47) into (3.44) we find

$$\varepsilon = (2/\omega_p \tau)(2x/r)^{1/2} \{[1+(x/r)^2]^{1/2} - x/r\}^{1/2} \tag{3.48}$$

or

$$\varepsilon = (2/\omega_p \tau) g(x, r), \tag{3.49}$$

where, so far, r is independent of frequency and temperature. The normalized frequency dependence of $g(x,r)$ in (3.49) for a particular r (namely 1.14) is represented by curve 1 in Fig. 3.5). The dashed line gives the Hagen–Rubens result for the same relaxation frequency. From the dc resistivity, the normalized relaxation frequency for Cu is $r = 0.72$. To represent this metal, curve 1 in Fig. 3.5 should be translated to the left from $r = 1.14$ to $r = 0.72$. The relative position of this curve with respect to the blackbody curve is determined simply by the position of the relaxation frequency r with respect to the blackbody centroid frequency m.

Unfortunately this simple expression for r does not work for other metals such as Pt, Pd, and Ta. For these metals, r decreases with increasing temperature. At the present time, it is not possible to predict the exact temperature dependance of r for the different metals, so we must determine r at each temperature from the dc resistivity and r_s, the radius of the free electron sphere, a quantity which is a measure of the electron density. They are related by [3.31]

$$1/\tau = \frac{4.545 \times 10^{14} \varrho_\mu(T)}{(r_s/a_0)^3}$$

in reciprocal seconds, so

$$r(T) = \frac{2411}{(r_s/a_0)^3} \frac{\varrho_\mu(T)}{T}, \tag{3.50}$$

where $\varrho_\mu(T)$ is the dc resistivity measured in $\mu\Omega$ cm, and a_0 is the Bohr radius. Values of r_s at room temperature have been tabulated for most metals [3.31, 32]. See also Appendix A [3.31] for numerical relations between τ, ϱ_μ, r_s/a_0, v_F, and ω_p.

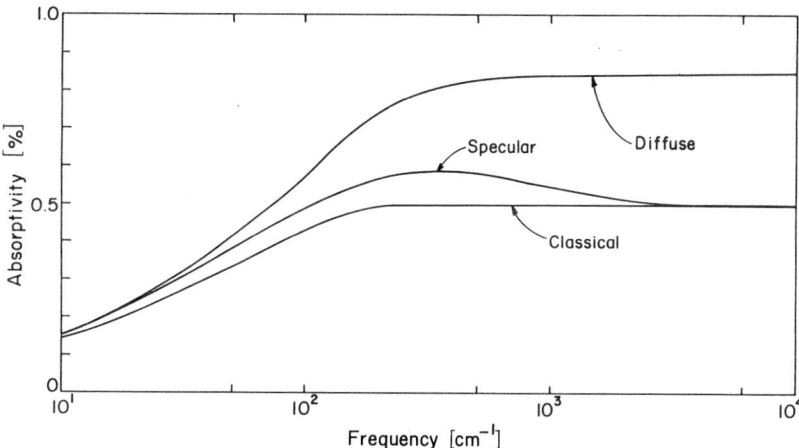

Fig. 3.6. Absorptivity of copper near room temperature. *Dingle*'s tables [3.35] have been used to estimate the diffuse and specular surface scattering limits. The classical skin effect is also shown for comparison

We complete our identification of the spectral emissivity for the Drude model by noting that

$$2/\omega_p\tau = 5.263 \times 10^{-6}(r_s/a_0)^{3/2}T \cdot r(T). \qquad (3.51)$$

c) Surface Scattering

The infrared properties of metals have been analyzed in some detail by *Reuter* and *Sondheimer* [3.33], *Holstein* [3.34], *Dingle* [3.35], *Pippard* [3.36], and others [3.37, 38]. The predicted infrared absorption is quite different depending on whether the conduction electrons colliding with the metal surface are specularly reflected or are diffusely reflected. The final expressions for the electric and magnetic fields involve complicated integrals which *Dingle* has computed and tabulated for a variety of cases. In Fig. 3.6 we present the absorptivity of copper near room temperature as obtained from *Dingle*'s tables [3.35]. Note that the diffuse absorptivity is always larger than the specular absorptivity. Also plotted is the predicted infrared absorption associated with the classical theory which we reviewed in the last section. It is clear that the classical theory is in poor agreement with the more complete *Dingle* results. The physical reason for this difference is that the classical theory does not take appropriate account of the surface. It is easiest to understand the surface absorption by focusing our attention on the region of interest, namely the high frequency relaxation region where $\omega\tau > 1$. From (3.46) the absorptivity is

$$\alpha = 2/\omega_p\tau_e,$$

where τ_e is now an effective relaxation time. We next assume that the electrons scatter diffusely from the surface but do not scatter in the bulk, so that there is one collision in the time needed to cross the electromagnetic skin depth ($\delta = c/\omega_p$) to the surface and back again. Thus

$$\tau_e \cong \frac{2c/\omega_p}{v_F}$$

so

$$\alpha_s \cong \frac{v_F}{c}.$$

A more accurate calculation of the absorptivity associated with diffuse surface scattering gives [3.34, 35]

$$\alpha_s = 3/4 \frac{v_F}{c}. \qquad (3.52)$$

We next include bulk scattering by assuming that the bulk and surface scattering mechanisms are independent. The effective relaxation time becomes

$$1/\tau_e = \sum_i (1/\tau_i), \qquad (3.53)$$

$$1/\tau_e = 1/\tau_b + (3/8)(\omega_p/c)v_F. \qquad (3.54)$$

In the region where $\omega\tau > 1$, (3.46, 54) give absorptivities for copper which are slightly larger than Dingle's exact results.

The calculated absorptivity for the specular reflection case with copper is also shown in Fig. 3.6. The absorptivity associated with specular reflection actually passes through a maximum at low frequencies [3.36]. This low-frequency anomalous-skin-effect absorption is not a surface absorption (in the sense of the one described above). When the radiation frequency $\omega < (\mathbf{k} \cdot \mathbf{v}_F)$ where $\mathbf{k}$ is the wavevector of the radiation and $\mathbf{v}_F$ the Fermi velocity vector, the electrons can absorb a photon directly and conserve both energy and momentum. In fact, the effective electrons for this process are those which do not hit the surface but "surf-ride" with the electromagnetic field through the skin depth. This anomalous-skin-effect absorption is relatively insensitive to boundary conditions (in contrast to the higher frequency case treated earlier) and it is not obvious that the absorption mechanisms are additive in this low-frequency region.

The results discussed above can be summarized with a frequency-relaxation time diagram which identifies the four different kinds of regions which we have specified [3.39, 40]. Such a diagram is given for copper in Fig. 3.7: region C is the transmission region where $\omega > \omega_p$; region A is the classical skin effect where $v_F\tau \ll \delta$ and $\omega\tau \ll 1$; region B is the classical relaxation region where $v_F\tau \ll \delta$ and

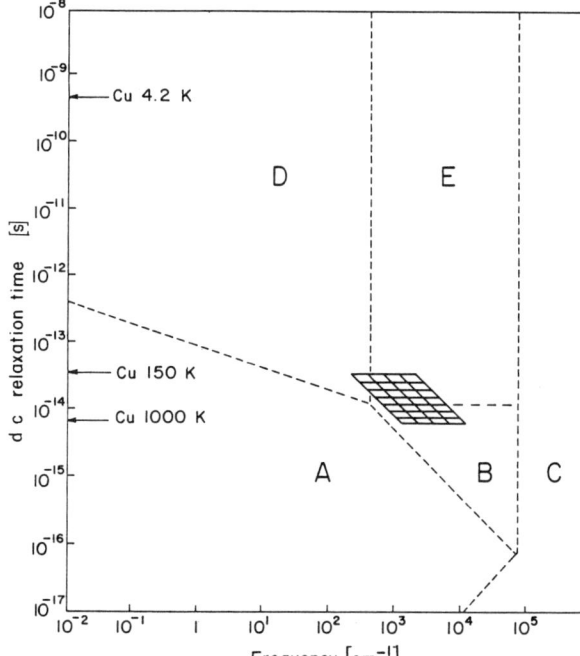

Fig. 3.7. Frequency-relaxation time diagram for copper [3.43]

$\omega\tau \gg 1$; region D the anomalous skin effect where $\delta \ll v_F\tau$ and $\delta \ll v_F/\omega$; and region E the surface scattering effect where $v_F/\omega \ll v_F\tau$. For copper at room temperature, $\langle\omega\rangle\tau > 1$ and $v_F/\langle\omega\rangle = 100\,\text{Å} < \delta = 180\,\text{Å} < v_F\tau = 210\,\text{Å}$. The cross-hatched parallelogram in Fig. 3.7 indicates the area of the frequency-relaxation time plane which contributes to the thermal emissivity of copper at high temperatures. A given horizontal line identifies the dc relaxation time at a particular temperature while the length of the line defines the most effective decade of thermal frequencies in the blackbody spectrum. Good conductors described by different dc relaxation times can be simulated to some extent by a vertical shift of the parallelogram in Fig. 3.7 to the new relaxation time. The positions of the boundaries change slightly from metal to metal since they depend on relaxation time and electron density; see [3.40] for a complete description of the boundary parameters. The diagram demonstrates that the relaxation region B, and the surface scattering region E of copper completely determine the amount of thermal radiation emitted. For metals with shorter relaxation times, the normal skin effect region A will play a secondary role. In all cases, for $T > \Theta$ the anomalous skin effect D can be neglected.

d) Electron-Phonon Scattering

Holstein has demonstrated that another relaxation mechanism which is purely a bulk process has been neglected in the classical theory [3.41]. He showed that, at 0 K, there is bulk absorption due to phonon generation by the excited electrons. *Gurzhi* [3.42] has given an explicit formula for this quantum effect which relates the effective τ_H (which includes the Holstein effect) to the dc relaxation time τ_b obtained from the dc resistivity,

$$\frac{1}{\tau_H} = \Phi(T)\frac{1}{\tau_b}, \qquad (3.55)$$

where [3.38]

$$\Phi(T) = 2\theta/5T + 4(T/\theta)^4 \int_0^{\theta/T} \frac{v^4 dv}{\exp(v) - 1}. \qquad (3.56)$$

For temperatures $T > \theta$, $\Phi(T)$ differs from 1 by less than 5%. Our final expression for the relaxation time in the diffuse scattering limit is

$$1/\tau_e = \Phi(T)/\tau_b + (3/8)(\omega_p/c)v_F. \qquad (3.57)$$

The corresponding normalized relaxation frequency is

$$r(T) = \frac{2411\varrho_\mu(T)}{(r_s/a_0)^3 T}\Phi(T) + \frac{1995}{(r_s/a_0)^{5/2}T}. \qquad (3.58)$$

A comparison of (3.58) with (3.50) indicates that the diffuse surface scattering term introduces a temperature dependence to the characteristic relaxation frequency of the metal. In general, r at high T will be smaller than r at room temperature, so that if a metal satisfies the condition $m/r > 1$ ($\langle\omega\rangle\tau > 1$) at room temperature then it is surely satisfied at high temperatures.

Equations (3.49, 51, 58) define the spectral emissivity of metals in the infrared. This emissivity depends mainly on two physical quantities, ϱ_μ and r_s. The third parameter, Θ, which determines $\Phi(T)$ plays a secondary role at room temperature. At high temperatures where $\Theta/T < 1$, $\Phi(T) \cong 1$ and the temperature dependence associated with the Holstein effect can be ignored. In any event, all three of these parameters are obtained from other than optical measurements.

3.2.2 Total Emissivities for Free-Electron Metals

a) Total Normal Emissivity

The total normal emissivity is defined by the equation

$$\varepsilon_N = \frac{\int_0^\infty \varepsilon L_\omega d\omega}{\int_0^\infty L_\omega d\omega}, \qquad (3.59)$$

where L_ω is given by (3.31) and ε by (3.38). In general, this integral is a function of $\omega_p, v_F,$ and τ (or, in our notation, r_s and ϱ_μ). But since $R/Z_0, X/Z_0 \ll 1$ for most metals, it is useful to introduce another integral, η, which approaches ε_N in this limit. We set

$$\eta = \frac{\int_0^\infty (4R/Z_0) L_\omega d\omega}{\int_0^\infty L_\omega d\omega}. \tag{3.60}$$

Substituting (3.49), the integral can be rewritten as

$$\eta = \frac{2}{\omega_p \tau} g_1(r), \tag{3.61}$$

where $g_1(r) = \int_0^\infty g(x,r) L_x dx \bigg/ \int_0^\infty L_x dx$ describes the spectral overlap between the two functions.

The quantities $2/\omega_p \tau$ and r are given in terms of ϱ_μ and r_s by (3.51), (3.58) respectively. The function $g_1(r)$ is graphed and tabulated in [3.43]. For $r < 2$, $g_1(r)$ differs from 1 by 10% or less. For good conductors, it is a reasonable approximation to set $g_1(r) \approx 1$. In addition, for $T > \Theta$, then $\Phi(T) = 1$ so (3.61) reduces to

$$\eta \approx \frac{1.269 \times 10^{-2}}{(r_s/a_0)^{3/2}} \varrho_\mu(T) + \frac{1.050 \times 10^{-2}}{(r_s/a_0)} \quad \text{(for } r<2, T>\Theta\text{)}. \tag{3.62}$$

Also, in this limit, $\eta \approx \varepsilon_N$, so the temperature dependence of ε_N is given by the temperature dependence of the dc resistivity.

b) Total Hemispherical Emissivity

The total hemispherical emissivity of a metal is defined as the total power radiated by the metal into a hemisphere compared with the total power radiated by a blackbody into the same solid angle, so [3.26]

$$\varepsilon_H = \frac{\int_0^\infty \int_0^{\pi/2} \varepsilon(\theta) L_\omega d(\sin^2\theta) d\omega}{\int_0^\infty L_\omega d\omega}, \tag{3.63}$$

where θ is the polar angle measured with respect to the normal to the metal, and

$$\varepsilon(\theta) = [\varepsilon_{TM}(\theta) + \varepsilon_{TE}(\theta)]/2 \tag{3.64}$$

is the emissivity for unpolarized radiation. $\varepsilon_{TM}(\theta)$ is the emissivity for plane polarized radiation with the plane of polarization parallel to the plane of incidence (transverse magnetic), and $\varepsilon_{TE}(\theta)$ is the emissivity for plane polarized radiation with the plane of polarization normal to the plane of incidence (transverse electric). The angular dependence of the emissivities is obtained from the Fresnel equations [3.44]. We find for metals where $R/Z_0, X/Z_0 \ll 1$ that

$$\varepsilon_{TM}(\theta) = \frac{(4R/Z_0)\cos\theta}{\cos^2\theta + (2R/Z_0)\cos\theta + (R^2+X^2)/Z_0^2} \tag{3.65}$$

and

$$\varepsilon_{TE}(\theta) = \frac{(4R/Z_0)\cos\theta}{1 + (2R/Z_0)\cos\theta}. \tag{3.66}$$

The angular dependence of emissivity for TM waves has a maximum at $\theta_m = \cos^{-1}(2R/Z_0)$ while the angular dependence of emissivity for TE waves has essentially a simple cosine dependence. Because of the different angular dependences, we calculate the contribution of the TM and TE waves to the total emissivity separately. We find that

$$\varepsilon_{TM}/2 = \frac{\int_0^\infty (4R/Z_0)\left(1 + (R/Z_0)\left\{(1/\zeta - \zeta)\left[\tan^{-1}\left(\frac{1+R}{\zeta R}\right) - \tan^{-1}(1/\zeta)\right] + \ln\left[\frac{R^2(1+\zeta^2)}{Z_0^2(1+2R/Z_0)}\right]\right\}\right)L_x dx}{\int_0^\infty L_x dx}, \tag{3.67}$$

where

$$\zeta^2 = \frac{[1+(x/r)^2]^{1/2} + x/r}{[1+(x/r)^2]^{1/2} - x/r},$$

and

$$\frac{\varepsilon_{TE}}{2} = \frac{\int_0^\infty (4R/3Z_0)[1 - (3/8)(4R/Z_0) + (3/20)(4R/Z_0)^2 - \ldots]L_x dx}{\int_0^\infty L_x dx}. \tag{3.68}$$

Both of these expressions simplify in the limit $r < m$ to

$$\frac{\varepsilon_{TM}}{2} \to \eta \tag{3.69}$$

and

$$\frac{\varepsilon_{TE}}{2} \to \frac{\eta}{3} \tag{3.70}$$

where η is given by (3.61). For good conductors, the hemispherical and normal emissivities are simply related. Combining (3.69) and (3.70), we have

$$\varepsilon_H = \frac{4}{3}\varepsilon_N \quad (r<m). \tag{3.71}$$

For large values of r, the simplification of (3.67) is not so obvious. The frequency dependences of the different terms in (3.67) are such that a near cancellation of some of the terms results. With a 10% loss in accuracy, we can replace (3.67) by

$$\frac{\varepsilon_{TM}}{2} = \frac{\int_0^\infty (4R/Z_0)\{1+(2R/Z_0)\ln(R/Z_0)\} L_x dx}{\int_0^\infty L_x dx}. \tag{3.72}$$

Neglecting the frequency dependence of the ln term, we find

$$\frac{\varepsilon_{TM}}{2} \approx \frac{2}{\omega_p \tau}[g_1(r) - (1/\omega_p \tau)g_2(r)\ln(\omega_p \tau)], \tag{3.73}$$

where $g_1(r)$ is defined after (3.61) and

$$g_2(r) = \frac{\int_0^\infty g^2(x,r) L_x dx}{\int_0^\infty L_x dx}.$$

$g_2(r)$ is graphed and tabulated in [3.43]. Note that $g_2(r) \approx g_1^2(r)$.

In a similar manner, (3.68) integrates to

$$\frac{\varepsilon_{TE}}{2} \approx \frac{2}{3\omega_p \tau}[g_1(r) - (3/4\omega_p \tau)g_2(r)]. \tag{3.74}$$

Together, (3.73, 74) give a good approximation to the hemispherical emissivity which is

$$\varepsilon_H \approx \frac{8}{3}\frac{g_1(r)}{\omega_p \tau} - \frac{g_2(r)}{(\omega_p \tau)^2}\left[\frac{1}{2} + 2\ln(\omega_p \tau)\right]. \tag{3.75}$$

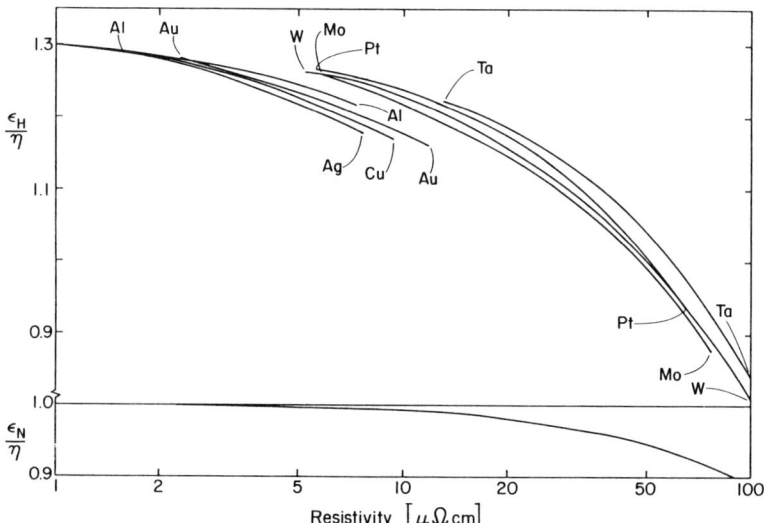

Fig. 3.8. Two emissivity ratios versus dc resistivity. ε_N is calculated from (3.59), η is calculated from (3.61), and ε_H is calculated from (3.63). [3.43]

Again, $2/\omega_p\tau$ and r are given in terms of ϱ_μ and r_s by (3.51, 58) respectively.

With (3.67, 68) we have also computed ε_H/η as a function of dc resistivity for a variety of metals[1]. The results are shown in the top part of Fig. 3.8. For $\varrho_\mu < 10\,\mu\Omega$ cm, ε_H can be set equal to $(4/3)\varepsilon_N$ with less than a 10% error. For $\varrho_\mu > 10\,\mu\Omega$ cm, a simple dependence on η does not occur [but see (3.75)]. The anomalous behavior comes about because the angle of maximum absorption for the TM waves also depends on the dc resistivity of the material. It is this dependence which produces the rapid variation in ε_H/η with dc resistivity.

In Fig. 3.9 the hemispherical emissivity for the same group of metals is plotted versus resistivity. Although the hemispherical emissivity does not vary as the first power of resistivity, a near universal dependence is still observed.

3.2.3 Spectral Selectivity Limits

Since the free-electron model gives the lowest emissivity for a given dc resistivity, it is now possible to determine a lower limit for the emissivity for any metal. The calculated hemispherical emissivities of aluminum, copper, and silver are compared with the experimental values [3.45–47] in Fig. 3.10. The calculated values for some transition elements are also shown. For these materials the measured emissivities are expected to be somewhat larger than the calculated values since we have ignored interband transitions.

---

[1] For Cu, Pt, Ag, $\varrho_\mu(T)$ was taken from [3.79], while for the rest of the elements, $\varrho_\mu(T)$ was taken from *Metals Reference Book*, Vol. III, edited by C.J. Smithells (Butterworths, London 1967). Values of r_s were taken from [3.31] or [3.32].

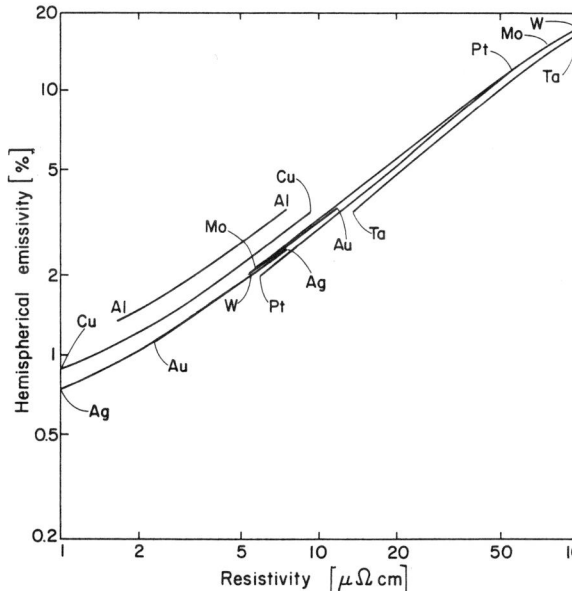

Fig. 3.9. Total hemispherical emissivity as a function of dc resistivity for a number of elements. These curves are calculated from (3.63) in the text. Since this calculation only counts the conduction electron contribution to the emissivity, these curves define lower bounds for real metals. [3.43]

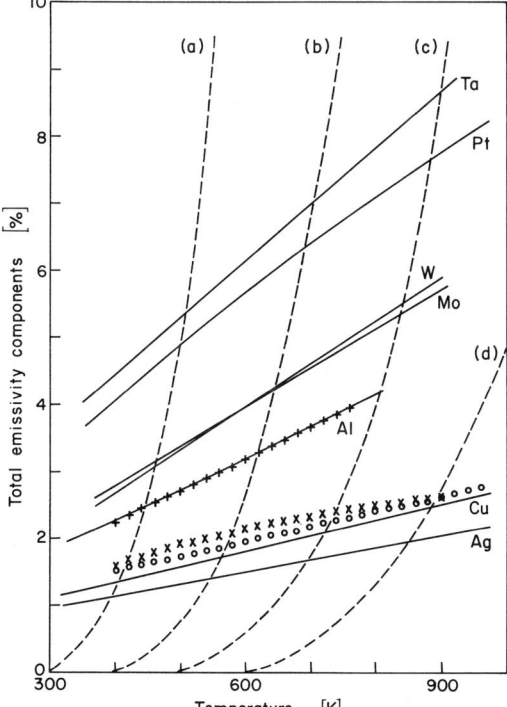

Fig. 3.10. Total emissivity components for an ideal selective surface versus temperature. The experimentally measured hemispherical emissivities are represented by discrete values: for copper ∘∘∘, for silver ×××, and for aluminum +++. The calculated hemispherical emissivity curves in the free-electron approximation are represented by the solid lines. The dotted curves indicate the contribution to the emissivity from a sharp step in the solar absorption spectrum. The edge frequency is: for (a), $3000\,\text{cm}^{-1}$; for (b), $4000\,\text{cm}^{-1}$; for (c), $5000\,\text{cm}^{-1}$; and for (d), $6000\,\text{cm}^{-1}$. The spectral shape is described by (3.76) in the text

To simulate the optical properties of an ideal selective surface, we coat the metal with a hypothetical material which does not change the infrared emissivity but causes the absorptivity to be 1 throughout the solar spectrum. The sharpest step function which satisfies this condition and still has a width of a few kT is

$$p = \exp[\hbar(\omega - \omega_0)/kT] \{\exp[\hbar(\omega - \omega_0)/kT] + 1\}^{-1}. \tag{3.76}$$

This function is 0 for $\omega \ll \omega_0$, 1/2 when $\omega = \omega_0$, and 1 when $\omega \gg \omega_0$.

The high frequency tail of the blackbody spectrum will overlap the low frequency end of p, the step function absorptivity, and produce a contribution to the emissivity of the composite structure. This contribution from the coating $\varepsilon_H(c)$ is

$$\varepsilon_H(c) = [1 - \varepsilon_H(m)] \frac{\int_0^\infty p L_\omega d\omega}{\int_0^\infty L_\omega d\omega}, \tag{3.77}$$

where $\varepsilon_H(m)$ identifies the emissivity for the bare metal. This contribution is also shown in Fig. 3.10 for different cutoff frequencies ω_0. To capture most of the solar spectrum, the cutoff should be set near $\omega_0 = 5000 \text{ cm}^{-1}$.

We have also plotted

$$[\varepsilon_H(\text{total})]^{-1} = [\varepsilon_H(c) + \varepsilon_H(m)]^{-1} \tag{3.78}$$

as a function of temperature in Fig. 3.11, since the figure of merit of selective surfaces is a/ε_H. At 900 K, the figure of merit is controlled by the emissivity of the step function edge, while at 600 K, the emissivity is controlled by the optical properties of the metal substrate itself. It is worth noting that a/ε_H must be less than 20 for temperatures above 700 K since these curves determine upper limits to the figure of merit for each substrate. If the coating does not have the ideal step function behavior, but turns on more slowly, say as a Lorentzian oscillator, then the figure of merit in the 700 K temperature region will be reduced accordingly. In this temperature region, the figure of merit is a sensitive function of the optical properties of the composite structure.

3.3 Composite Coatings

3.3.1 The Dipole Approximation

The general problem of the electromagnetic response of an isolated conductor of arbitrary size, shape, and frequency-dependent dielectric constant has not been solved. For the special case of a homogeneous sphere embedded in a

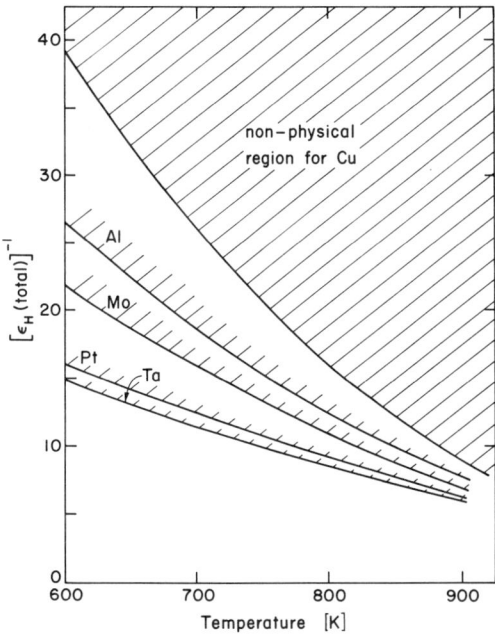

Fig. 3.11. A fundamental limit to the figure of merit versus temperature. The dc electrical resistivity of the metal substrate and the kT width of the best solar absorption edge determine a natural upper limit to the figure of merit for selective surfaces

homogeneous medium, however, the theory of *Mie* [3.48] and *Debye* (see [3.49]) is a complete description, provided the frequency dependence of the index of refraction of the material composing the sphere is known.

Physically, one imagines that the electric and magnetic fields of the incident light waves induce currents to flow within the conducting sphere. At large distances from the particles, these currents may be viewed as giving rise to oscillating electric and magnetic dipole radiation fields, electric and magnetic quadrupole radiation fields, etc. The solution to the Mie problem, therefore, takes the form of an expansion of the electromagnetic fields reradiated by the sphere in spherical harmonics, i.e., a multipole expansion in partial waves, the coefficients of which depend only on the indices of refraction of the sphere and host material and the ratio of the sphere radius to the wavelength of the incident light.

As might be expected, certain simplifications arise under the special circumstances where the sphere is much larger or much smaller than the incident wavelength. We will consider only the latter case – spheres (or in general, particles of ellipsoidal shape) small in comparison with the wavelength of radiation used. The coefficients of Mie's solution then take the form of a power series in the small quantity $kd/2$, where k is the wavevector of the light incident on the sphere (of radius $d/2$), in the medium surrounding the sphere.

A feature typical of multipole expansions of radiation fields is present in this case: as long as the conductivity of the sphere remains finite, the amplitude of the *l*th *magnetic* partial wave is comparable in magnitude to the amplitude of the $(l-1)$st *electric* partial wave. This observation is the key to the simplicity of the theory of the optical properties of small metal particles. In particular, one expects the amplitude of the magnetic *dipole* radiation field terms to be comparable to that of the electric *quadrupole* amplitude, which is smaller in order of magnitude than the electric *dipole* term (the lowest order term) by a factor of $(kd/2)^2$. Since the small particles we consider here are of size $d \sim 50$ Å, and optical wavelengths are of order 5000 Å, $(kd/2)^2 \sim 10^{-4} \ll 1$. We may, therefore, confine our attention, at least as far as solar radiation and thermal reemission are concerned, to the electric dipole part of the Mie solution.

The results can be written in terms of the scattering and absorption cross sections, Q_{sca} and Q_{abs}, which are defined by

$$\frac{I_{sca}}{I_0} = \pi \frac{d^2}{4} Q_{sca} \tag{3.79}$$

and

$$\frac{I_{abo}}{I_0} = \pi \frac{d^2}{4} Q_{abs}, \tag{3.80}$$

where I_{sca} is the total scattered intensity, I_{abs} the total absorbed intensity, and I_0 the incident intensity.

For small metal spheres such that $x = \pi d/\lambda = kd/2 \ll 1$, where d is the diameter of the sphere and λ is the wavelength in the medium, the absorption and scattering cross sections take particularly simple forms. These cross sections are [3.50]

$$Q_{sca} = \frac{8}{3} x^4 \operatorname{Re} \left\{ \frac{\varepsilon_1 - \varepsilon_0}{\varepsilon_1 + 2\varepsilon_0} \right\}^2 \tag{3.81}$$

and

$$Q_{abs} = 4 \operatorname{Im} \left\{ x \left[\frac{\varepsilon_1 - \varepsilon_0}{\varepsilon_1 + 2\varepsilon_0} \right] \right\}, \tag{3.82}$$

where ε_1 is the dielectric constant of the particle, and ε_0 is the dielectric constant of the host medium.

When the size of a metal particle is much less than the wavelength of the impinging radiation, the scattering cross section, which varies as $(kd/2)^4$, is much smaller than the absorption cross section, which varies as $kd/2$. This condition is readily satisfied in the visible region for metal particles 50–100 Å in diameter. We shall limit our study to such small particles, so scattering processes will be ignored in this paper.

To begin our investigation, let us consider an electromagnetic wave incident on a metal particle. For the very small particles which we have in mind, the E_0

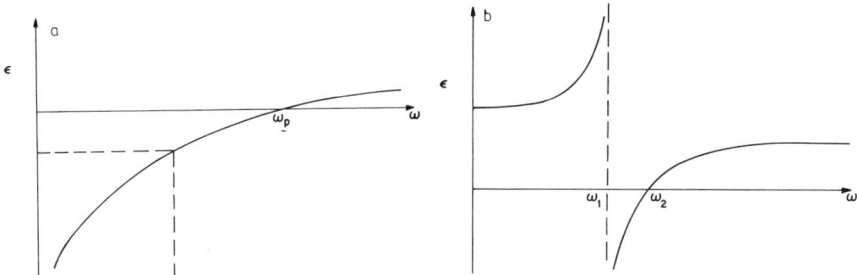

Fig. 3.12a. Real part of the dielectric function for a free electronlike metal. **(b)** Real part of the response function for a small particle constructed of this metal

field outside the particle can be taken as uniform. Let the dielectric function of the metal of which the sphere is composed by $\varepsilon(\omega)$ and the field inside the particle E_1; then by matching boundary conditions, we obtain the well-known result that

$$E_1 = \left[\frac{3}{\varepsilon(\omega)+2}\right] E_0. \tag{3.83}$$

Now in Sect. 3.1.2, we found that the real part of the dielectric function of a simple metal is negative for frequencies less than the plasma frequency. At some frequency in this region, $\varepsilon(\omega_1) = -2$ and a very large internal electric field is produced by a modest external electric field, characteristic behavior of a resonance process. The response function for a small metal particle is very different then from the response function of the bulk material from which the particle is constructed. The contrast is illustrated in Fig. 3.12. In (a), we show the real part of the dielectric function of a simple metal. In (b), we show the response function for a small particle made up of the same metal as in (a). For the bulk case (a), there is no restoring force for transverse waves, and the dielectric function has a pole at $\omega = 0$; while for a small particle, the geometric boundary itself provides the restoring force and a pole occurs at a finite frequency. In short, the boundary has changed the optical properties of free electrons into those appropriate to bound electrons.

3.3.2 Composite Dielectric Function for a Drude Metal

The electromagnetic properties of composite media have attracted considerable interest during the last century. A complete history of work in heterogeneous media in the nineteenth century has been given by *Landauer* [3.51]. He also traces the history of the so-called Maxwell–Garnett optical theory of composite media [3.52] and shows that it is another description of the Clausius–Mossotti relation [3.53]. This relation is still used to describe the optical properties of

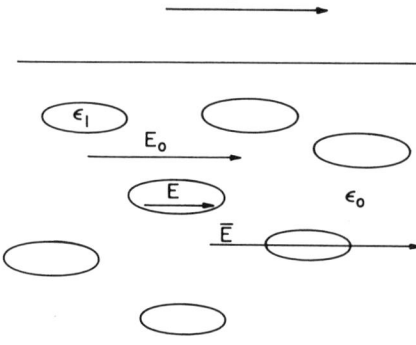

Fig. 3.13. Metallic ellipsoids with dielectric function ε, embedded in a nonabsorbing medium with dielectric constant ε_0. The electric fields inside the ellipsoids are denoted by E_1 and in the nonabsorbing medium by E_0. $\bar{E}$ is the value of the electric field averaged over a volume containing many particles

composite media although only recently has it become apparent through the work of *Smith* [3.54] and of *Lamb* et al. [3.55] why the Maxwell–Garnett results should follow. They show that, for the usual composite medium, the distribution of metal and dielectric are not topologically equivalent; thus theories which impose topological symmetry on the two components (such as the *Bruggeman* effective medium theory [3.56]) cannot hold.

A vast literature of experiment and theory has grown up around the Maxwell–Garnett expressions. The application to discontinuous films has been described in some detail by *Heavens* [3.57]. Recent studies on composite media include [3.58–76].

We now describe the optical properties of a composite medium in which one of the two components consists of ultrafine Drude metal particles and the medium is described by the Maxwell–Garnett equation. We start with a layer of nonabsorbing material with dielectric constant ε_0 which contains a volume fraction f of small absorbing ellipsoids with dielectric function $\varepsilon_1(\omega)$ as shown in Fig. 3.13.

This composite medium is polarized by an external field E. The polarization in the medium can be described in terms of an average dielectric function $\bar{\varepsilon}$. If the field inside the ellipsoids, E_1, and the field outside the ellipsoids, E_0, are nearly homogeneous, then $\bar{\varepsilon}$ is easily determined by using the Maxwell–Garnett approximation. For this case, it is necessary that the metal composite should have a density less than 25% of the density of the bulk metal. This condition comes about because the averaging process for the fields must be carried out over dimensions comparable to the wavelength of the radiation, i.e., the sample must contain many particles within λ. Therefore, for ellipsoids of revolution with axes a, a and b, and for wavelength λ,

$$f \geq a^2 b/\lambda^3 . \tag{3.84}$$

To ensure that the fields E_1 and E_0 have small inhomogeneities, the ellipsoids should be separated by a distance large compared to the mean diameter, so [3.62]

$$a^2 b/\lambda^3 \ll f \ll 0.5 . \tag{3.85}$$

The average electric field, $\bar{E}$, inside the composite medium is assumed to be a volume average of the uniform fields E_1 and E_0, so

$$\bar{E} = fE_1 + (1-f)E_0. \tag{3.86}$$

In addition, the average polarization inside the composite medium is defined as

$$4\pi\bar{P} = (\bar{\varepsilon} - 1)\bar{E}. \tag{3.87}$$

For particles with an ellipsoidal shape, the average dielectric constant is [3.77]

$$\frac{\bar{\varepsilon}}{\varepsilon_0} = \frac{\varepsilon_0(1-f)(1-L) + \varepsilon_1[(1-f)L+f]}{\varepsilon_0[1-(1-f)L] + \varepsilon_1(1-f)L}, \tag{3.88}$$

where L is the depolarizing coefficient which is a geometrical constant depending on the ratios of the axes of the ellipsoid. For any ellipsoid, $L_x + L_y + L_z = 1$. For our case, with the ellipsoid axes of revolution perpendicular to the surface, $2L_x + L_z = 1$.

For small metallic ellipsoids, the dielectric function of the bulk metal can be written as [3.62]

$$\varepsilon_1(\omega) = \varepsilon_\infty(1 - \omega_p^2/\tilde{\omega}^2), \tag{3.89}$$

where

$$\omega_p^2 = 4\pi N_e^2/m^*\varepsilon_\infty \tag{3.90}$$

and

$$\tilde{\omega}^2 = \omega^2 + i\omega/\tau. \tag{3.91}$$

The expression for $\tilde{\varepsilon}$ is obtained by inserting (3.89) into (3.88). We find that

$$\frac{\bar{\varepsilon}}{\varepsilon_0} = \frac{\bar{\varepsilon}_\infty}{\varepsilon_0}\left(\frac{\omega_2^2 - \tilde{\omega}^2}{\omega_1^2 - \tilde{\omega}^2}\right), \tag{3.92}$$

where

$$\frac{\bar{\varepsilon}_\infty}{\varepsilon_0} = \frac{\varepsilon_0(1-f)(1-L) + \varepsilon_\infty[(1-f)L+f]}{\varepsilon_0[1-(1-f)L] + \varepsilon_\infty(1-f)L}; \tag{3.93}$$

$$\omega_2^2 = \frac{\varepsilon_\infty[(1-f)L+f]\omega_p^2}{\varepsilon_0(1-f)(1-L) + \varepsilon_\infty[(1-f)L+f]} \tag{3.94}$$

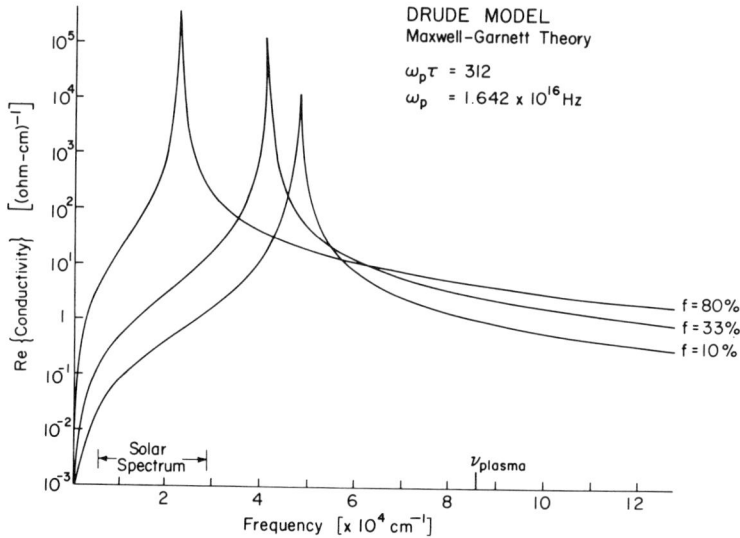

Fig. 3.14 Real conductivity for a Drude metal mixed with vacuum ($\varepsilon_0 = 1.0$) as a function of frequency for several metal sphere volume filling fractions f [3.75]

and

$$\omega_1^2 = \frac{\varepsilon_\infty(1-f)L\omega_p^2}{\varepsilon_0[1-(1-f)L]+\varepsilon_\infty(1-f)L}. \tag{3.95}$$

Note that (3.92) has a resonance when $\tilde{\omega}^2 = \omega_1^2$. For small damping, $\omega_p\tau \gg 1$ and the resonance occurs at the frequency given by (3.95). This feature is displayed in Fig. 3.12b.

The damping of the resonance is determined by the mean free path of the carriers in the bulk for particle sizes greater than 200 Å at room temperature. For smaller metallic particles ≤ 100 Å, the mean free path is determined by the particle size itself and should be essentially temperature independent. The dependence of the resonance frequency on filling fraction is shown in Fig. 3.14. For convenience, $\mathrm{Re}\{\bar{\sigma}\} = \dfrac{\omega}{4\pi}\,\mathrm{Im}\{\bar{\varepsilon}\}$ is plotted.

a) A Calculation with Spherical Particles

To estimate the frequency associated with the sphere mode we rewrite (3.95) as

$$\omega_1^2 = \frac{\omega_p^2}{1+\dfrac{\varepsilon_0}{\varepsilon_\infty}\left[\dfrac{1}{(1-f)L}-1\right]}. \tag{3.96}$$

The plasma frequency for metals occurs in the uv part of the spectrum. To shift the sphere mode to the center of the solar radiation spectrum (1.8 eV), we want a material with $\varepsilon_0/\varepsilon_\infty$ as large as possible; a filling factor, f, as large as possible; and L as small as possible. L is minimized for spherical particles where $L_x = L_y = L_z = 1/3$, so we restrict ourselves to this geometry.

The bulk plasmon frequency, ω_p, for copper occurs at 7 eV [3.78]. For $\varepsilon_0/\varepsilon_\infty = 3$, $f = 1/4$, and $L = 1/3$, we have $\omega_1^2 = \omega_p^2/10$ so $\omega_1 = 2.2$ eV. To decrease the frequency still further to $\omega_1 = 1.8$ eV (the center of the solar spectrum) would require a filling factor $f = 1/2$ in (3.95). It should be noted that our mean field calculation is not expected to be very accurate for such large f values, and it may be necessary to start with a metal with a somewhat smaller ω_p.

3.3.3 Composite Dielectric Function for Transition Metals

In this section, we calculate from the experimental optical constants of bulk metals the optical properties of composite structures consisting of submicroscopic metal spheres embedded in a dielectric medium. A variety of metallic elements is investigated. As expected, sharp sphere resonances are observed for the noble metals in agreement with the results of the last section, while near-resonant behavior is obtained for many of the transition elements in the visible part of the spectrum. To illustrate how effective this composite medium would be as a solar collector, we have also calculated the reflectivity of a layer of this composite material on a copper substrate.

The complexities of the optical properties of a real metal can be incorporated into (3.88) by noting that

$$\varepsilon_1(\omega) = (n_1 + ik_1)^2, \qquad (3.97)$$

where n_1 and k_1 are the optical constants of the metal. These quantities can be obtained for many elements from [3.79]. The relations between the optical constants of the composite medium $\bar{n}$ and $\bar{k}$ and the experimental optical constants are given in Appendix A.

The absorption coefficient of the composite medium is

$$\bar{\alpha} = (2\omega/c)\bar{k} \qquad (3.98)$$

so we first investigate the wavelength dependence of the extinction coefficient, k, in order to maximize the absorption of the solar spectrum on the one hand and minimize the infrared absorption on the other.

The extinction coefficients for small particles for Al, Cu, Au, and Co are plotted as a function of wavelength in Fig. 3.15. The spheres are embedded in a medium of dielectric constant $n_0 = 3^{1/2}$. For the noble metals, a resonance occurs in the visible part of the electromagnetic spectrum; while for Al, the resonance occurs at too small a wavelength to be useful for solar energy applications. For Co particles, no resonance is observed, although the extinction coefficient does go through a maximum and is fairly constant over the

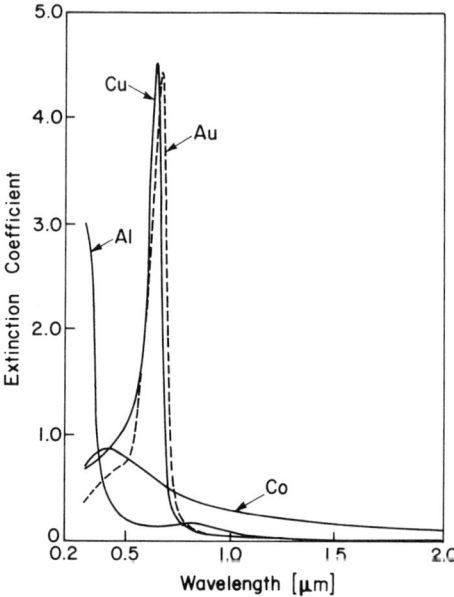

Fig. 3.15. The extinction coefficient of a composite medium containing metallic particles. The dielectric medium has an index $n_0 = \sqrt{3}$ and the particles have a filling factor of 0.45. (Calculated)

visible part of the spectrum. But to obtain the same attenuation as for copper, the composite medium containing Co would have to be about a factor of 5 thicker.

Experimental studies of the transmission spectrum of gold smokes have been made by *Harris* et al. [3.80]. Using the bulk constants of gold [3.79] and (3.98) we have been able to fit the resonance peak satisfactorily with the parameters given in Fig. 3.16. The difference in the peak widths may be due to a shorter electron relaxation time in the metal particles than the bulk value (which we have assumed) or to the particle distribution being nonuniform and consequently influencing the f factor. The decrease in the experimental transmission spectrum near 12 μm has been identified with WO_2 [3.80] which was produced during the metal particle formation and is not a property of the metallic particles. For a complete study of the optical properties of gold particles, the reader is referred to the work by *Granquist* and *Hunderi* [3.81].

3.3.4 Metallic Particulate Coatings on Copper Surfaces

To illustrate how metal particles on a copper substrate would behave optically, we have simulated such a medium on a computer. With $\bar{n}$ and $\bar{k}$ determined from the preceding section, the intensity reflected at normal incidence can be obtained from a simple variation of the standard reflectivity formula which is given in Appendix B.

For the first example, the particles are made of copper and the same relaxation time as for bulk copper is assumed. The calculated reflectivity for

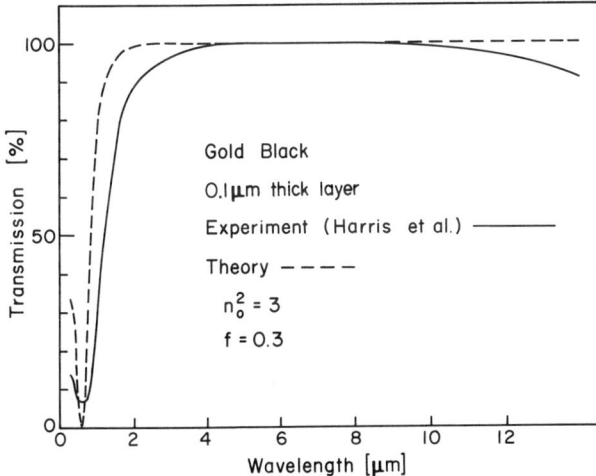

Fig. 3.16. Transmission spectrum of gold smoke. The experimental measurements on a 0.1 μm thick layer are represented by solid line and were obtained by *Harris* et al. [3.80]. The dashed line is calculated from (3.88) using $n_0 = \sqrt{3}$, $f = 0.3$ and the published optical constants of bulk gold

three different thicknesses of the composite coating is shown in Fig. 3.17a. The variation in reflectivity with thickness demonstrates two things: (1) no interference process is involved in the rapid wavelength dependence of the reflectivity at 0.6 μm, since the edge does not shift with thickness; and (2) a very thin layer ∼1000 Å produces complete absorption over the visible region of the electromagnetic spectrum for a volume fraction of 0.15 spheres. More of the solar spectrum would be absorbed than is shown in Fig. 3.17a because of the shorter relaxation times associated with small particles, but not enough to make this combination an efficient solar collector.

There are two other parameters available in the model besides the layer thickness d, the index of the medium in which the spheres are placed, n_0, and the packing fraction, f. In Fig. 3.17b, the reflectivity is given for three different values of n_0^2. Although a shoulder does develop on the reflectivity edge for the largest value of n_0^2, the edge itself remains essentially unchanged.

The largest changes in the wavelength dependence of the reflectivity are associated with changes in the filling factor, f, and these cases are shown in Fig. 3.17c. The unusual curve associated with $f = 0.45$ comes about because the absorption over a small wavelength interval is too strong, hence much of the shorter wavelength region is reflected by the layer itself.

The two parameters can be traded off against each other to some extent. From (3.96), identical resonance frequencies can be obtained for n_{01}^2, f_1 and n_{02}^2, f_2 when

$$n_{01}^2 \left(\frac{2+f_1}{1-f_1}\right) = n_{02}^2 \left(\frac{2+f_2}{1-f_2}\right). \tag{3.99}$$

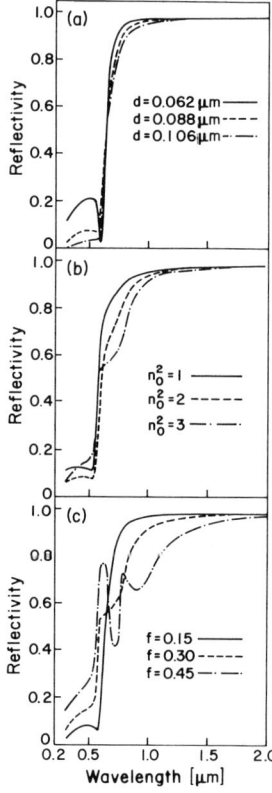

Fig. 3.17a. Reflectivity vs wavelength for copper particles on a copper substrate for different d. d is the thickness of the composite structure, $n_0 = \sqrt{3}$, and $f = 0.15$. (Calculated)

b Reflectivity vs wavelength for copper particles on a copper substrate for different n_0. n_0 is the index of refraction, $f = 0.3$, and $d = 0.088\,\mu\text{m}$. (Calculated)

c Reflectivity vs wavelength for copper particles on a copper substrate for different f. f is the filling factor, $n_0 = \sqrt{3}$, and $d = 0.88\,\mu\text{m}$. (Calculated)

But because the extinction coefficient varies as

$$\bar{k} \propto n_0^2 f/(1-f)^2, \tag{3.100}$$

then, for a given resonance frequency, increasing n_0^2 and decreasing f produces a smaller $\bar{k}$ at a given wavelength. This can be compensated for to some extent by increasing the layer thickness d.

For copper, no combination of the three parameters d, n_0^2, and f has been found which produces an efficient solar collector. This conclusion should apply to all the noble metals since the particle resonances for free-electron-like metals are much too narrow in wavelength for these to be effective absorbers for the entire solar spectrum.

The transition metal elements (which are not free-electron-like) should be better candidates for the particular application of matching to the solar spectrum with a composite layer. For example, the extinction coefficient of Co particles (shown in Fig. 3.15) has a weaker wavelength dependence in the visible, producing a better match to the solar spectrum.

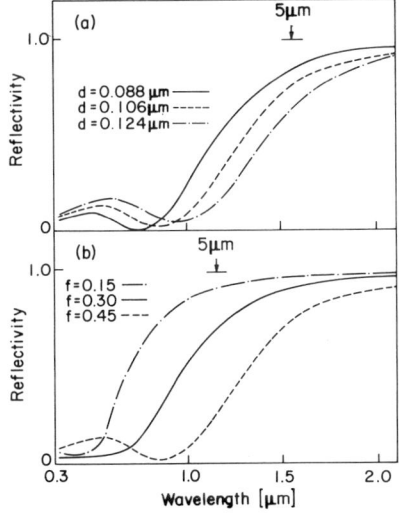

Fig. 3.18a. Reflectivity vs wavelength for nickel particles on a copper substrate for different d. $n_0 = \sqrt{3}$ and $f = 0.45$. (Calculated)
b Reflectivity vs wavelength for nickel particles on a copper substrate for different f. $n_0 = \sqrt{3}$ and $d = 0.106\,\mu m$. (Calculated)

For transition elements, we find that all three of the parameters d, n_0^2, and f markedly alter the wavelength dependence of the reflectivity of nickel particles [3.82] on copper. The reflectivity versus wavelength for three different layer thicknesses is calculated in Fig. 3.18a. As the thickness is increased, the minimum in the reflectivity moves from $0.75\,\mu m$ to $1\,\mu m$ indicating that an interference effect is present here. In addition, at shorter wavelengths the absorptivity in the layer itself is large enough to control the reflectivity of the system.

As the filling factor is increased, the near resonance moves to larger wavelengths, as shown in Fig. 3.18b. For $f = 0.3$, the surface layer provides an almost perfect match between the copper and free space. When f is increased to 0.45, the resonance becomes stronger (3.100), and moves to larger wavelengths providing a better match at $0.85\,\mu m$; however, the absorptivity of the layer is too large near $0.5\,\mu m$, hence the reflectivity increases in this region.

From these calculations, we conclude that a good match to the solar spectrum can be obtained if the extinction coefficient increases with increasing wavelength over the solar spectrum interval and then is zero for larger wavelengths. No single physical process produces such an effect, but we shall see that the composite extinction coefficient of some other transition metal elements come fairly close to this ideal behavior.

The $\bar{k}$ values of the transition elements in the fifth column of the periodic table are shown in Fig. 3.19. From these calculations, we conclude that Ta has too sharp a sphere resonance to be an efficient solar collector, while with V, no resonance is observed at all. Nb looks like a promising candidate because of the double resonant character of $\bar{k}$ in which the longer wavelength resonance is stronger than the shorter wavelength one.

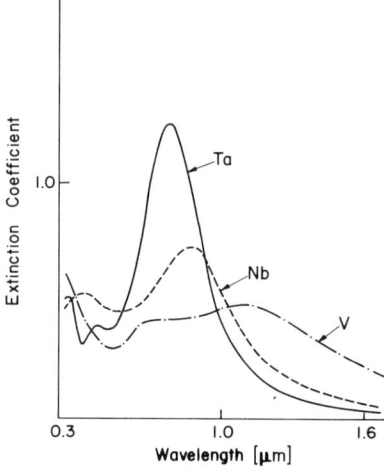

Fig. 3.19. Extinction coefficient of a composite medium containing transition metal elements. $n_0 = \sqrt{3}$ and $f = 0.45$. (Calculated)

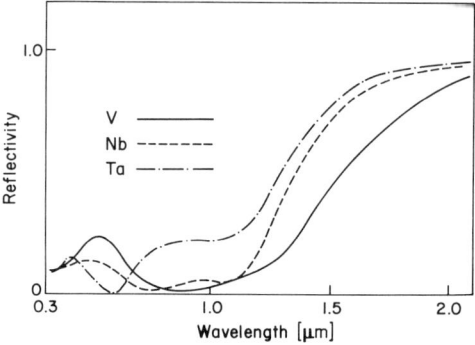

Fig. 3.20. Reflectivity vs wavelength for transition metal particles from the fifth column of the periodic table. $n_0 = \sqrt{3}$, $f = 0.45$, and $d = 0.124\,\mu\text{m}$. (Calculated)

In Fig. 3.20 the reflectivity for a Nb composite layer [3.83] is compared with one of Ta and V. As we anticipated, the Ta layer does not match particularly well because of the strong resonance near 0.8 μm. For the particular parameters given here, the resonance is much too strong and contributes to the reflectivity.

At first sight, it is surprising that the V composite layer [3.82] matches so well to the solar spectrum over such a large wavelength interval. The unfortunate reflectivity peak at 0.5 μm is due to the small extinction coefficient in this region (see Fig. 3.20). This flaw may not be as significant as it appears, since higher order sphere modes which we have neglected in our analysis and which occur in this 0.5 μm wavelength region should eliminate this problem.

In Fig. 3.21, the solar spectrum has been superimposed over the reflectivity curve for two particulate coatings on copper. The Ta parameters have been decreased, so a better match is obtained here than previously shown in Fig. 3.20. A tungsten particulate coating appears to be almost as effective as does vanadium.

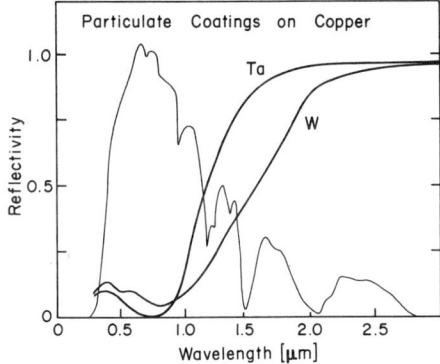

Fig. 3.21. Solar spectrum superimposed on the reflectivity of particulate metal coatings on copper. (Calculated)

Because of the shorter relaxation times associated with small particles, the experimental absorptions should extend over a somewhat larger wavelength interval than given by the computer simulation which used the experimental bulk optical constants for all elements.

3.4 Composite Metals

3.4.1 Experimental Parameters for Copper

Another way to obtain solar selectivity is to change the optical properties of the metal substrate itself. If a metal such copper could be made dark in color in the visible but still highly reflecting in the infrared, then more lenient tolerances could be permitted on the spectral selectivity of the composite coating overlayer described in Sect. 3.3.4.

Copper co-sputtered with small quantities of SiC or Al_2O_3 has been used for CO_2 laser mirror applications. This dispersion-hardened copper produces a stable fine-grained polishable material exhibiting a surface roughness of less than 15 Å rms when mechanically polished. At 10.6 μm, the percentage absorption is observed to vary monotonically with the dc electrical resistivity, but the measured absorption values are 5 to 10% smaller than those given by the Drude theory.

In this section, we analyze the optical properties of submicroscopic SiC or Al_2O_3 particles in a copper matrix by means of the Maxwell–Garnett equation. We show that the emissivity of the metal composite is less than would be calculated from the experimental dc conductivity. In addition, the reststrahl absorption associated with SiC or Al_2O_3 which occurs in the infrared does not increase the emissivity appreciably over that associated with the copper matrix itself. The largest change in the optical properties of dispersion-hardened metals is expected to occur in the visible and ultraviolet region of the electromagnetic spectrum [3.84] – a region in which these materials have not been systematically measured and a region in which the Maxwell–Garnett approach cannot be used.

We begin our analysis of the infrared properties by first identifying the experimental parameters for dispersion-hardened copper and then using these quantities to find plasmalike sphere modes associated with the dispersoid material. For a free-electron model, resonances in the uv region are found. Next, we introduce the effects of lattice vibrations into the model and calculate the absorption in the reststrahl region. Finally, a simple rule is provided for estimating the optical properties of dispersion hardened metals in the infrared.

For copper co-sputtered with 0–5 vol % SiC or Al_2O_3, *Stewart* [3.85] has found the following properties: (a) the typical diameter of a dispersoid particle in copper is less than 100 Å and may reside preferentially at grain boundaries; (b) the dc resistivity of the hardened metal varies between 2 and 11 $\mu\Omega$ cm, while the corresponding absorption at 10.6 µm varies between 0.8 and 2.5% as the volume percentage of the dispersoid is increased to 5%. From this information, the parameters appropriate for a Drude model calculation of the optical properties can be obtained. The effective relaxation time for the conduction electrons is $\tau \leq 6 \times 10^{-15}$ s (hardened copper). At 10 µm, $\omega\tau < 1$. In this limit, the penetration depth of the electromagnetic radiation is $\delta \geq 200$ Å, and the mean free path is

$$l = v_F \tau \leq 100 \text{ Å} \quad \text{(hardened copper)}.$$

For the 10 µm wavelength region, then $\omega\tau \leq 1$ and $l/\delta < 1$; the classical skin effect regime should describe the infrared properties to a first approximation.

3.4.2 Free-Electron Model Composite

To see how the optical properties of hardened copper will be changed from the properties of the matrix itself we start with an isolated dielectric sphere embedded in a metal matrix. The theory for scattering and absorption by isolated spheres has been described in Sect. 3.3.1. The scattering and absorption cross-sections are given by (3.81, 82) respectively. For the problem at hand, we set $\varepsilon_1 = \varepsilon_d$, the dielectric constant of the sphere which is embedded in a metal matrix with dielectric function $\varepsilon_0 = \varepsilon_m$. We now have the added complication that the k vector of the radiation in the medium is complex [see (3.16)] with

$$k^2(\omega) = i4\pi\omega\sigma(\omega)/c^2 \tag{3.101}$$

so

$$x = (\text{Re}\{k\} + i\,\text{Im}\{k\})\frac{d}{2}. \tag{3.102}$$

In the region of the solar spectrum $x \sim 1$, and Q_{sca} is more important than Q_{abs}. The effective medium approach will not give an accurate description of the optical properties in this frequency region.

To determine which extinction mechanism (absorption or scattering), is more important in the infrared, we investigate the ratio of Q_{sca} to Q_{abs} in the limit where $\omega\tau < 1$, $\varepsilon_{m1} \ll 0$, and $\varepsilon_{m2} \gg 1$.

In this limit,

$$\frac{Q_{sca}}{Q_{abs}} = \frac{(2/3)(d/2\delta)^4}{2[1 + 3\varepsilon_{m2}\varepsilon_d/2(\varepsilon_{m1}^2 + \varepsilon_{m2}^2)](d/2\delta)}, \qquad (3.103)$$

where $\delta = (c/2\pi\omega\sigma_0)^{1/2}$ is the classical skin depth. The absorption mechanism dominates as long as

$$Q_{sca}/Q_{abs} \ll 1$$

which, from (3.103), implies that $d \ll 3\delta$. This condition is easily satisfied for SiC hardened copper.

To estimate the dielectric function including finite concentration effects, we must calculate the average electric field over dimensions comparable with the wavelength of the radiation. In the metal, the effective wavelength is of the order of the skin depth, $\delta \gg 200$ Å. To ensure that many particles are averaged over, the fraction of volume filled by the particles should be larger than the ratio of the particle volume to the wavelength. We can now estimate an upper bound on the dispersoid size which satisfies the condition

$$f > [d/\delta]^3.$$

For $f = 0.05$, then $d < (0.05\delta)^{1/2}$ or the particle diameter $d < 70$ Å. This upper bound is consistent with the bound on the experimental dispersoid particles [3.85].

Consequently, for these small particles the average electric field inside the composite medium can be calculated as a volume average of the appropriate uniform fields in each of the two components. The general expression for the average dielectric function in this limit has already been described in Appendix A. This dielectric function has a resonance when

$$\varepsilon_m = -\frac{L(1-f)\varepsilon_d}{[1-(1-f)L]}. \qquad (3.104)$$

The resonance frequency increases with increasing f, e.g., the larger the density of the dispersoid, the larger is the frequency associated with the uniform polarization mode in the dielectric particles. But for the low concentrations we are treating, the finite concentration effect is only of minor importance.

In order to estimate the frequency associated with the particle resonance, we treat a specific model. Let us assume that the optical properties of the dielectric particle can be represented by a real dielectric constant ε_d while the dielectric function, $\varepsilon_m = \varepsilon_{m1} + i\varepsilon_{m2}$, of the copper matrix is represented by a Drude model. The real part of this function is

$$\varepsilon_{m1} = \varepsilon_{m\infty}\left(1 - \frac{\omega_p^2 \tau^2}{1 + \omega^2 \tau^2}\right) \tag{3.105}$$

and the imaginary part is

$$\varepsilon_{m2} = \varepsilon_{m\infty} \frac{(\omega_p/\omega)\omega_p \tau}{1 + \omega^2 \tau^2}, \tag{3.106}$$

where $\omega_p^2 = 4\pi N e^2/m$ is the plasma frequency.

From (3.104), the frequency ω_d associated with the dielectric sphere resonance is

$$\frac{\omega_d}{\omega_p} = \left(\frac{1-(1-f)L}{1+(1-f)L(\varepsilon_d/\varepsilon_{m\infty}-1)} - \frac{1}{\omega_p^2 \tau^2}\right)^{1/2}. \tag{3.107}$$

To estimate the frequency associated with ω_d we note that for SiC hardened copper, $\omega_p \tau \cong 100$ and $\varepsilon_d = 7$ [3.86].

If we assume $\varepsilon_{m\infty} = 2$, and that we have a low concentration of particles, (3.107) reduces to

$$\frac{\omega_d}{\omega_p} = \left(\frac{1-L}{1+2.5L}\right)^{1/2}. \tag{3.108}$$

For spherical particles $\omega_d/\omega_p \cong 0.60$; while for needles with E perpendicular to the needle axis $\omega_d/\omega_p \cong 0.5$. As $\omega_p \cong 7\,\text{eV}$ for copper, the particle resonance for SiC in copper occurs in the ultraviolet region of the electromagnetic spectrum.

3.4.3 Infrared Emissivity

In the infrared region, $\omega \sim \omega_p/100 \ll \omega_d$ and these resonances do not participate directly in the emissivity in this long wavelength region. To estimate the emissivity, we must examine the dielectric function when it contains the Drude model in the limit that $\varepsilon_{m1} \ll 0$, $\varepsilon_{m2} \gg 1$. The real and imaginary parts of the dielectric function for the composite medium are given by (C.2, 3) in Appendix

C. In the low frequency region where $\omega\tau \ll 1$, (C.1) reduces to

$$\bar{\varepsilon}_2 = \varepsilon_{m2}\left(1 - \frac{f}{[1-(1-f)L]}\right) \quad \text{in the limit } \omega \to 0.$$

For $f \ll 1$,

$$\bar{\varepsilon}_2 \cong \varepsilon_{m2}\left(1 - \frac{f}{1-L}\right). \tag{3.109}$$

In the limit of zero frequency, this quantity can be related to the dc conductivity of the composite medium, since

$$\bar{\sigma}_1(\text{dc}) = (\omega/4\pi)\bar{\varepsilon}_2 = \varepsilon_{m\infty}\frac{\omega_p^2 \tau_c}{4\pi}\left(1 - \frac{f}{1-L}\right), \tag{3.110}$$

where τ_c is the relaxation time for copper in the composite structure. It is instructive to compare (3.110) with the appropriate expression for a Drude metal, namely

$$\sigma_1(\text{dc}) = \varepsilon_{m\infty}\frac{\omega_p^2 \tau_s}{4\pi}. \tag{3.111}$$

We are now in a position to understand how the emissivity in dispersion-hardened copper can be less than that calculated from the Drude model with the experimental conductivity. In the relaxation region where $\omega\tau > 1$, the normal emissivity depends both on $\bar{\sigma}_1$ and on τ_c so that

$$\varepsilon_c = \frac{2}{\omega_p \tau_c} = \frac{2}{(4\pi\bar{\sigma}_1 \tau_c)^{1/2}}, \quad \omega\tau > 1. \tag{3.112}$$

Since the experimental conductivity is given and ω_p is assumed constant, we can relate the two emissivities ε_s and ε_c by setting (3.110) equal to (3.111). In this case

$$\tau_s = \tau_c\left(1 - \frac{f}{1-L}\right) \tag{3.113}$$

or

$$(\varepsilon_c - \varepsilon_s)/\varepsilon_s = -\frac{1}{2}\left(\frac{f}{1-L}\right) \tag{3.114}$$

so,

$$\Delta\varepsilon/\varepsilon_s < 0.$$

For spherical particles,

$$(\varepsilon_c - \varepsilon_s)/\varepsilon_s = -(3/4)f, \tag{3.115}$$

while for aligned anisotropic shapes, the difference can be even greater. We conclude that the emissivity of dispersion-hardened copper in the relaxation regime, $\omega\tau > 1$, is always less than the emissivity calculated from the experimental dc conductivity by using the Drude model and neglecting the composite nature of the material.

3.4.4 Influence of Lattice Vibrations

The transverse optic mode, ω_T, in SiC is at 800 cm^{-1}, while the longitudinal optic mode is at 970 cm^{-1} ($\sim 10\,\mu$m). The optic modes of Al$_2$O$_3$ occur at slightly larger frequencies, In order to calculate the emissivity of the composite medium in the 10 μm region, the lattice vibrations in the SiC particles must be taken account of. The frequency-dependent dielectric function which includes the lattice response can be written as [3.62]

$$\varepsilon_d(\omega) = \varepsilon_{d\infty} \frac{(\omega_L^2 - \omega^2 - i\omega\gamma)}{(\omega_T^2 - \omega^2 - i\omega\gamma)}, \tag{3.116}$$

where γ is the linewidth of the lattice resonance which occurs at $\omega = \omega_T$. At this frequency, the frequency function has an ac conductivity

$$\sigma_{d1}(\omega_T) = (\omega_T/4\pi)\varepsilon_{d2}(\omega_T) = \varepsilon_{d\infty} \frac{(\omega_L^2 - \omega_T^2)}{\gamma}. \tag{3.117}$$

For SiC, $\varepsilon_{d\infty} = 7$.

The importance of the frequency dependence of the dielectric function can be estimated by comparing the ac conductivity at the lattice resonance to the conductivity associated with the free electrons as given by the Drude model at the same frequency. The ratio of these two conductivities at the transverse optic mode frequency is

$$\frac{\sigma_{d1}(\omega_T)}{\sigma_{m1}(\omega_T)} = \frac{\varepsilon_{d\infty}[(\omega_L^2 - \omega_T^2)/\gamma]}{\varepsilon_{m\infty}\omega_p^2\tau} \simeq \frac{(\omega_L/\gamma)\,\omega_L}{(\omega_p\tau)\,\omega_p}. \tag{3.118}$$

If we now set $\omega_p\tau = \omega_L/\gamma = 100$, then

$$\frac{\sigma_{d1}(\omega_T)}{\sigma_{m1}(\omega_T)} = [m/M]^{1/2} \simeq 1/120, \tag{3.119}$$

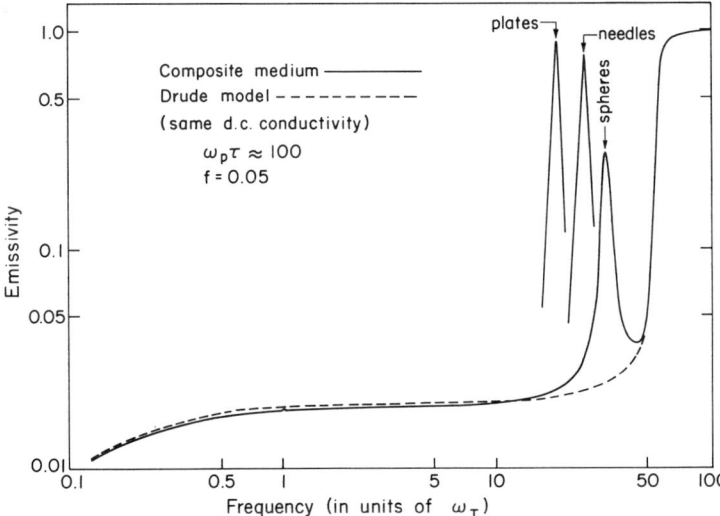

Fig. 3.22. Calculated emissivity for dispersion-hardened copper vs frequency. The calculated infrared emissivity includes the influence of lattice vibrations in the dispersion material. The parameters for the calculation described in the text are $\omega_p\tau=97.5$, the filling factor $f=0.05$, the transverse optic mode linewidth $\gamma=0.01\,\omega_T$, the longitudinal optic mode frequency $\omega_L=1.2\,\omega_T$ and $\varepsilon_{d\infty}/\varepsilon_{m\infty}=3.5$. The frequency axis is normalized in terms of the transverse optic mode frequency, ω_T, of the dielectric particles embedded in the copper. (For SiC or Al_2O_3, $\omega_T \sim 1000\,cm^{-1}$). The uv resonance at 33 on the horizontal axis is associated with uniform polarization sphere modes in the composite medium. The resonance is stronger and occurs at lower frequencies for two other particle shapes appropriate to a depolarizing factor $L=1/2$ or $L=2/3$. The dashed curve is calculated for a Drude model with the same dc conductivity

where M ($=8.42\,amu$) is the reduced mass associated with the SiC ions. Equation (3.111) shows that, even at resonance, the lattice conductivity is less than one percent of the electron conductivity. No major changes can occur in the results which were derived in the last section in which a frequency independent dielectric constant was used for the SiC particle.

Our calculations so far have placed some bounds on the frequency dependence of the emissivity of dispersion-hardened copper, but it is of some interest to have the complete frequency dependence for a few representative cases eventhough the scattering contribution is not treated properly.

We have used (3.105, 106) together with (C.2, 3) to calculate the frequency dependence of the absorptivity for spherical SiC particles. The results are shown in Fig. 3.22. The abscissa is normalized in terms of the transverse optic mode frequency $\omega_T(\sim 1000\,cm^{-1})$ and the plasma frequency ω_p ($\omega_p\tau=97.5$), with a filling factor of $f=0.05$. For comparison, the absorptivity is also shown for a Drude model with the same dc conductivity. Equation (3.113) was used to find the appropriate relaxation time. Notice that the composite absorptivity is smaller than that given by the Drude model over the entire infrared region, but

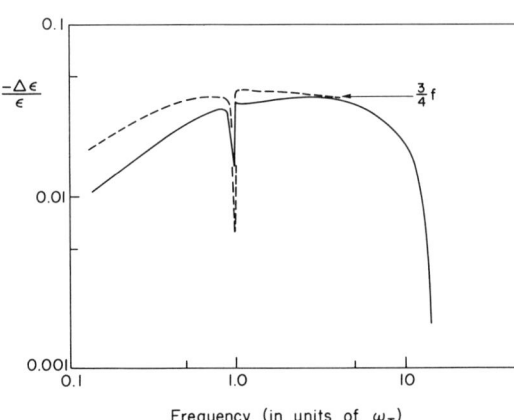

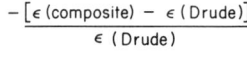

Fig. 3.23. Relative change in emissivity vs frequency. For a given dc conductivity of the composite medium, the calculated emissivity is less than that calculated using a Drude model which ignores the composite structure. The larger $\omega_p\tau$, the larger is the relative change in the emissivity at all infrared frequencies. For spherical particles in copper, the relative change in emissivity does reach a limiting value of $-(\Delta\varepsilon/\varepsilon) = 3f/4$. The rapid change at $\omega/\omega_T = 1$ is associated with lattice vibration in the dispersoid material

larger in the visible and uv parts of the electromagnetic spectrum. In particular, an additional sphere mode occurs in the uv below the bulk plasma frequency for $L = 1/3$. The depolarizing factor L has a strong influence on the exact frequency of the particle resonance as previously shown in (3.108). In Fig. 3.22, we also show the strengths and frequencies for two other particle shapes corresponding to $L = 1/2$ and $L = 2/3$. Although we can identify resonance, the absorptivity as given by Fig. 3.22 does not tell the whole story. A resonance also occurs in the scattering cross section at the same frequency [see (3.81)] and this contribution which has been neglected for the infrared calculation should now be included. Our emissivity calculation is only accurate in magnitude for $\omega/\omega_T < 5$; we underestimate the effect at large frequencies.

The relative change in emissivity, $\Delta\varepsilon/\varepsilon$, between the composite medium and a Drude metal with the same dc conductivity is illustrated more clearly in Fig. 3.23. The emissivity is calculated for two different relaxation times. $\Delta\varepsilon/\varepsilon < 0$ over the entire region. Below the transverse optic mode frequency, the magnitude of $\Delta\varepsilon/\varepsilon$ increases with increasing relaxation time of the copper matrix. At the transverse optic mode frequency where $\omega/\omega_T = 1$, the magnitude of $\Delta\varepsilon/\varepsilon$ decreases, but the emissivity of the composite is still less than for a Drude metal with the same dc conductivity. Finally, in the region where $\omega\tau > 1$ (e.g., $\omega/\omega_T > 1$), $\Delta\varepsilon/\varepsilon$ is frequency independent and approaches the value $-(3/4)f$ which was previously obtained in (3.115) where lattice vibrations were neglected completely.

In conclusion, the introduction of submicroscopic dielectric particles into a metal such as copper changes the optical properties of the metal from those represented by a simple plasma to those of a plasma plus discrete electronic transitions. The detailed spectra of the descrete transitions have not been explored experimentally.

3.5 Transforming the Solar Spectrum with Selective Surfaces

3.5.1 Selective Surface Configuration

Of the total solar energy falling on a solar cell, only the fraction of photons with energy greater than the energy gap of the material can be used to create electron-hole pairs. Under ideal conditions, a photon of energy $\hbar\omega$ which creates an electron-hole pair cannot use all of this energy in useful work. The fraction of the energy which is greater than the energy gap is irreversibly lost as heat. Due to the wide frequency distribution of the solar spectrum, the maximum efficiency for an optimum gap material is reduced to 42% by this effect.

In this section, we show how the solar spectrum could be transformed to a narrow frequency distribution if ideal selective surfaces could be fabricated. For a given solar energy flux with a sharp photon distribution at the gap energy of the material, the number of electron-hole pairs could be enhanced by a factor of 4–8 over that given by a wide distribution of photon energies. Incorporating the transformed solar spectrum, together with the experimental parameters of a particular semiconductor, we estimate maximum possible efficiency of a PbS photovoltaic cell.

To transform the solar frequency spectrum, we use two selective surfaces attached to a thin metal sheet placed above the photocell as shown in Fig. 3.24a. The entire assembly is in vacuum so that energy can be transmitted only

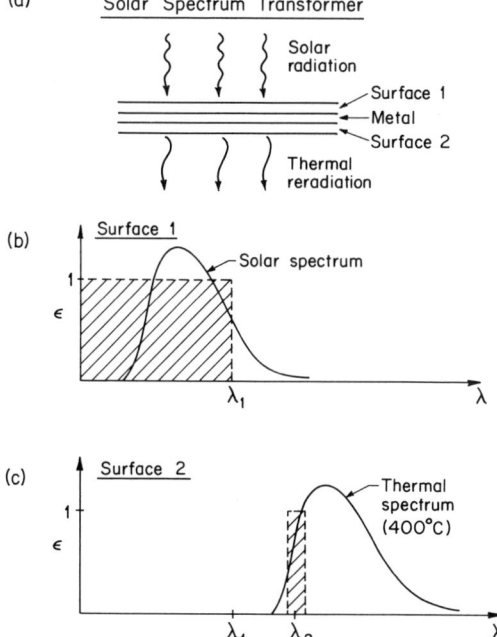

Fig. 3.24a. Solar spectrum transformer. A thick metal plate coated with two different selective surfaces and isolated in a vacuum transforms the broad band of frequencies associated with the solar spectrum to a narrow band of frequencies in the infrared

b Wavelength dependence of the emissivity of the two selective surfaces. For a PbS solar cell, surface 1 should have an edge at $\lambda_1 = 2\,\mu m$

c Surface 2 should have a notch at $3\,\mu m$ with a width of $1\,\mu m$. The metal plate would reach a steady state temperature of about 400 °C for ideal surfaces by radiation to and from the metal sheet. The transformer works as follows: the solar spectrum is absorbed by selective surface 1 which has an emissivity as shown in Fig. 3.24b. Ideally, at wavelength λ_1 the emissivity changes from 0 to 1. Surface 2, on the other hand, has a low emissivity everywhere except at wavelength λ_2, as shown in Fig. 3.24c.

With solar energy incident on surface 1, the metal heats up until the thermal reradiation at wavelength λ_2 is just sufficient to carry away the incident energy absorbed at surface 1. Very little thermal energy will be radiated from surface 1 because of the low emissivity of this ideal surface for wavelengths larger than λ_1. Because the blackbody distribution function varies exponentially when $hc/\lambda \gg kT$, it is possible to choose $\lambda_1 < \lambda_2$ such that most of the radiation absorbed by surface 1 is reradiated by surface 2.

Let us call the total solar energy absorbed U_s, and the number of incident solar photon N_s, and define the average energy of a solar photon $\hbar\bar{\omega}_s$, so that

$$U_s = N_s \hbar\bar{\omega}_s.$$

Similarly, the total energy radiated by surface 2 is

$$U_2 = N_2 \hbar\omega_2,$$

with N_2 the number of photons radiated by surface 2, and $\hbar\omega_2$ the energy of each of these. For our condition of steady state $U_s = U_2$, so

$$N_2/N_s = \bar{\omega}_s/\omega_2 > 1. \tag{3.120}$$

The inequality means that the number of electron-hole pairs produced by the narrow-band thermal reradiation is larger than that obtained if each photon in the solar spectrum produced one electron-hole pair.

We would now like to estimate the maximum possible efficiency of this system for some real material. For selective surface 1, we assume a coating which has low emissivity for wavelengths larger than 2 µm. Selective surface 2 is assumed to have a peak at 3 µm with a full width of 1 µm.

A PbS photocell has a gap at $\hbar\omega_g = 0.37$ eV which matches the 3 µm radiation quite closely [3.87]. The resultant configuration is shown in Fig. 3.25.

Putting numbers into (3.120), we can now calculate the enhancement factor. The average energy of a photon in the solar spectrum is $\hbar\bar{\omega}_s = 1.32$ eV [3.88] while for the reradiation, $\hbar\omega_2 = 0.41$ eV; hence

$$N_2/N_s = 3.2.$$

From the literature, the maximum voltage V_{max} developed by PbS at room temperature is $V_{max} = 65$ mV [3.89]. From *Loferski*'s paper [3.88], we estimate that the corresponding voltage at maximum power transfer $V_{mp} = 30$ mV.

Hybrid Thermal Photovoltaic Receiver

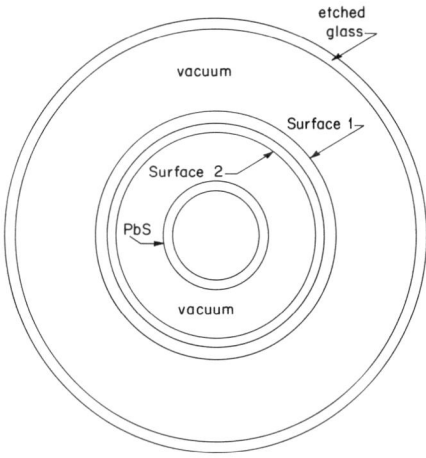

Fig. 3.25. Hybrid thermal-photovoltaic receiver. The assembly has a cylindrical geometry

Neglecting reflectivity losses at the surface of PbS, the maximum efficiency, η_{max}, which is the ratio of the maximum electrical power generated to the solar power absorbed per unit area, is [3.88]

$\eta_{max} = 14\%$.

It is worth noting that for small band gap materials, η_{max} varies rapidly with temperature, essentially because the dark current varies exponentially with temperature.

The configuration shown in Fig. 3.25 is actually the ideal arrangement for minimizing heat transfer to the central chamber. The narrow band of radiation centered around 3 µm is used mainly to generate electron-hole pairs and does little to heat the central pipe. All other wavelengths emitted from surface 2 come from a material with a very small emissivity (see Fig. 3.24c).

A novel feature of this device is that its efficiency is controlled by the selectivity of the two transformer surfaces. The two surfaces operate at the stagnation temperature where the power in is equal to the power out; hence the figure of merit is the solar absorptivity divided by the thermal emissivity. Concentrating the sunlight cannot be used in this case to make up for inadequate selectivity of the surfaces.

3.5.2 High Temperature Selective Surfaces

a) Smokes

Although it has long been known that semiconductor and metal blacks are transparent in the infrared [3.90, 91], only recently has this knowledge been transferred to solar energy collection techniques. The first efforts were on low-temperature selective coatings. The chemical conversion oxide coatings on copper and steel and one- and two-layer NiS–ZnS electrodeposited coatings were discovered to possess a fair degree of selectivity, but were found to be unstable above 220 °C. These efforts have been reviewed by *Mattox* [3.92].

The first experimental study of the selective radiation properties of particulate coatings for solar energy applications was by *Williams* et al. [3.93]. They measured the optical properties of PbS particulate coatings on aluminum substrates. They also demonstrated the feasibility of selective paints made of PbS in a silicone binder.

The original work on gold black by *Harris* et al. [3.80] has been extended by *McKenzie* [3.94], *O'Neill* et al. [3.95], and *Granquist* and *Hunderi* [3.81]. The Maxwell–Garnett model is found to agree fairly well with the optical measurements. Unfortunately, the gold-black system can only be used for modeling small particle effects, as it sinters above 100 °C and looses its selective behavior.

The use of black-chrome electroplated coatings as solar selective surfaces was first suggested by *Tabor* in 1959 [3.96]. The selective coatings produced by standard black chrome baths are unstable under uv irradiation in air at room temperature, but the Harshaw Chemical Co. black chrome electroplated coatings[2] appear to be stable to 300 °C. A detailed description of this coating has been given by *McDonald* [3.97], and *Pettit* and *Sowell* [3.98].

Evaporated chrome blacks have been studied by *Harding* [3.99], and *Granquist* and *Niklasson* [3.100]. *Harding* showed that the optical properties were selective and that the material was stable to 200 °C in vacuum. *Granquist* and *Niklasson* interpreted their optical data within the Maxwell–Garnett theory by generalizing it to include dipole-dipole coupling among aggregated spheres, cubelike shapes, and oxide pellicles. Agreement between theory and experiment was achieved when it was recognized the spherical particles were aggregated into linear chains. None of the smokes which have been investigated so far appear to be stable above 300 °C.

b) Cermet Films

The filtering properties of cermets were first described by *Zeller* and *Kuse* [3.101]. The applicability of semiconductor cermets to the photothermal conversion of solar energy was shown by *Gittleman* [3.102]. The first demonstration of the high temperature stability of cermets was made by *Fan* and *Zavracky* [3.103]. They found that MgO/Au-metal composites were stable at different temperatures depending on the metal backing. The cermet-copper samples were stable to 200 °C, while the optical properties of those prepared with stainless steel and aluminum began to degrade above about 300 °C. The optical properties obtained by depositing the cermet films on stainless steel coated with a 1000 Å molybdenum layer were stable to 400 °C. They found a solar absorptivity of over 0.9 and an infrared emissivity of less than 0.1 at room

---

2 Harshaw Chrom Onyx Black Chrome Plating Process CRX-0270, Harshaw Chemical Company.

temperature. We should note that the emissivity which they calculated was the normal emissivity. The hemispherical emissivity for this cermet metal composite should be 0.12 [see (3.71)].

Fan and *Spura* [3.104] have made selective surfaces from rf-sputtered Cr_2O_3/Cr cermet films. They found that the room temperature solar absorptivity was 0.93 and the normal emissivity was 0.08. (The appropriate hemispherical emissivity is 0.11.) The composite structure consisted of a 350 Å thick Cr_2O_3/Cr cermet on top of a 1500 Å Ni layer which was on top of stainless steel. The structure was stable to 300 °C.

Gittleman et al. [3.105] studied the optical properties of cermets of Au–MgO and W–MgO. They also measured the optical properties of Si–CaF_2 and Si–MgO. They found that the optical properties of Au–MgO did not agree with the Maxwell–Garnett equations. On the other hand for the W–MgO films the observed dielectric constants were in good agreement with the Maxwell–Garnett theory. They estimated the photothermal conversion efficiency of the solar energy of these materials but they assumed that the hemispherical reflectivity in the infrared is equal to the normal reflectivity. From the derivation of hemispherical emissivity in Sect. 3.2.2, this assumption underestimates the size of the hemispherical emissivity by a factor as large as 1/3.

Craighead and *Buhrman* [3.106, 107] have produced selectively absorbing Ni/Al_2O_3 cermet films which are stable to 500 °C. This high temperature stability compares favorably with the earlier work on the stability of A–M–A interference films by *Schmidt* (reported in [3.108]). The composite films were produced by simultaneous evaporation of the metal and dielectric with a dual electron-beam evaporator. The optical properties of the composite agreed fairly well with those calculated from the Maxwell–Garnett theory. *Craighead* and *Buhrman* also showed as did *Bastien* [3.109], *Richie* and *Window* [3.110], and *Stephens* and *Cody* [3.111] that greater efficiencies could be obtained with graded composite films. For reference, we list some of the earlier papers on graded media [3.112–119]. The graded Ni/Al_2O_3 surface produced by *Craighead* and *Buhrman* gave a solar absorptivity of 0.94 and an emissivity of 0.1. Again, the normal emissivity was used instead of the hemispherical emissivity.

3.5.3 Outlook

Two kinds of results have been found with composite coatings. For smokes, a good figure of merit can be obtained, but the coatings are not stable above 300 °C; while for cermets, stability can be obtained, but the figure of merit is less than 10. This latter result is surprising. Because of the precise tailoring of the optical properties in the visible, cermet films can currently be produced with solar absorptivities of 0.94. Even higher values can be expected in the future. The relatively poor figure of merit for the cermets described here stems from a large infrared emissivity. This result is unexpected. According to the Maxwell–

Garnett theory, the infrared emission from the coating should be extremely small in the infrared spectral region.

A figure of merit larger than 20 has been produced with a PbS particulate film on silver [3.120], but this coating is not expected to be stable at high temperatures. A figure of merit larger than 20 at 300 °C has also been produced by reactively sputtering hafnium carbide onto silver in the presence of N_2 gas [3.121, 122]. Unfortunately the solar absorptivity was less than 0.9.

In conclusion, the selective composite coatings described in the text do not approach the limiting figure of merit described in Sect. 3.2.3. Whether or not any coating can reach this limit and maintain its integrity through many high temperature cycles will remain an open question for some time to come.

Acknowledgments. This work has been supported by the Department of Energy under Contract No. EG-77-S-03-1456. Additional support was received from the National Science Foundation under Grant DMR-76-81083 through Cornell Materials Science Center, Report No. 3046. Some of this material was prepared while the author was an Erskine Fellow at the University of Canterbury, Christchurch, New Zealand in 1976.

Appendix A

The optical constants $\bar{n}$ and $\bar{k}$ of the composite medium containing spherical metal particles which have bulk optical constants n_1 and k_1 and the parameters of the medium are related as follows:

Given n_1 and k_1, then

$$\bar{n}^2 - \bar{k}^2 = \frac{ac+bd}{c^2+d^2} \tag{A.1}$$

and

$$2\bar{n}\bar{k} = \frac{bc-ad}{c^2+d^2}, \tag{A.2}$$

where

$$\begin{aligned} a &= (n_1^2 - k_1^2)(1+2f) + 2n_0^2(1-f) \\ b &= 2n_1 k_1(1+2f) \\ c &= (n_1^2 - k_1^2)(1-f) + n_0^2(2+f) \\ d &= 2n_1 k_1(1-f) \end{aligned} \tag{A.3}$$

and f is the volume fraction of metal particles embedded in the medium of index n_0.

Appendix B

Using the values of $\bar{n}_i$ and $\bar{k}_i$ obtained from Appendix A for the ith element, and the optical constants of copper, n_2 and k_2, for the substrate, the reflectivity at normal incidence from a medium consisting of a composite particulate layer on the copper substrate is

$$R_i(\lambda) = \frac{(g_{1i}^2 + h_{1i}^2)\exp\bar{\alpha}_i + (g_{2i}^2 + h_{2i}^2)\exp(-\bar{\alpha}_i) + A_i\cos\bar{\gamma}_i + B_i\sin\bar{\gamma}_i}{\exp\bar{\alpha}_i + (g_{1i}^2 + h_{1i}^2)(g_{2i}^2 + h_{2i}^2)\exp(-\bar{\alpha}_i) + C_i\cos\bar{\gamma}_i + D_i\sin\bar{\gamma}_i} \quad (B.1)$$

with

$$\begin{aligned} A_i &= 2(g_{1i}g_{2i} + h_{1i}h_{2i}) \\ B_i &= 2(g_{1i}h_{2i} - g_{2i}h_{1i}) \\ C_i &= 2(g_{1i}h_{2i} - h_{1i}h_{2i}) \\ D_i &= 2(g_{1i}h_{2i} + g_{2i}h_{1i}), \end{aligned} \quad (B.2)$$

where

$$g_{1i} = \frac{1 - \bar{n}_i^2 - \bar{k}_i^2}{(1+\bar{n}_i)^2 + \bar{k}_i^2}, \qquad h_{1i} = \frac{2\bar{k}_i}{(1+\bar{n}_i)^2 + \bar{k}_i^2},$$

$$g_{2i} = \frac{\bar{n}_i^2 - n_2^2 + \bar{k}_i^2 - k_2^2}{(\bar{n}_i + n_2)^2 + (\bar{k}_i + k_2)^2}, \qquad h_{2i} = \frac{2(\bar{n}_i k_2 - n_2 \bar{k}_i)}{(\bar{n}_i + n_2)^2 + (\bar{k}_i + k_2)^2} \quad (B.3)$$

and

$$\bar{\alpha}_i = 4\pi \bar{k}_i d/\lambda \quad \text{and} \quad \bar{\gamma}_i = 4\pi \bar{n}_i d/\lambda. \quad (B.4)$$

Appendix C

For particles with an ellipsoidal shape, the average dielectric function is [3.77]

$$\frac{\bar{\varepsilon}}{\varepsilon_m} = \frac{\varepsilon_m(1-f)(1-L) + \varepsilon_d[(1-f)L + f]}{\varepsilon_m[1-(1-f)L] + \varepsilon_d(1-f)L}. \quad (C.1)$$

The real part of this function is

$$\bar{\varepsilon}_1 = \varepsilon_{m1} + f\left\{\frac{C(\varepsilon_{m1}A + \varepsilon_{m2}B) + D(\varepsilon_{m1}B - \varepsilon_{m2}A)}{A^2 + B^2}\right\} \quad (C.2)$$

while the imaginary part is

$$\bar{\varepsilon}_2 = \varepsilon_{m2} + f\left\{\frac{C(\varepsilon_{m2}A - \varepsilon_{m1}B) + D(\varepsilon_{m1}A + \varepsilon_{m2}B)}{A^2 + B^2}\right\}, \quad (C.3)$$

where

$$A = \varepsilon_{m1}[1-(1-f)L] + \varepsilon_{d1}(1-f)L$$
$$B = \varepsilon_{m1}[1-(1-f)L] + \varepsilon_{d1}(1-f)L$$
$$C = \varepsilon_{d1} - \varepsilon_{m1}$$
$$D = \varepsilon_{d2} - \varepsilon_{m2}.$$

References

3.1 H. Tabor: Bull. Res. Counc. Isr. **54**, 28 (1956)
3.2 L.H. Shaffer: Sol. Energy **2**, 21 (1958)
3.3 J.T. Gier, R.V. Dunkle: *Trans. Conf. on the Use of Solar Energy*, Vol. II (University of Arizona Press, Tucson 1958) p. 41
3.4 B.O. Seraphin, A.B. Meinel: In *Optical Properties of Solids – New Developments*, ed. by B.O. Seraphin (North-Holland, Amsterdam 1975) Chap. 17, p. 927
3.5 H. Tabor: Proc. Nat. Acad. Sci. U.S.A. **47**, 1271 (1961)
3.6 F. Daniels: *Direct Use of the Sun's Energy* (Yale University Press, New Haven 1969)
3.7 D.K. Edwards, J.T. Gier, K.E. Nelson, R.D. Roddick: Sol. Energy **6**, 1 (1962)
3.8 R. Schmidt, R. Park: Appl. Opt. **4**, 917 (1965)
3.9 T.F. Irvine, J.P. Hartnett, E.R.G. Eckert: Sol. Energy **2**, 12 (1958)
3.10 M. Born, E. Wolf: *Principles of Optics* (Pergamon, Oxford 1970) p. 633
3.11 W.T. Doyle: Phys. Rev. **111**, 1067 (1958)
3.12 R.H. Doremus: J. Chem. Phys. **40**, 2389 (1964)
3.13 R.H. Doremus: J. Chem. Phys. **42**, 414 (1965)
3.14 D.C. Skillman, C.R. Berry: J. Chem. Phys. **48**, 3297 (1968)
3.15 U. Kreibig, P. Zacharias: Z. Phys. **231**, 128 (1970)
3.16 K. Kimoto, Y. Kamiya, M. Nonoyama, R. Uyeda: Jpn. J. Appl. Phys. **2**, 702 (1963)
3.17 N. Wada: Jpn. J. Appl. Phys. **6**, 553 (1967)
3.18 K. Kimoto, I. Nishada: Jpn. J. Appl. Phys. **6**, 1047 (1967)
3.19 C.G. Granquist, R.A. Buhrman: J. Appl. Phys. **47**, 2200 (1976)
     R.A. Buhrman, C.G. Granquist: J. Appl. Phys. **47**, 2220 (1976)
3.20 D. Bohm, D. Pines: Phys. Rev. **82**, 625 (1952)
3.21 F. Wooten: *Optical Properties of Solids* (Academic Press, New York 1972) p. 220
3.22 R. Ruppin: Surf. Sci. **34**, 20 (1973)
3.23 J. Stone: *Radiation and Optics* (McGraw-Hill, New York 1963) p. 374
3.24 R.H. Ritchie: Surf. Sci. **34**, 1 (1973)
3.25 J.M. Elson, R.H. Ritchie: Phys. Rev. B **4**, 4129 (1971)
3.26 Y.S. Toulovkian (ed.): *Thermophysical Properties of Matter*, Vol. 7 (Plenum, New York 1970) p. 24a
3.27 B. Donovan: *Elementary Theory of Metals* (Pergamon, Oxford 1967) p. 215
3.28 L.D. Landau, E.M. Lifshitz: *Electrodynamics of Continuous Media* (Addison-Wesley, Reading, Mass. 1960) p. 280
3.29 A.B. Pippard: In *Optical Properties and Electronic Structure of Metals and Alloys*, ed. by F. Abeles (North-Holland, Amsterdam 1966) p. 622
3.30 F.J. Blatt: *Physics of Electronic Conduction in Solids* (McGraw-Hill, New York 1968) p. 194
3.31 N.W. Ashcroft, N.D. Mermin: *Solid State Physics* (Holt, Rinehart, and Winston, New York 1976) p. 5
3.32 N.F. Mott, H. Jones: *The Theory of the Properties of Metals and Alloys* (Dover, New York 1936), Appendix II, p. 318

3.33 G.E.H. Reuter, E.H. Sondheimer: Proc. R. Soc. London A **195**, 336 (1948)
3.34 T. Holstein: Phys. Rev. **88**, 1427 (1952)
3.35 R.B. Dingle: Physica **19**, 311 (1953)
3.36 A.B. Pippard: In *Advances in Electronics and Electron Physics*, ed. by L. Marton (Academic Press, New York 1954) p. 1
3.37 J.A. McKay, J.A. Rayne: Phys. Rev. B **13**, 673 (1976)
3.38 N.W. Ashcroft, K. Sturm: Phys. Rev. B **3**, 1898 (1971)
3.39 H.B.G. Casimir, J. Ubbink: Philips Tech. Rev. **28**, 300 (1967)
3.40 F. Wooten: *Optical Properties of Solids* (Academic Press, New York 1972) p. 93
3.41 T. Holstein: Ann. Phys. (N.Y.) **29**, 410 (1964); Phys. Rev. **96**, 535 (1954)
3.42 R.N. Gurzhi: Zh. Eksp. Teor. Fiz. **33**, 660 (1957) [Sov. Phys.-JETP **6**, 506 (1958)]; Zh. Eksp. Teor. Fiz. **35**, 965 (1958) [Sov. Phys.-JETP **8**, 673 (1959)]
3.43 A. Sievers: J. Opt. Soc. Am. **68**, 1505 (1978)
3.44 D.S. Jones: *The Theory of Electromagnetism* (Pergamon, Oxford 1964) p. 322
3.45 K.G. Ramanathan, S.H. Yen: J. Opt. Soc. Am. **67**, 32 (1977)
3.46 E.A. Estalote, K.G. Ramanathan: J. Opt. Soc. Am. **67**, 39 (1977)
3.47 K.G. Ramanathan, S.H. Yen, E.A. Estalote: Appl. Opt. **16**, 2810 (1977)
3.48 G. Mie: Ann. Phys. (Leipzig) **25**, 377 (1908)
3.49 M. Kerker: *The Scattering of Light and Other Electromagnetic Radiation* (Academic Press, New York 1969)
3.50 H.C. van de Hulst: *Light Scattering by Small Particles* (Wiley, New York 1957)
3.51 R. Landauer: In *Electrical Transport and Optical Properties of Inhomogeneous Media*, ed. by J.C. Garland, D.B. Tanner (American Institute of Physics, New York 1978) p. 2
3.52 J.C.M. Garnett: Philos. Trans. R. Soc. London **203**, 385 (1904); **205**, 237 (1906)
3.53 N.W. Ashcroft, N.D. Mermin: *Solid State Physics* (Holt, Rinehart, and Winston, New York 1976) p. 542
3.54 G.B. Smith: J. Phys. D **10**, L39 (1977)
3.55 W. Lamb, D.M. Wood, N.W. Ashcroft: In *Electrical Transport and Optical Properties of Inhomogeneous Media*, ed. by J.C. Garland, D.B. Tanner (American Institute of Physics, New York 1978) p. 240
3.56 D.A.G. Bruggeman: Ann. Phys. (Leipzig) **24**, 636 (1935)
3.57 O.S. Heavens: *Optical Properties of Thin Solid Films* (Butterworths, London 1955)
3.58 O. Hunderi, D. Beaglehole: Phys. Rev. B **2**, 321 (1970)
3.59 G. Zacschmar, A. Nedoluha: J. Opt. Soc. Am. **62**, 348 (1972)
3.60 J.I. Gittleman, Y. Goldstein, S. Bozowski: Phys. Rev. B **5**, 3609 (1972)
3.61 R.W. Cohen, G.D. Cody, M.D. Coutts, B. Abeles: Phys. Rev. B **8**, 3689 (1973)
3.62 L. Genzel, T.P. Martin: Surf. Sci. **34**, 33 (1973)
3.63 P.H. Lissberger, R.G. Nelson: Thin Solid Films **21**, 159 (1974)
3.64 T. Yamaguchi, S. Yoshida, A. Kinbara: Thin Solid Films **21**, 173 (1974)
3.65 B. Abeles, Ping Sheng, M.D. Coutts, Y. Arie: Adv. Phys. **24**, 407 (1975)
3.66 L. Genzel, T.P. Martin, U. Kreibig: Z. Phys. B **21**, 339 (1976)
3.67 E.B. Priestley, B. Abeles, R.W. Cohen: Phys. Rev. B **12**, 2121 (1975)
3.68 D.B. Tanner, A.J. Sievers, R.A. Buhrman: Phys. Rev. B **11**, 1330 (1975)
3.69 B. Abeles: In *Applied Solid State Physics*, ed. by R. Wolfe (Academic Press, New York 1976)
3.70 C.G. Granquist, R.A. Buhrman, J. Wyns, A.J. Sievers: Phys. Rev. Lett. **37**, 625 (1976)
3.71 D.R. McKenzie: J. Opt. Soc. Am. **66**, 249 (1976)
3.72 J.D. Eversole, H.P. Broida: Phys. Rev. B **15**, 1644 (1977)
3.73 M. Rasigni, G. Rasigni: J. Opt. Soc. Am. **67**, 510 (1977)
3.74 V.V. Truong, G.D. Scott: J. Opt. Soc. Am. **66**, 124 (1976); **67**, 502 (1977)
3.75 D.M. Wood, N.W. Ashcroft: Philos. Mag. **35**, 269 (1977)
3.76 C.G. Granquist: In *Electrical Transport and Optical Properties of Inhomogeneous Media*, ed. by J.C. Garland, D.B. Tanner (American Institute of Physics, New York 1978) p. 196
3.77 W.L. Bragg, A.B. Pippard: Acta Crystallogr. **6**, 865 (1953)
3.78 D. Pines: *Elementary Excitations in Solids* (Benjamin, New York 1964)
3.79 American Institute of Physics Handbook, 3rd ed. (American Institute of Physics, New York)

3.80 L. Harris, R.T. McGinnies, B.M. Siegel: J. Opt. Soc. Am. **38**, 582 (1948)
L. Harris, D. Jeffries, B.M. Siegel: J. Appl. Phys. **19**, 791 (1948)
L. Harris, J.K. Beasley: J. Opt. Soc. Am. **42**, 134 (1952)
L. Harris: *The Optical Properties of Metal Blacks and Carbon Blacks* (Eppley Foundation for Research, Monograph Ser. No. 1, Newport, Rhode Island 1967)
3.81 C.G. Granquist, O. Hunderi: Phys. Rev. B **16**, 3513 (1977)
3.82 P.B. Johnson, R.W. Christy: Phys. Rev. B **9**, 5056 (1974)
3.83 J.H. Werner, D.W. Lynch, C.G. Olsen: Phys. Rev. B **7**, 4311 (1973)
3.84 A.A. Lucas: Phys. Rev. B **7**, 3527 (1973)
3.85 R.W. Stewart: Semiannual Technical Report, Dec. 1972, Battelle, Pacific Northwest Lab., Richland, Washington 99352
3.86 E. Burstein: In *Phonons and Phonon Interactions*, ed. by T.A. Bak (Benjamin, New York 1964) p. 296
3.87 C. Kittel: *Introduction to Solid State Physics* (Wiley, New York 1953) p. 302
3.88 J.J. Loferski: J. Appl. Phys. **27**, 777 (1956)
3.89 J. Bloem: Appl. Sci. Res. B **6**, 92 (1955)
3.90 A.H. Pfund: J. Opt. Soc. Am. **23**, 275 (1933)
3.91 W.E. Forsythe: *Measurement of Radiant Energy* (McGraw-Hill, New York 1937)
3.92 D.M. Mattox: J. Vac. Sci. Technol. **13**, 127 (1976)
3.93 D.A. Williams, T.A. Lappin, J.A. Duffie: J. Eng. Power, Trans. ASME, p. 213 (July 1963)
3.94 D.R. McKenzie: J. Opt. Soc. Am. **66**, 249 (1976)
3.95 P. O'Neill, C. Doland, A. Ignatiev: Appl. Opt. **16**, 2822 (1977)
3.96 H. Tabor: "Research on Optics of Selective Surfaces", Final Report AF 61 (052-279) (May, 1959)
3.97 G.E. McDonald: Sol. Energy **17**, 119 (1975)
3.98 R.B. Pettit, R.R. Sowell: J. Vac. Sci. Technol. **13**, 596 (1976)
3.99 G.L. Harding: Thin Solid Films **38**, 109 (1976)
3.100 C.G. Granquist, G.A. Niklasson: J. Appl. Phys. **49**, 3512 (1978)
3.101 H.R. Zeller, D. Kuse: J. Appl. Phys. **44**, 2763 (1973)
3.102 J.J. Gittleman: Appl. Phys. Lett. **28**, 370 (1976)
3.103 J.C.C. Fan, P.M. Zavracky: Appl. Phys. Lett. **29**, 478 (1976)
3.104 J.C.C. Fan, S.A. Spura: Appl. Phys. Lett. **30**, 511 (1977)
3.105 J.I. Gittleman, B. Abeles, P. Zanzucchi, Y. Arie: Thin Solid Films **45**, 9 (1977)
3.106 H.G. Craighead, R.A. Buhrman: Appl. Phys. Lett. **31**, 423 (1977)
3.107 H.G. Craighead, R.A. Buhrman: J. Vac. Sci. Technol. **15**, 269 (1978)
3.108 J. Jurisson, R.E. Peterson, H.Y.B. Mar: J. Vac. Sci. Technol. **12**, 1010 (1975)
3.109 A.C. Bastien: "The Use of Inhomogeneous Metal Dielectric Films as Selective Absorbing Surfaces for Solar Conversion Panels", Perkin Elmer, Norwalk, Conn, Final Report No. 13223 (Nov. 1976) NSF Grant No. AER 74-16330
3.110 I.T. Richie, B. Window: Appl. Opt. **16**, 1438 (1977)
3.111 R.B. Stephens, G.D. Cody: Thin Solid Films **45**, 19 (1977)
3.112 J. Wallot: Ann. Phys. (Leipzig) **60**, 734 (1919)
3.113 J.A. Stratton: *Electromagnetic Theory* (McGraw-Hill, New York 1941) p. 589, prob. 8
3.114 V.A. Bailey: Phys. Rev. **96**, 865 (1954)
3.115 H. Osterberg: J. Opt. Soc. Am. **58**, 513 (1958)
3.116 S.F. Monaco: J. Opt. Soc. Am. **51**, 280 (1961)
3.117 R. Jacobson: Opt. Acta **10**, 309 (1963)
3.118 R. Jacobsson, J.O. Martensson: Appl. Opt. **5**, 29 (1966)
3.119 G. Tyres: *Radiation and Propagation of Electromagnetic Waves* (Academic Press, New York 1969)
3.120 T.J. McMahon, S.N. Jasperson: Appl. Opt. **13**, 2750 (1974)
3.121 R. Blickensderfer, D.K. Deardorff, R.L. Lincoln: Sol. Energy **19**, 429 (1977)
3.122 R. Blickensderfer: *Proceedings D.O.E./D.S.T. Thermal Power System Workshop on Selective Absorber Coatings* (Solar Energy Research Institute, Golden, CO 1977) p. 359

4. Solar Photoelectrolysis with Semiconductor Electrodes

H. Gerischer

With 35 Figures

With the exception of water power and wind energy, all our traditional energy resources are the product of photosynthesis. Since the energy conversion in photosynthesis is an electrochemical process, namely a redox reaction of excited chlorophyll molecules, many attempts have been made to develop devices for solar energy conversion along similar lines [4.1–3]. It turns out, however, that the efficiency of artificial systems of this kind is very limited, and the prospects of developing more efficient and durable systems appear to be very low. The principal disadvantage is the limited light absorption of a thin layer with photoactive dye molecules such as the thylacoid membrane of a plant. However, a thin layer is required to prevent large energy losses in electron transmission through the membrane. Nature has compensated for this disadvantage first by using multilayers which successively absorb the sunlight and second by adding antenna molecules which transport the absorbed light energy to chlorophyll molecules at the reaction centers by resonating energy transfer [4.4–6]. Abandoning the membrane would mean using homogeneous photoredox reactions. However, the decisive benefit of the membrane is then lost, namely the separation of the products which prevents the occurrence of back reactions. The latter processes restrict the exploitation of all homogeneous photoreactions to an unsatisfactorily low yield of energy conversion.

The only alternative which preserves the advantages of the photosynthetic process regarding energy conversion efficiency and still exploits photoredox reactions is a photoelectrolytic cell with semiconductor electrodes as the energy converters. Figure 4.1 schematically indicates the relation between photosynthesis and photoelectrolysis in terms of the electron energies used for the redox reactions which provide the chemical energy for consecutive use. One sees that photosynthesis can be considered as a kind of internal electrolysis where both the electronic and the ionic currents pass the membrane in parallel. Photoelectrolysis, however, needs a counterelectrode which must be in electronic contact with the semiconductor and can pick up one of the mobile charge carriers generated in the semiconductor by light absorption. To make this charge separation efficient, one further needs a space-charge layer in the semiconductor beneath the interface as indicated in this figure. Such systems are closely related to solid-state semiconductor solar cells. Photoelectrolysis has attracted wide interest in recent years as a possible alternative for harvesting solar energy [4.7–14]. In this article, the basic principles of such photoelectrochemical devices will be explained and the various material problems will be discussed and analyzed.

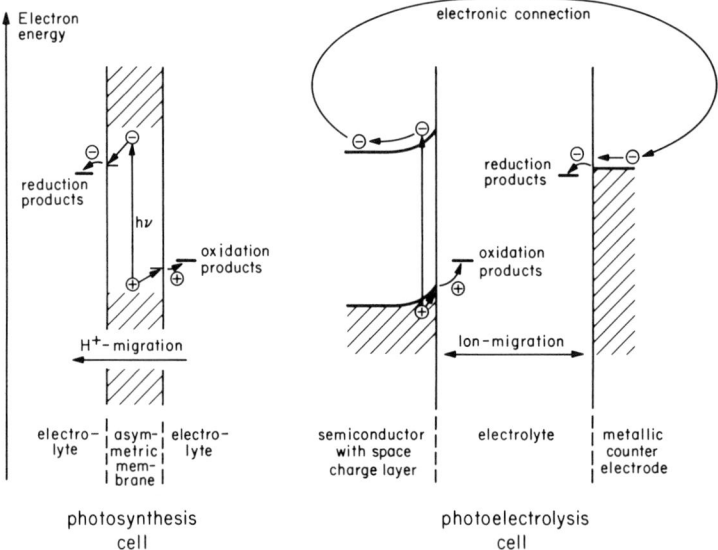

Fig. 4.1. Principal schemes of photosynthesis and photoelectrolysis in terms of electron transfer reactions

4.1 Principles of Photoelectrolysis

4.1.1 The Semiconductor-Electrolyte Interface

Like any other contact between an electronic and an electrolytic conductor, the semiconductor-electrolyte interface forms a barrier to the transfer of electric current. As a consequence, excess electric charges of opposite sign can be accumulated at both sides of this barrier creating electric fields which serve as a driving force for charged species to overcome this barrier. The charge carriers can be electrons, if suitable electron acceptors or donors are present in the electrolyte, or ions. Since semiconductors can rarely be electrolytically deposited from ions in solution, we shall discuss only such ion-transfer reactions where the components of a semiconductor pass the interface as ions formed by interaction with species from the electrolyte. This describes the electrolytic decomposition of a semiconductor.

The accumulated excess charge at such a contact forms an electric double layer which has a particular structure depending mainly on the concentration of the mobile and immobile charge carriers at both sides of the interface [4.15–18]. In photoelectrochemical cells, one has to use concentrated electrolyte solutions to minimize internal resistances, which means ion concentrations in the order of 10^{21} cm^{-3} (about 1 molar). The carrier density in the semiconductor, however, must be kept much lower to create the extended

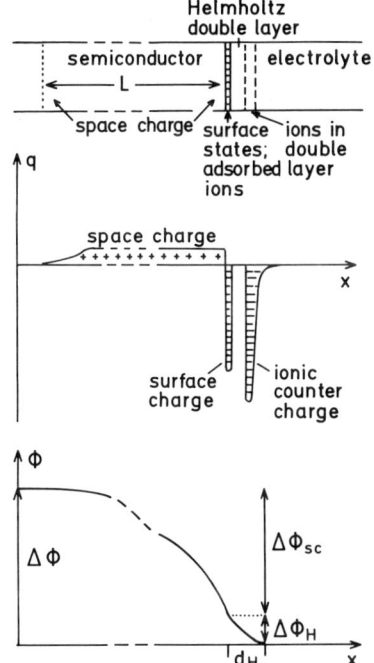

Fig. 4.2. Electric double layer at the semiconductor-electrolyte interface for an n-type semiconductor with positive excess charge. Spatial charge (q) and potential (ϕ) distribution

electric field in a space-charge layer underneath the semiconductor surface needed for the charge separation mechanism during illumination. Fortunately, this does not increase the resistance to intolerably high values owing to the much higher mobility of electrons and holes in comparison to ions.

If there were no specific interaction between charged species and the surface, a very simple double-layer structure would result, with a diffuse space charge on the side of the semiconductor and a condensed ionic countercharge arranged in a plate-like fashion on the electrolyte side. However, specific interaction can be caused by the existence of electronic surface states on the semiconductor or by chemical bonding between ionic components of the electrolyte and the semiconductor surface atoms. With some simplifying assumptions, a double-layer model can be derived which takes into account this special role of the surface for the attachment of electric charge. The result is shown in Fig. 4.2 for the situation of an n-type semiconductor with enough positive excess charge such that a depletion layer is formed with a constant density of positive charge of ionized donors in the bulk of the space-charge region.

It can be shown that the amount of adsorbed ionic charge at the surface depends mainly on the composition of the electrolyte but very little on the excess charge on the semiconductor [4.19, 20]. Consequently, only the distribution of the electronic charge in the semiconductor varies with the applied electrode potential and this controls the differential capacity of the interface. In an ideal depletion layer where electronic equilibrium is fully established during a periodic cycle of the voltage, and in the absence of a charge variation in surface states, the measured capacity depends only on the potential drop in the space-charge layer. The following equations represent the results for this ideal case [4.21]:

$$C_{\text{diff}}(\Delta\phi_{\text{sc}}) = \left(\frac{\varepsilon\varepsilon_0 e N_d}{2}\right)^{1/2} (\Delta\phi_{\text{sc}} - kT/e)^{-1/2}, \quad (4.1)$$

$$\Delta\phi_{\text{sc}} + \Delta\phi_{\text{H}} = \Delta\phi, \quad (4.2)$$

where $\Delta\phi_{\text{sc}}$ is the potential drop in space-charge layer; $\Delta\phi_{\text{H}}$, the potential drop in Helmholtz double layer; ε_0, the permittivity of free space; ε, the dielectric constant relative to vacuum; e, the charge of electron; N_D the concentration of donors per cubic centimeter.

Equation (4.1) predicts that a plot of C^{-2} versus $\Delta\phi$ should give a straight line if $\Delta\phi_{\text{H}}$ remains constant. $\Delta\phi$, however, cannot be measured directly between two different phases. Electrode potentials are measured against a reference electrode which means that the sum of two or more electric potential differences is obtained. Since the potential difference between the reference electrode and the electrolyte is kept constant, the measured potential difference V corresponds to $\Delta\phi$ plus a constant. We shall denote this value V as the electrode potential versus the standard hydrogen electrode. Unless otherwise stated the hydrogen electrode will be used as the reference system, i.e.,

$$V = \Delta\phi - \Delta\phi_{\text{NHE}}. \quad (4.3)$$

Part of $\Delta\phi$ can, however, be measured absolutely if (4.1) or another relation between C_{diff} and $\Delta\phi_{\text{sc}}$ is valid. For semiconductor electrodes, a situation of particular interest is when $\Delta\phi_{\text{sc}} = 0$. In this case, no electric field is present inside the semiconductor and the potential energy of the electrons, which is represented by the position of the band edges, is constant from the bulk up to the surface. This is the so-called flat-band situation, and the corresponding electrode potential is called the flat-band potential V_{fb}.

Combining (1)–(3) gives the so-called Mott–Schottky relation

$$C^{-2} = \frac{V - V_{\text{fb}} - kT/e}{\frac{1}{2}\varepsilon\varepsilon_0 e N_D} = \frac{1.41 \times 10^{32}}{\varepsilon N_D}(V - V_{\text{fb}} - kT/e) \quad [\text{F}^{-2}\,\text{cm}] \quad (4.4)$$

showing that the flat-band potential can be determined from capacity measurements if a depletion layer is formed. This implies that no variation of surface charge occurs in the range of the experiment.

Our knowledge of the energy position of band edges in relation to electrolytes results from such experiments. The distance between the band edges and the Fermi level in the bulk of a semiconductor can be derived from

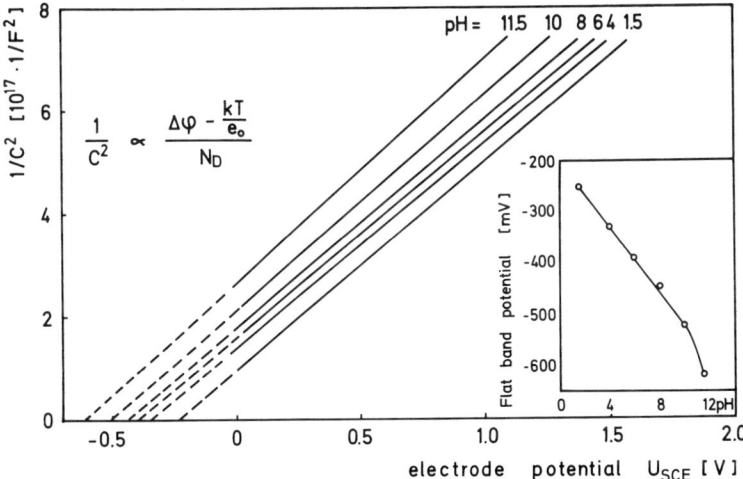

Fig. 4.3. Mott-Schottky plots of differential capacity versus electrode potential for ZnO electrodes in aqueous electrolytes of different pH; flat-band potential dependence on pH

the electron or hole concentration at equilibrium conditions, n_0 or p_0, if the band gap and the effective densities of states for the conduction band, N_c, and the valence band, N_v, are known.

For nondegenerate semiconductors this relation is

$$E_c = E_F - kT \ln(n_0/N_c), \tag{4.5}$$

or

$$E_v = -E_F + kT \ln(p_0/N_v), \tag{4.6}$$

To connect this relation with the electrode potentials V measured versus a reference electrode, one only has to convert electrode potentials into the free energy of electrons at the same electrostatic potential. This is obtained from the equation

$$E_F = -eV + \text{const}_{\text{ref}}, \tag{4.7}$$

where the constant defines the free energy of the electron in the reference electrode. For the standard hydrogen electrode, this constant has a value between -4.5 and $-4.7\,\text{eV}$ [4.22–24]. The energy of the Fermi level at the flat-band potential is immediately obtained by inserting V_{fb} into (4.7). One then can obtain the position of the band edges with (4.5) and (4.6) either in terms of Fermi energies, or of the corresponding electrode potentials.

An example of such measurements is shown in Fig. 4.3 for a zinc oxide electrode. In this case the position of the flat band potential is pH-dependent as it is for all other semiconducting oxides. This pH-dependence can be explained by the basic or acidic character of the oxide surface which can interact with water by binding protons or hydroxyl ions. The following equilibria have to be considered:

$$M-O + H_2O \rightleftarrows M^+-OH + OH^- \qquad (4.8)$$

and

$$M-O + H_2O \rightleftarrows M\begin{smallmatrix}OH\\OH\end{smallmatrix} \rightleftarrows M\begin{smallmatrix}O^-\\OH\end{smallmatrix} + H^+. \qquad (4.9)$$

This corresponds to a charge separation, since one part of the electric charge is localized at the surface while the counter-charge remains in the electrolyte. Such equilibria are potential dependent, and the potential drop in the Helmholtz double layer varies therefore with the pH. Neglecting complications like the chemical activity of the surface in dependence on the surface composition, one can derive the following equilibrium conditions at room temperature for the dependence of $\Delta\phi_H$ on the hydrogen ion activity as given by the pH value of the solution [4.19]:

$$\Delta\phi_H = \text{const} + 0.059\,\text{pH} \quad [V]. \qquad (4.10)$$

If $\Delta\phi_H$ remains unaffected by an electronic excess charge on the semiconductor, Mott–Schottky plots of the capacity should be shifted in parallel by a variation of the pH, as indeed has been found in many cases. This is demonstrated in Fig. 4.3, where the intersection of the straight lines with the abscissa shows the position of the flatband potential.

The contribution of electronic surface states to $\Delta\phi_H$ is much more complex and cannot be discussed here in any detail. If the charge of the surface states varies considerably, deviations from linearity in Mott–Schottky plots are observed. Such deviations can, however, also be caused by different phenomena, like ionization of deep donor levels. A distinction can only be made from case to case with the help of other types of experiments. Another possible way to measure quickly the approximate position of the flat-band potential is the observation of photocurrents which decrease to zero in the absence of a band bending.

Figure 4.4 gives a summary of band-edge positions for oxides and other compound semiconductors obtained from capacity or photovoltage measurements by various authors. Due to the different electrolytes used and the variability of the surface properties of the materials, these data can only give the general trend. The measured values show a large scatter. This is not surprising, for the flat-band potential is not an absolutely defined thermodynamic quantity. It corresponds to a Volta potential which depends essentially on

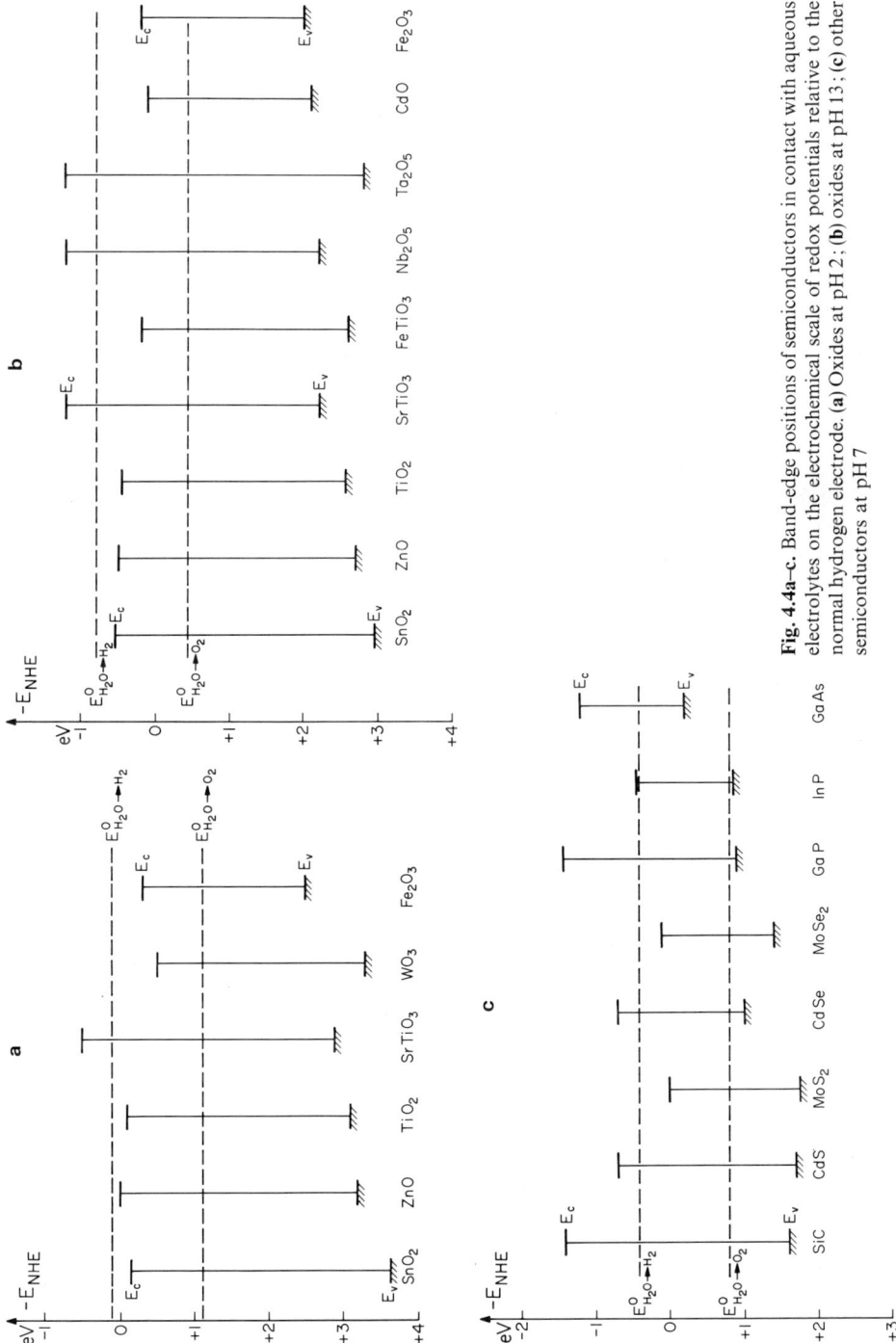

Fig. 4.4a–c. Band-edge positions of semiconductors in contact with aqueous electrolytes on the electrochemical scale of redox potentials relative to the normal hydrogen electrode. (**a**) Oxides at pH 2; (**b**) oxides at pH 13; (**c**) other semiconductors at pH 7

crystal surface orientation and structure as well as on impurities and adsorption. Nevertheless, it is an important quantity, in particular, for kinetic considerations.

Some attempts have been made to predict the position of the flat-band potentials or band edges from physical properties of the semiconductor materials. The close correspondence to the Volta potentials in vacuo suggests the use of work functions and electron affinities for this purpose as in solid-state heterojunctions. In some cases, close parallelism has been observed between the band edge position relative to the Fermi level of a solid metal (Au) and the position on the electrochemical energy scale [4.25, 26].

Butler and *Ginley* [4.27, 28] have used the electronegativities of the components of a compound semiconductor to predict the electron affinity of the semiconductor which they define as the mid-energy of the band gap. They have recognized that such comparisons suffer from the fact that the voltage drop in the Helmholtz double layer will be different for different materials even in the same electrolyte. For example, oxides have different affinities for the reactions (4.8) and (4.9) with water. To avoid this problem they have proposed to use the point of zero surface charge of the materials for a comparison with theoretical predictions. This situation can be obtained from charge distribution between collodial particles and electrolyte solutions [4.29], or from electrokinetic measurements [4.30]. This approach could give satisfactory results within the uncertainty of the variation of surface properties, but many more experimental results are needed. It does not take into account the possibility that chemisorbed species or a layer of adsorbed water molecules, preferentially oriented, may create a potential step between the electrolyte and the semiconductor. Such dipole layers can cause drastic shifts of Volta potentials, as is well known at interfaces with gases.

4.1.2 Electron Transfer Reactions of Semiconductor Electrodes

Electron transfer is a fast process compared with motions of atoms. The theoretical description of electron transfer reactions is therefore based on the Franck–Condon principle analogous to electronic excitation by light absorption [4.31–34]. As a consequence, electron transfer between weakly interacting systems can be described as an exchange of electrons between quantum states of equal energy. Therefore, the energy distribution of occupied and vacant quantum states of electrons in an interacting system is decisive for the rate of electron transfer.

Due to the presence of a band gap in a semiconductor, there is a distinction between electron transfer reactions according to whether the conduction band or the valence band is involved. Which process is preferred depends on the nature of the redox reactants in the electrolyte, particularly whether its electronic states available for electron transfer are located in the energy range of the conduction or valence band [4.33, 35]. The energy levels of a redox couple in an electrolyte are controlled by the ionization energy of the reduced species, and the electron affinity of the oxidized species in solution. These fluctuate over a considerable energy range due to the varying interaction with the surrounding electrolyte. The standard redox potential is an average of the ionization energy and the electron affinity of a redox couple. This energy level is reached with equal probability by fluctuations of the ionization energy of the reduced, and of the electron affinity of the oxidized, species. These fluctuations of the energy levels averaged over time give a Gaussian distribution of the energy levels around characteristic mean values for each component of a redox system [4.36].

These characteristic energies represent the ionization energy of the reduced species, E^0_{red}, and the electron affinity of the oxidized species, E^0_{ox}, in their most probable state of solvation. The width of the distribution function is controlled by the so-called reorganization energy λ_{redox}. This is the energy released due to alterations of the solvation structure after the oxidation state of a redox species is changed suddenly under Franck–Condon conditions. In other words, λ_{redox} is the energy needed to bring the solvation shell (including the vibrational frequencies) of one redox species from its most probable state into the most probable solvation structure of its redox counterpart.

The time-average distribution of the redox energy levels for each single redox species, as a statistical average over the whole ensemble, can be described as follows:

$$W_{red}(E)dE = (4\pi kT\lambda)^{-1/2} \exp[-(E-E^0_{red})^2/4\pi kT\lambda]dE, \qquad (4.11a)$$

$$W_{ox}(E)dE = (4\pi kT\lambda)^{-1/2} \exp[-(E-E^0_{ox})^2/4\pi kT\lambda]dE. \qquad (4.11b)$$

The most probable energy levels E^0_{red} and E^0_{ox} are connected with the standard redox potential or its equivalent, the standard redox Fermi level $E^0_{F,redox}$, by the symmetrical relations

$$E^0_{red} = E^0_{F,redox} - \lambda; \quad E^0_{ox} = E^0_{F,redox} + \lambda. \qquad (4.12)$$

This model leads to two different cases of the energy-state distribution at the contact of a semiconductor with a redox electrolyte.

These are shown in Fig. 4.5 for the equilibrium situation. The decisive parameter which controls which situation is found, is the position of the Fermi level of the redox system in relation to the position of the band edges. The Fermi level of the redox system, $E_{F,redox}$, is equivalent to the redox potential V_{redox}, as given by (4.7). Figure 4.5 shows the Fermi level for the standard state of the redox system with equal concentrations of the oxidized and reduced species. If $E_{F,redox}$ is closer to the conduction band edge E_c than to the valence band edge E_v, case (a) is found at equilibrium with a predominant electron exchange in the conduction band. In the converse situation with $E_{F,redox}$ closer to E_v, electron exchange predominates in the valence band [case (b)].

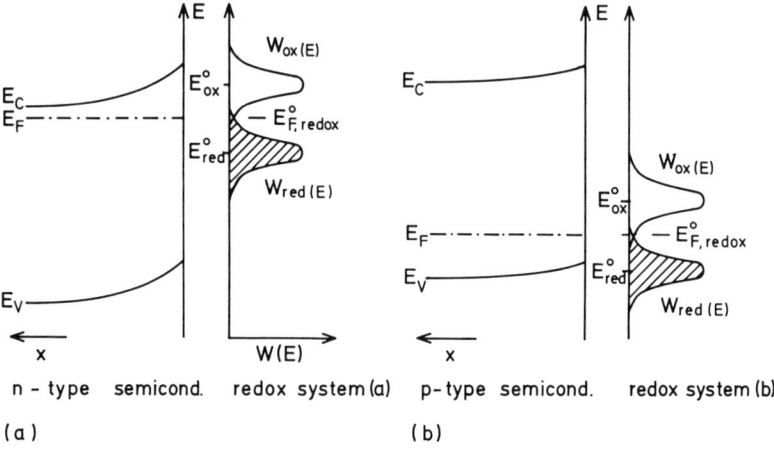

Fig. 4.5a and b. Electron energy distribution at the contact between a semiconductor and a redox electrolyte for two different redox systems at equilibrium

At equilibrium, the Fermi level in the semiconductor surface adjusts itself to the position of the redox Fermi level by the appropriate charging of the double layer. If a variation of the double-layer charge does not considerably change the potential drop in the Helmholtz part, the energy position of the band edges relative to the energy levels of the redox couple in solution does not change either. However, what will drastically vary with the double-layer charge is the surface concentration of electrons and holes (equilibrium between bulk and surface presumed). The equilibrium for electron transfer between a semiconductor electrode and a redox electrolyte is established by the adjustment of electron and hole concentrations at the surface to the applied voltage. We see that in thermodynamic terms, it is the entropy factor of the free energy which adjusts to the equilibrium condition at the semiconductor electrode. This is in contrast to the situation at metal electrodes where it is the potential energy term which makes the adjustment by double layer charging.

At semiconductor electrodes, this situation leads to particularly simple kinetic equations for electron transfer reactions, since electrons and holes appear directly as reactants in the kinetic formulas. One obtains for the rate of electron transfer in the conduction band, j_c:

$$j_c = j_{c,0}\left(\frac{C_{red}}{C_{red,0}} - \frac{n_s}{n_{s,0}}\frac{C_{ox}}{C_{ox,0}}\right), \qquad (4.13)$$

where $j_{c,0}$ is the exchange current density at equilibrium; $C_{red,0}$ is the concentration of the reduced component at equilibrium; $C_{ox,0}$ is the concentration of the oxidized component at equilibrium; and $n_{s,0}$ is the surface concentration of electrons at equilibrium.

The terms C_{red} and C_{ox} take into account that the concentration of the reactants may vary if a current flows (concentration polarization). The surface concentration of electrons is related to the bulk concentration n_0 by the potential drop $\Delta\phi_{sc}$ in the space-charge layer

$$n_s = n_0 \exp(-e\Delta\phi_{sc}/kT). \tag{4.14}$$

The corresponding relations for the rate of electron transfer in the valence band are:

$$j_v = j_{v,0}\left(\frac{p_s}{p_{s,0}}\frac{C_{red}}{C_{red,0}} - \frac{C_{ox}}{C_{ox,0}}\right), \tag{4.15}$$

where $j_{v,0}$ is the exchange current density at equilibrium; and $p_{s,0}$ is the surface concentration of holes at equilibrium.

The relation between the surface concentration and bulk concentration of holes is

$$p_s = p_0 \exp(e\Delta\phi_{sc}/kT). \tag{4.16}$$

The theory of electron transfer, described above, gives the following relation between the exchange current densities in the conduction and the valence band:

$$\ln\left(\frac{j_{c,0}}{j_{v,0}}\right) = \ln\left(\frac{N_c}{N_v}\right) - \frac{E_{gap}+2\lambda}{2\lambda kT}\cdot(E_i - E_{F,redox}), \tag{4.17}$$

where N_c, N_v are the effective densities of states in the conduction or valence band, respectively; λ is the reorganization energy of the redox system; $E_{F,redox}$ is the Fermi energy of the redox system [see (4.7)]; and $E_i = (E_c + E_v)/2$ is the energy of the middle of the band gap.

Neglecting concentration polarization ($C_{red} = C_{red,0}$ and $C_{ox} = C_{ox,0}$) and assuming $\Delta\phi_H$ = constant, so that $dV = d\Delta\phi_{sc}$, one obtains from (4.13–16) the following simple current-voltage relations:

$$j_c = j_{c,0}[1 - \exp(e\eta/kT)], \tag{4.18a}$$

$$j_v = j_{v,0}[\exp(e\eta/kT) - 1], \tag{4.18b}$$

where $\eta = V - V_0$ is the overvoltage. If $\Delta\phi_H$ is not constant, but varies approximately linearly with V in the form

$$d\Delta\phi_H = \beta dV; \quad d\Delta\phi_{sc} = (1-\beta)dV; \quad 0 < \beta < 1 \tag{4.19}$$

the following equations for the current-voltage curves are obtained:

$$j_c = j_{c,0}\{\exp(\alpha_c\beta e\eta/kT) - \exp[-(1-\alpha_c\beta)e\eta/kT]\}, \tag{4.20a}$$

$$j_v = j_{v,0}\{\exp[[1-(1-\alpha_v)\beta]e\eta/kT] - \exp[-(1-\alpha_v)\beta e\eta/kT]\}, \tag{4.20b}$$

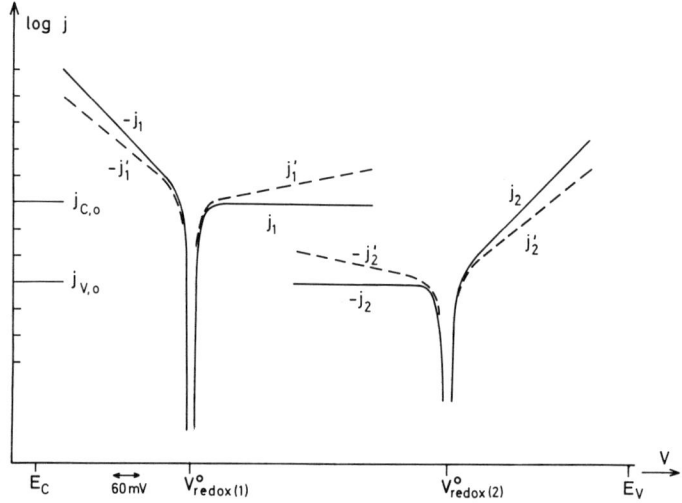

Fig. 4.6. Typical current voltage curves for redox reactions at semiconductor electrodes for electron exchange with conduction band (redox 1) or with valence band (redox 2). Full lines: constant $\Delta\phi_H$; broken lines: $\beta=0.4$, $\alpha=0.5$

where α_c is the charge transfer coefficient for the anodic process in the conduction band, and α_v in the charge transfer coefficient for the anodic process in the valence band. α_c and α_v are between 0 and 1; for simple one-electron transfer reactions, their values are close to 1/2.

Figure 4.6 shows current voltage curves for the ideal case ($\beta=0$) and a case with $\beta=0.4$, $\alpha=0.5$ for the two redox systems, one with electron exchange in the conduction band, the other in the valence band.

For both bands to contribute significantly to the current, the semiconductors must have a small band gap and a high rate of minority carrier generation. Semiconductors for solar energy conversion, however, must have a band gap >1.2 V and a high electric conductivity in the bulk. This means that only highly n-type or p-type materials can be employed. In the absence of illumination, minority carriers will not be available at the surface of these materials, and reactions which consume minority carriers will be limited to the extent of the generation rate. On the other hand, reactions with majority carriers can proceed at a high rate, and will be controlled by their surface concentration.

If surface states are present to a large extent and the amount of charge due to them varies drastically with the applied voltage, the linearity of the logarithmic current-voltage curves of Fig. 4.6 is lost, and the curves become rather complex. Another complication which is introduced by the presence of surface states is electron transfer via such surface states [4.33, 37] which modifies the kinetic equations considerably. We shall come back to this modification later.

4.1.3 Photocurrents and Photovoltages

The origin of photoeffects at semiconductor electrodes is the same as in solid-state p–n junctions or Schottky barriers. If light with more than the band-gap energy is absorbed in a space-charge layer, the generated electron-hole pairs are separated by the electric field present there. A Schottky barrier can be formed at the semiconductor-electrolyte contact if the electrolyte contains a suitable redox system which consumes majority carriers in reaching equilibrium for electron exchange. It may also be formed by applying an external voltage which removes the majorities from the interface. In both cases a depletion layer is formed beneath the surface of the semiconductor.

Generation of minority carriers by illumination will cause a photocurrent if the minority carriers can react at the interface with species of the electrolyte. Otherwise, charge separation will create a countervoltage which will compensate the voltage drop in the space-charge layer such that electron-hole pair generation is compensated for by recombination. In most systems, minorities will react with the electrolyte or cause electrolytic decomposition of the semiconductor.

The current at constant illumination intensity depends on $\Delta\phi_{sc}$ and will reach a saturation limit when the space-charge layer extends so far that all light is absorbed within it. This should result in a quantum yield close to one, which in some cases is indeed obtained. However, more often recombination at the surface or in the space-charge layer causes considerable losses, especially if the minority carriers are trapped at recombination centers in the space-charge layer.

A model can easily be treated [4.38, 39] in which all minority carriers generated in the space-charge region of length L contribute to the photocurrent. The photocurrent-voltage curve for monochromatic light which is absorbed in the semiconductor with an absorption coefficient a then has the following form for an n-type semiconductor:

$$i = eI_0[1 - (1 + \alpha L_p)\exp(-\alpha L)] \tag{4.21}$$

with $L = L_D[(V - V_{fb})/kT]^{1/2}$, where $L_D = (2\varepsilon\varepsilon_0 kT/eN_D)^{1/2}$ is the Debye length of the space-charge layer, and $L_p = (D_p\tau_p)^{1/2}$ is the diffusion length of holes.

Wilson [4.40] has taken surface recombination into account to explain the S-shaped appearance of the measured photocurrent-voltage curves. With adjustable parameters for the rate of surface recombination and the rate of charge transfer at the interface, he simulated the measured current-voltage curve to a good approximation. However, as *Wilson* himself has pointed out, this model neglects serious complications like the influence of the photogenerated carriers on the charge distribution in the depletion layer, an effect which should be very important if minority carriers are trapped in the depletion layer. A digital simulation of the processes in the depletion layer has been worked out by *Laser* and *Bard* [4.41] and a very detailed analysis of the

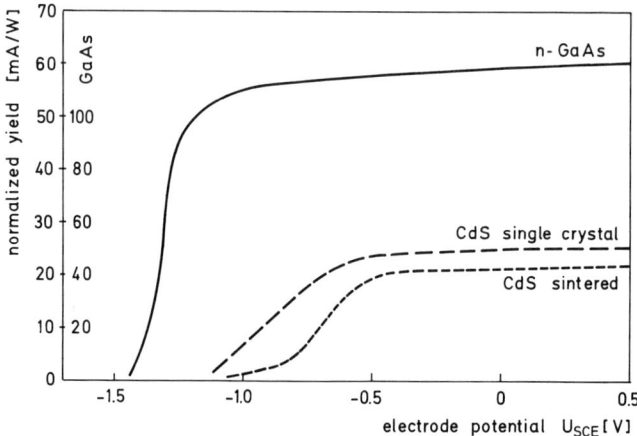

Fig. 4.7. Photocurrent-voltage curves of n-type semiconductors in contact with redox-electrolytes: n-GaAs in S^{2-}/S_n^{2-}-aqueous solution; CdS in $[Fe(CN)_6]^{4-}/[(Fe(CN)_6]^{3-}$-aqueous solution. Illumination intensity $5\,\text{mW}\,\text{cm}^{-2}$

problem including the influence of the rate of the electrode reaction has recently been published by *Reiss* [4.42].

Since the theoretical situation is rather unsatisfactory, individually measured photocurrent-voltage curves are at present the only reliable information of the material properties regarding the efficiency of charge separation in the depletion layer. In the best systems, the photocurrent approaches saturation at voltages in the order of 0.3 V above the flat-band potential. Figure 4.7 gives three examples of photocurrent-voltage curves for different materials under illumination with light quanta in the energy range above the band gap energy. The voltage range between the onset of the photocurrent and the value where saturation is reached gives some indication about the quality of the barrier with respect to charge separation. We see in Fig. 4.7 that the GaAs crystal used in this experiment has a much better quality than the CdS samples. Figure 4.8 demonstrates the influence of the absorption coefficient on the photocurrent-voltage curves.

Light with energy below that of the band gap, which penetrates much deeper into the crystal, gives a far smaller photocurrent and does not saturate before such a high field strength is reached near the surface that the dark current begins to increase significantly.

Photovoltages are even more difficult to analyze or to calculate because the simplifying condition, that all minority carriers which are generated in the depletion layer or enter it also reach the surface, cannot be applied in this case. Recombination in the depletion layer and at the surface in balance with the generation rate of carriers determine the photovoltage reached in the steady state. One should also not forget that the photovoltage is only well defined if the semiconductor electrode is in electronic equilibrium with the electrolyte in

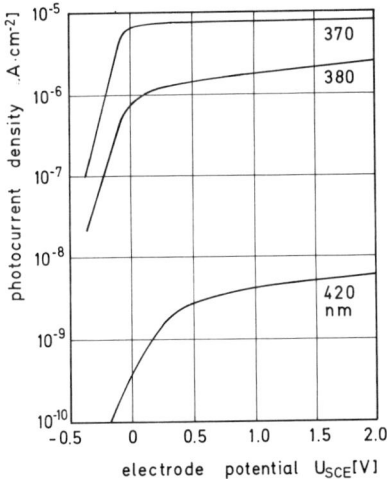

Fig. 4.8. Photocurrent voltage curves of ZnO electrodes in aqueous electrolytes for irradiation with light of different wavelengths. Above band gap energy: $\lambda = 370$ nm; at band gap energy: $\lambda = 380$ nm; below band gap energy: $\lambda = 420$ nm

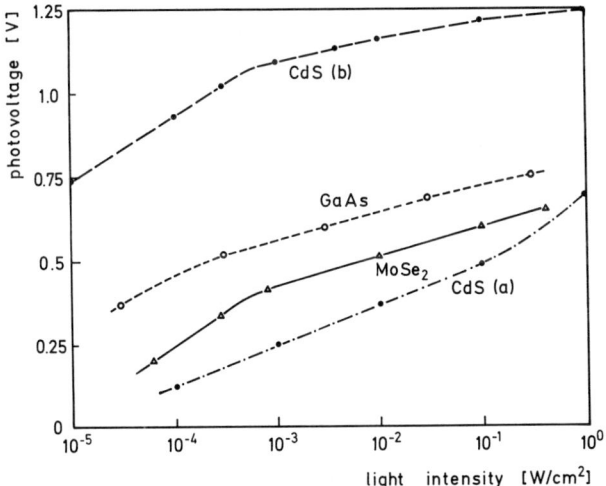

Fig. 4.9. Photovoltages at open circuit vs light intensity at "above-band-gap" illumination for three semiconductors in contact with redox electrolytes. GaAs in S^{2-}/S_n^{2-} aqueous solution; MoSe in I^-/I_3^- aqueous solution; CdS in S^{2-}/S_2^{2-} aqueous solution (a); CdS in $[Fe(CN)_6]^{4-}/[Fe(CN)_6]^{3-}$ aqueous solution (b)

darkness. Equilibrium requires the presence of a redox couple and necessitates relating the photovoltage to the equilibrium potential of this redox system.

Figure 4.9 shows some examples of the relation between photovoltage and light intensity with simulated sunlight illumination. Photovoltage spectra are not as characteristic for the semiconductor properties as photocurrent spectra since the approximate, logarithmic relation between absorbed light intensity

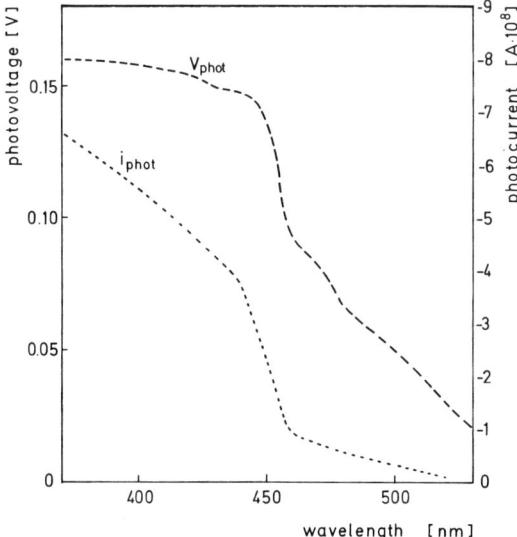

Fig. 4.10. Photovoltage and photocurrent spectrum of a p-GaP electrode in the redox-electrolyte Na_2S/Na_2S_2 (1 M)

and photovoltage effaces a great deal of the differences. To demonstrate this, Fig. 4.10 shows a comparison of a photocurrent spectrum with a photovoltage spectrum at similar conditions.

4.1.4 The Driving Force of Photoelectrolysis

The illuminated semiconductor electrode of a photoelectrolytic cell is connected to a counterelectrode of high conductivity to close the electric circuit with a minimum of electric resistance. If no corrosion process occurs at either one of the two electrodes, then in the nonilluminated state, with the circuit closed, the Fermi level of the semiconductor is equal to that of the counterelectrode. We shall assume here that both electrodes are inert and act only as electron donors or acceptors. In order to obtain electronic equilibrium with the electrolyte, a redox system must be present which exchanges electrons rapidly with at least one of the electrodes of this galvanic cell. It will usually be the counterelectrode with metallic properties which will have the highest rate of electron exchange.

For the purpose of photoelectrolysis, a Schottky barrier must be formed at the semiconductor electrolyte contact by the adjustment to equilibrium with the redox couple in the electrolyte. The initial amount of band bending $\Delta\phi_{sd,0}$ between the bulk and the surface at equilibrium is given by the difference between the redox potential and the flat-band potential

$$\Delta\phi_{sc,0} = V_{redox} - V_{fb}. \tag{4.22}$$

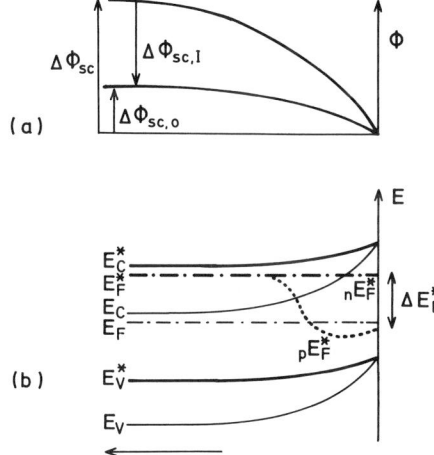

Fig. 4.11a and b. Curves of electrical potential (**a**) and electron energies (**b**) at band edges in the dark (thin lines, index 0) and under illumination (thick lines, index I)

This situation is shown in Fig. 4.11a. This figure represents also schematically the situation with illumination of the depletion layer. $\Delta\phi_{sc,I}$ depends on the resistance of the circuit and on the illumination intensity I. The difference $(\Delta\phi_{sc,I} - \Delta\phi_{sc,0})$ can be considered as the driving force of electrolysis for the majority carriers (electrons in Fig. 4.11a) since, in the bulk of the n-type semiconductor, they have this excess energy with respect to the counterelectrode.

Usually, at solid-state Schottky barriers, only the difference of the electronic free energy of the majority carriers between the metal contact and the bulk of the semiconductor is considered as the photovoltage. A driving force for minority carriers which they might need to overcome any barrier for charge transfer at the interface is neglected. This is based on the assumption that electrons or holes can pass the interface very easily. This approximation which even for solid/solid contacts is not always justified, can generally not be maintained for semiconductor-electrolyte interfaces because electron transfer between redox systems and semiconductors is often quite slow at the equilibrium potential. Either activation energy barriers or chemical barriers for the formation of energy rich intermediates must be overcome. This causes the well-known overvoltages in electrochemical reactions.

If the electrochemical reaction in which the minorities are consumed is slow, an additional driving force is needed for the photoelectrolytic cell. In such cases – and generally for all photoreactions at a semiconductor-electrolyte interface, as we shall see later – it is useful to characterize the free energies of the majority and the minority carriers individually. This idea has been very successfully used for semiconductors to describe the behavior of electrons and holes in the steady state of illumination by their so-called quasi-Fermi energies [4.43]. This concept is applicable as long as the distribution of electrons and holes over the quantum states of their respective energy bands corresponds very closely to thermal equilibrium, although the electron distribution between the conduction and the valence band can be very far from this. Since energy dissipation within the bands is very fast compared with recombination between bands, this is certainly justified for illuminations of solar light intensities and for boundary layers with modest band bending. Recently, *Nozik* and *Williams* have discussed the possibility that holes or electrons reach the interface as hot charge carriers and can in this way initiate reactions which would need more energy than available at the band edge [4.44]. There is, however, up to now no proof for such a mechanism playing any considerable role in useful solar photoelectrolysis.

We define the individual Fermi energies of electrons and holes in the absence of degeneracy by:

$$_nE_F = E_c + kT\ln(n/N_c); \quad n \ll N_c, \tag{4.23}$$

$$_pE_F = E_v - kT\ln(p/N_v); \quad p \ll N_v. \tag{4.24}$$

At thermal equilibrium, both Fermi energies coincide as the equilibrium condition between electrons and holes indicates,

$$n_0 p_0 = n_i^2 = N_c N_v \exp[(E_c - E_v)/kT] \tag{4.25}$$

$$_{n_0}E_F = {_{p_0}}E_F = E_F.$$

If in the steady state of illumination, the local concentration of electrons and holes differs from the equilibrium value by Δn and Δp respectively, the definitions above give the quasi-Fermi levels E_F^* as follows:

$$_nE_F^* = E_c + kT\ln[(n_0 + \Delta n)/N_c] = E_F + kT\ln(1 + \Delta n/n_0), \tag{4.26}$$

$$_nE_F^* = E_v - kT\ln[(p_0 + \Delta p)/N_v] = E_F - kT\ln(1 + \Delta p/p_0). \tag{4.27}$$

One sees that only for the minority carriers can a large deviation of the quasi-Fermi energy from its equilibrium value be achieved by illumination.

With the help of this definition, we can now take into account, at least in principle, the driving forces for both electronic carriers which is needed by the majority carriers for their passage through the external circuit, and by the minority carriers for their transfer through the interface. The whole driving force of a photoelectrolytic cell in this sense is the difference of the quasi-Fermi levels ΔE_F^* of the majority carriers outside the space-charge layer and the minority carriers at the surface of the semiconductor in contact with the electrolyte. The illuminated depletion layer between these two points acts as the generator of this force. This is also indicated in Fig. 4.11b together with a hypothetical sketch of the two quasi-Fermi levels in the boundary layer of the semiconductor.

4.2 Photodecomposition of Semiconductor Electrodes

The most serious problem in applying semiconductors in photoelectrochemical cells is their susceptibility to photodecomposition. This is a fundamental materials problem and deserves particular attention. If we understand the conditions and mechanisms of photodecomposition, we have a better chance to select the optimal materials and to find means for stabilization of such systems. This section is therefore exclusively concerned with a thorough analysis of this problem.

4.2.1 Energetic and Thermodynamic Aspects

Unlike the photodecomposition of solid compounds in vacuo or in contact with a gas where the products are either the components of the material in their elemental state (like the photodecomposition of silver halides) or newly formed electroneutral compounds with the surrounding gas, the result of photodecomposition in contact with an electrolyte is usually an ionic oxidation or reduction product of the semiconductor. The electric balance comes from simultaneous reduction or oxidation of a component in the electrolyte. Such decomposition reactions are electrochemical reactions connected with the consumption or liberation of electronic charge at the interface. Their rate can be measured as a current and depends on the free energy of the electrons at the surface of the semiconductor.

It has been found that the electrolytic oxidation of semiconductors is always connected with the holes of the valence band as electronic reactants, while the electrolytic reduction of semiconductors is connected with electrons of the conduction band [4.45]. We shall therefore formulate the decomposition of a binary compound semiconductor MX as follows

$$MX + zh^+ + zY^- \cdot solv + solv \rightarrow M^{z+} \cdot solv + zY^- \cdot solv + X, \tag{4.28}$$

$$MX + ze^- + solv + A^{z+} \cdot solv \rightarrow M + X^{z-} \cdot solv + A^{z+} \cdot solv. \tag{4.29}$$

These are only the simplest types of decomposition reactions. In more complex systems, both components can be transformed to higher oxidation states with additional hole consumption or, in some cases, coupled with electron injection. Similarly, reduction can proceed further for both components, if such valency states are possible, via more electron consumption or eventually via hole injection. The steps which are rate determining in all such processes involve holes as reactants in case of oxidation and electrons in case of reduction.

This role of the hole or the electron in such electrolytic processes is quite understandable. The presence of holes at the surface is equivalent to missing electrons in the bonding states of the valence band. This must weaken the bonds between neighboring atoms and will make them more susceptible to interaction with the electrons of nucleophilic reactants. To what degree this bond weakening initially occurs will depend on the localization of the wave functions. All surface atoms are statistically involved, but atoms at some particular surface sites, e.g., kink sites, dislocations, or other lattice defects, will be especially favoured for "transitorial hole capturing". Interaction with electron-donating atoms or molecules from the electrolyte will then cause final localization of holes by forming new chemical bonds.

The same but electrically opposite role can be attributed to electrons of the conduction band. If they occupy antibonding states, their localization causes equal bond weakening by making this site susceptible to interaction with electrophilic reactants of the electrolyte. By forming new bonds with such partners, reduced states are stabilized and decomposition can proceed.

Reactions (4.28) and (4.29) represent redox reactions, which occur irreversibly. However, they can be treated as reversible and a thermodynamic redox potential can be attributed to them [4.46, 47]. Formally, this redox potential can be derived from thermodynamic quantities of all species involved in these reactions. Redox potentials are measured against a reference electrode and thermodynamic quantities are well defined only for electroneutral reactions. One has therefore to add the corresponding electrode reaction of the reference system to the reaction in question to get an electroneutral cell reaction for which free energy data can be found in tables. Using the hydrogen electrode as a reference system, the following reactions have to be used for completion of the galvanic cell reaction:

$$z H^+ \cdot \text{solv} \rightleftharpoons (z/2) H_2 + \text{solv} + zh^+ \tag{4.30}$$

$$z H^+ \cdot \text{solv} + ze^- \rightleftharpoons (z/2) H_2 + \text{solv}. \tag{4.31}$$

Both reactions are fully equivalent. However, we use both formulations, one with holes, and the other with electrons, to make these reactions consistent with the formulation of the decomposition reactions in (4.28, 29).

The combination of reactions (4.28) and (4.30) gives

$$MX + zH^+ \cdot \text{solv} + zY^- \cdot \text{solv} \rightleftharpoons M^{z+} \cdot \text{solv} + zY^- \cdot \text{solv} + H_2 \tag{4.32}$$

with a free energy difference of ΔG_{32}. From this thermodynamic value one obtains the standard redox potential of reaction (4.28) versus the standard hydrogen electrode. We shall denote this as the thermodynamic decomposition potential for oxidation by holes, $_pV_{\text{decomp}}$.

$$_pV_{\text{decomp}} = \frac{\Delta G_{32}}{zF} \tag{4.33}$$

with F the Faraday constant.

Combining (4.29, 31) results in

$$MX + (z/2)H_2 + solv \rightleftharpoons M + zH^+ \cdot solv + X^{z-} \cdot solv \qquad (4.34)$$

with ΔG_{34}. The thermodynamic decomposition potential for reduction by electrons is

$$_nV_{decomp} = -\Delta G_{34}/zF. \qquad (4.35)$$

We have introduced in the previous section individual Fermi energies of electrons and holes [cf. (4.23, 24)] and we have shown before that redox potentials and Fermi levels correspond to each other [see (4.7)]. Using these relations, we find that the decomposition potentials given by (4.33, 35) correspond to definite positions of the Fermi levels of holes or electrons. These can be transformed into critical Fermi energies for decomposition of the semiconductor by holes ($_pE_{decomp}$) or electrons ($_nE_{decomp}$) on the scale of electronic energies which represent the properties of the solid. We denote these quantities as "decomposition Fermi levels". Particularly instructive is the position of the decomposition Fermi levels relative to the band edges at the interface [4.46]. This can be obtained if the flat-band potential and the distance between the Fermi level in the bulk and the band edges are known [see (4.5–7)]:

$$E_{decomp} = E_{fb} - e(V_{decomp} - V_{fb}). \qquad (4.36)$$

This equation is valid for anodic and cathodic decomposition Fermi levels.

If both decomposition Fermi levels are known, or can be derived from thermodynamic data, one can predict whether a semiconductor in a particular electrolyte is thermodynamically stable at a given redox potential. The criteria are obvious:

For stability:

$$_pV_{decomp} > V_{redox} > _nV_{decomp}; \quad \text{equivalent to}$$
$$_pE_{decomp} < E_{redox} < _nE_{decomp}.$$

For instability:

$$V_{redox} > _pV_{decomp} \quad \text{or} \quad V_{redox} < _nV_{decomp}; \quad \text{equivalent to}$$
$$E_{redox} < _pE_{decomp} \quad \text{or} \quad E_{redox} > _nE_{decomp}.$$

Since the decomposition potentials depend on the composition of the electrolyte (solvents, counter ions, complexing ligands) these criteria will vary for each individual semiconductor to some extent with the particular electrolyte. However, the decomposition Fermi level in some way reflects the bond strength between the components of the semiconductor and therefore one can expect to find a general trend for different materials of similar character.

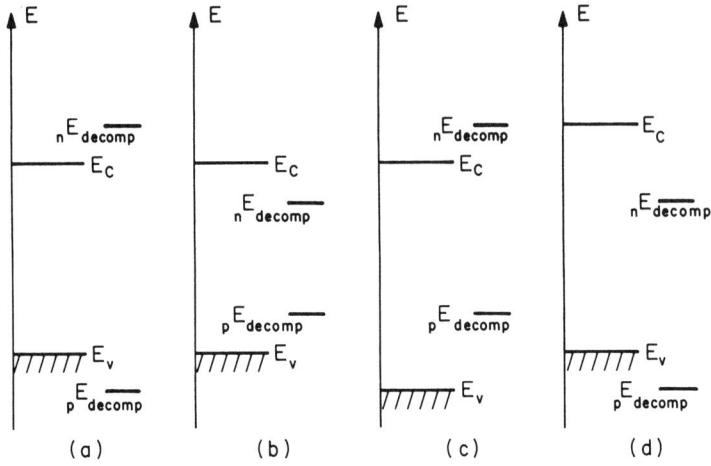

Fig. 4.12a–d. Thermodynamics of photodecomposition of semiconductors. Four typical cases: (**a**) stable, (**b**) anodically and cathodically unstable, (**c**) cathodically stable, (**d**) anodically stable

We turn now to the illuminated semiconductor. The only difference here is that we have to consider quasi-Fermi levels in the steady state instead of the real equilibrium values, although the conditions for anodic or cathodic decomposition are the same as above. We have previously defined the quasi-Fermi levels of electrons and holes in relation to the energy position of the band edges (4.26, 27). We have also seen that the only systems of any interest for photoelectrolysis are those which form Schottky barriers at equilibrium. Since the maximum splitting of the two quasi-Fermi levels cannot exceed the band gap, the range of the quasi-Fermi levels which can be reached under illumination is limited to the range between the position of the band edges at the semiconductor electrolyte contact. The position of the decomposition Fermi levels in relation to the position of the band edges is therefore decisive for the stability of a semiconductor electrode against photodecomposition.

Consequently, four different cases for semiconductor electrodes with regard to their photodecomposition behavior can be envisaged. They are represented in Fig. 4.12. If the decomposition Fermi level is located outside the band gap the semiconductor is thermodynamically stable; if one or both are positioned inside the gap, it is unstable for decomposition in one or both directions. In this case, the farther the distance of the decomposition Fermi level from the respective band edge, the easier it will be to start the decomposition reaction. Unfortunately, no real semiconductor electrode seems to fall into category (a) of Fig. 4.12, where one would be absolutely safe against photodecomposition. All presently known semiconductors belong to categories (b) or (c) and therefore are at least susceptible to anodic photodecomposition. A semiconductor falling under category (b) can even decompose under illumination anodically and cathodically at the same time. This mechanism of photodecomposition without any external current flow corresponds to photodecomposition in vacuo.

4.2.2 Kinetic Aspects

Whether or not photodecomposition occurs really depends to a large extent on kinetics. If activation barriers are high, a semiconductor can be stable enough for practical use even under conditions where thermodynamics predict instability. There is good reason why kinetics can help to prevent decomposition of semiconductor electrodes, at least to some extent. This is connected with the mechanism of crystal decomposition [4.48].

It is generally accepted that formation and decomposition of a crystal proceeds via kink sites, where the atoms average half of the binding energy they would have inside the crystal. Kink sites are present in steps on the surface or at the intersections of dislocations. If they are scarce or blocked by inhibitors, their formation needs extra energy, thus slowing down the rate of decomposition. Even if enough active kink sites are present, complete removal of an atom or a molecule from a kink site at a crystal lattice can only occur in a series of steps. Since interaction with a number of next nearest neighbors has to be interrupted this never occurs in one stage. The thermodynamic free energy change for the process of crystal decomposition is therefore the sum over various reaction steps. At equilibrium, ΔG between the species in kink sites and the products in solution is zero. This means for a complex reaction occurring in steps that one or more of the steps need excess energy, others release energy. The free energy difference is still zero for each of the individual steps due to the compensation of the differences in the energy terms by the entropy terms which contain the concentrations (or activities) of the intermediates. There is usually one step which leads to the most energetic intermediate and will therefore be rate determining. Consequently, the activation energy needed to perform this reaction step will be the controlling factor for decomposition. This means that the electronic free energy needed as a driving force for the decomposition reaction can differ considerably from the thermodynamic standard free energy of the overall reaction which we have derived in the first part of this section.

To give an example, a model surface is shown in Fig. 4.13 in a two-dimensional representation of a compound semiconductor MX. The drawing shows the stepwise oxidation of the two molecules forming the kink site and demonstrates the successive association of holes and of ligands, L, to the surface. The filled and vacant electronic orbitals of the surface atoms are also indicated. This model approximates the decomposition of CdS or ZnO, where the electronegative component is oxidized no further than to the elemental state. In the case of GaAs, the oxidation of both components would go on to higher valence states and the number of single steps for the oxidation of a kink site molecule would multiply.

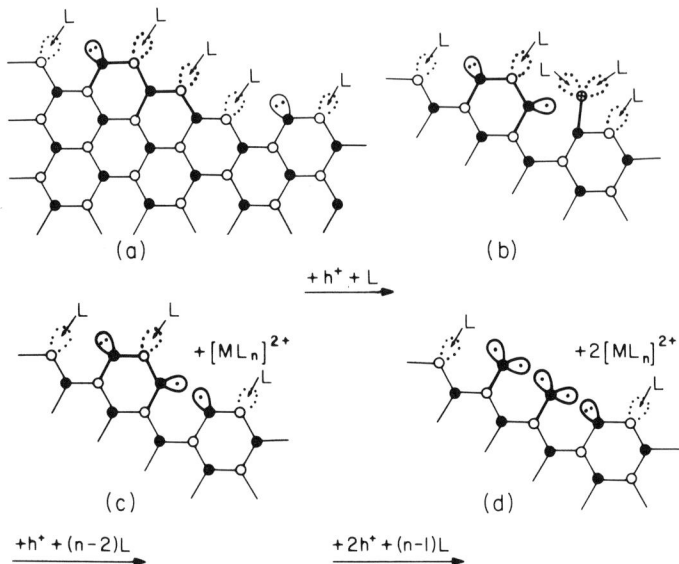

Fig. 4.13a–d. Two-dimensional model for oxidation by photogenerated holes of kink-site atoms with electronic orbitals which are exposed to the electrolyte

For this relatively simple model system, we can divide the overall reaction into three steps, two electrochemical ones and one chemical

$$MX + h^+ + L \to M_{ad}^+ L + X_{ad}, \tag{4.37a}$$

$$M_{ad}^+ L + h^+ + L \to [ML_2]^{2+}, \tag{4.37b}$$

$$X_{ad} + X_{ad} \to X_2, \tag{4.37c}$$

where the index ad denotes an adsorbed or chemisorbed state at the surface. The net reaction consists of two times each reaction (a) and (b) plus once reaction (c):

$$2MX + 4h^+ + 4L \to 2[ML_2]^{2+} + X_2. \tag{4.38}$$

The decomposition Fermi level calculated from the thermodynamic data for reaction (4.38) tells us nothing about the energetics of reaction (4.37a) which generates the intermediates with the highest energy and will therefore be rate determining. We can attribute to this step its own individual redox potential for equilibrium at standard conditions. This will exceed the overall decomposition potential by some amount, and possibly a very large one. Consequently at the equilibrium potential of the overall reaction, the concentration of the intermediates will be very small or even negligible with the result that the reaction rate can be extremely low.

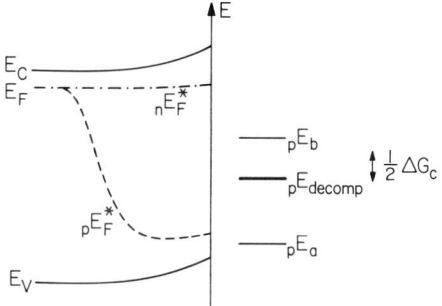

Fig. 4.14. Thermodynamics of photodecomposition with intermediate steps [cf. (4.37a–c, 38)]. $_pE_a$ and $_pE_b$ are decomposition Fermi energies for the individual redox reactions within the sequence of the overall process

A better indicator of the free energy of holes (or electrons) required to cause decomposition of the semiconductor in such a case is the redox potential, or the Fermi energy for the slowest redox step in such a sequence. This is shown schematically in Fig. 4.14 for the sequence of reactions (4.37a–c) with the net reaction (4.38). The distance of the standard Fermi level of the redox step, needing the highest free energy of holes, from the Fermi level of the net reaction gives some idea how much the quasi-Fermi energy of holes has to be increased to start photodecomposition. However, one should not forget that such a reaction will already proceed at a much smaller concentration of the intermediates than needed for the standard state. Therefore, the reaction will reach a considerable rate at quasi-Fermi energies intermediate between this individual free-energy level and the real equilibrium Fermi energy for this net reaction. How early the decomposition reaction begins and how fast it proceeds depends on other kinetic factors and cannot be predicted from general principles.

The model of Fig. 4.13 gives some indication of the conditions necessary for the intermediates to be formed in crystal decomposition which will control the height of activation barriers. To interrupt a bond to neighboring atoms of the crystal, it is not enough to remove electrons from bonding orbitals. One also needs reaction partners which can interact with the weakened bond site. This interaction depends mainly on the overlapping between the electronic wave function of this activated atom and the reactant. If the electron orbitals of surface atoms available for interaction extend into the electrolyte, interaction is strong and the activation barrier to form the intermediate will be low. If there is little overlap between the wave functions of the reactants the activation barrier will be very high. We shall discuss in the next part of this section a class of crystals, namely layer crystals, where the rate of decomposition is drastically decreased by the shielding of electronic orbitals of the crystal against overlapping with the wave functions of the reactants from solution.

There is another process which can help prevent photodecomposition. It is the competition of other redox reactions if they are better scavengers for the photogenerated minorities. This is usually the case for one-electron transfer reactions if they involve no kinetic complications. They have high rate constants and are usually much faster than electrochemical decomposition reactions. This is unfortunately not true for most of the redox reactions in which the solvent itself is oxidized or reduced, e.g., for the redox reactions of water. These reactions are complex and proceed equally through a series of steps as does the crystal decomposition. They undergo similar kinetic restrictions and will therefore only protect a semiconductor against decomposition if their thermodynamic decomposition potential is reached at much lower concentration of the respective minorities than needed for photodecomposition. In Fig. 4.15, the critical conditions for such competitive reactions which might protect a semiconductor against photodecomposition are schematically indicated.

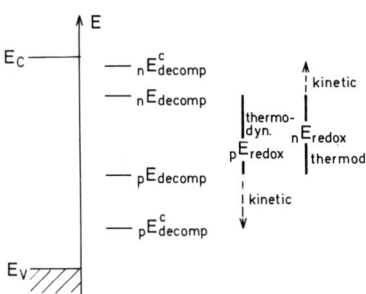

Fig. 4.15. Ranges of thermodynamic and kinetic protection of semiconductors against photodecomposition by redox systems. E^c_{decomp} are the critical Fermi levels with respect to kinetics

4.2.3 Materials for Electrochemical Solar Cells

Various semiconductors which can be doped highly enough to get sufficient conductivity have been used for solar photoelectrolysis. A thermodynamic analysis of most systems for which the required free-energy data are available (sometimes only in an approximate form for reaction products in particular solutions) has shown that all these semiconductors are, in principle, unstable against anodic photodecomposition in aqueous solutions. In spite of this, some materials have resisted decomposition very well for reasons we shall analyze below.

Figure 4.16 gives a summary of the position of band edges and decomposition Fermi levels for a series of semiconductors which have been studied in aqueous electrolytes. The reactions for which the decomposition Fermi levels are calculated are:

$$SnO_2 + 4HCl \cdot aq \rightarrow SnCl_4 \cdot aq + O_2 + 2H_2$$
$$SnO_2 + 2H_2 \rightarrow Sn + 2H_2O$$
$$WO_3 + H_2O \rightarrow WO_2 + (1/2)O_2 + H_2$$
$$WO_3 + 3H_2 \rightarrow W + 3H_2O$$
$$ZnO + 2HCl \cdot aq \rightarrow ZnCl_2 \cdot aq + (1/2)O_2 + H_2$$
$$ZnO + H_2 \rightarrow Zn + H_2O$$

$$TiO_2 + 4HCl \cdot aq \rightarrow TiCl_4 \cdot aq + O_2 + 2H_2$$
$$TiO_2 + 2H_2 \rightarrow Ti + 2H_2O$$
$$Cu_2O + H_2O \rightarrow 2CuO + H_2$$
$$Cu_2O + H_2 \rightarrow 2Cu + H_2O$$
$$CdS + 2HCl \cdot aq \rightarrow CdCl_2 \cdot aq + 2S$$
$$CdS + H_2 \rightarrow Cd + H_2S \cdot aq$$
$$MoS_2 + 2H_2O \rightarrow MoO_2 + 2S + 2H_2(_pE_{dec})$$
$$MoS_2 + 2H_2 \rightarrow Mo + 2H_2S \cdot aq$$
$$GaP + 6H_2O \rightarrow Ga(OH)_3 + H_3PO_3 \cdot aq + 3H_2$$
$$GaP + (3/2)H_2 \rightarrow Ga + PH_3$$
$$GaAs + 5H_2O \rightarrow Ga(OH)_3 + H_3AsO_3 \cdot aq + eH_2$$
$$GaAs + (3/2)H_2 \rightarrow Ga + AsH_3.$$

This figure contains also the decomposition Fermi levels for water ($_pE_{H_2O}$ for oxidation to O_2 and H^+, $_nE_{H_2O}$ for reduction to H_2 and OH^-) for the same solution in which the band-edge positions have been measured. Since these quantities depend on pH and the composition of the electrolyte they can differ for different solutions, even for the same material.

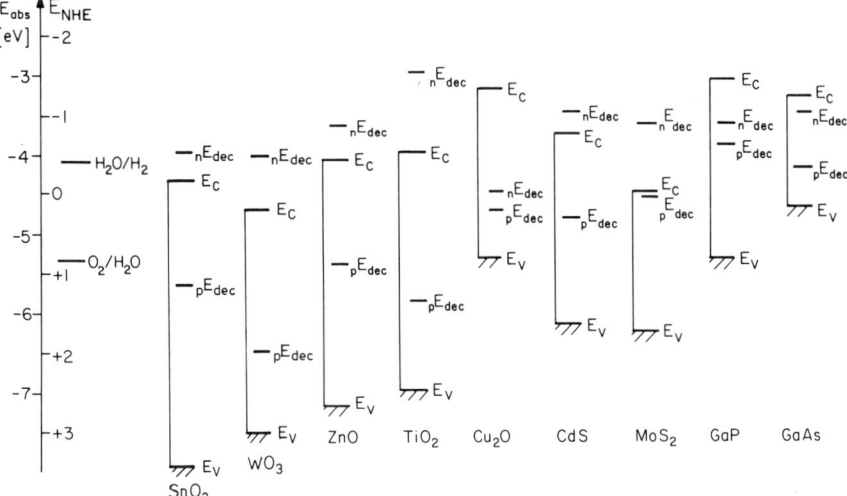

Fig. 4.16. Position of band edges and decomposition Fermi levels for various semiconductors.

The anodic decomposition of oxides competes in aqueous electrolytes with the oxidation of water. The oxidation Fermi level of water is shifted in a positive direction by 0.059 V per pH unit at normal temperature. Since the flat-band potential of oxides is shifted with pH by about the same amount, the stability of oxides against photodecomposition can be influenced little by a change in pH values. At least, this is the thermodynamic conclusion. If, despite this, there is a change, it must be due to kinetic reasons.

The experimentally found stability of a number of oxides like TiO_2, titanates, SnO_2, WO_3 and others against anodic photodecomposition obviously has kinetic reasons. This is clearly indicated by the thermodynamic data. The decomposition Fermi level of all these oxides is located far above the valence band edge. The quasi-Fermi level of holes can therefore be shifted below this level at relatively modest illumination intensity. We see, however, that the decomposition Fermi level of water is above this decomposition Fermi level of the semiconductor in most oxides. This indicates that there is a good chance that the water oxidation will proceed fast enough to keep the concentration of holes at the surface below the critical value needed to trigger the decomposition process of the crystal. In the case of ZnO, the decomposition Fermi level of the crystal and of the solvent H_2O are rather close. Consequently, both processes occur in parallel with variable rate ratios, as has been found in some experiments [4.49]. The ratio depends on crystal orientation, light intensity and the pH of the electrolyte.

Cu_2O is very unstable because it can easily be oxidized to CuO. The decomposition potentials for the formation of Cu^+ ions and O_2 or Cu^{2+} ions and O_2 are closer to the valence band edge and these reactions are therefore more difficult to reach. On the other hand, Cu_2O can also easily be reduced by electrons, since the Fermi level for cathodic decomposition is far below the position of the conduction band edge. It is therefore not surprising that Cu_2O crystals are reported to disproportionate into Cu and Cu^{2+} under illumination [4.50]. That even the stability of very stable oxides might be only a relative one has recently been demonstrated for TiO_2, which showed clear indications of photocorrosion if a high light intensity was employed [4.51].

The situation for the nonoxidic semiconductors which are represented in Fig. 4.16 shows an even more unfavorable picture. All these semiconductors are susceptible to anodic photodecomposition. Since the decomposition Fermi level is located above, often far above, the valence band edge and is above the Fermi level for water decomposition, the latter normally cannot compete with the photodecomposition of the crystal. Here the situation can be somewhat different with varying pH since the flat band potential of these materials does not necessarily depend on pH. If this is nevertheless the case, it is an indication that the surface has reacted with water to form an oxidic surface compound like $\geqslant$P–OH or $\geqslant$As–OH for phosphides or arsenides.

It has been proposed to use semiconductors in their saturated solution as photoelectrodes, like CdS in sulfide electrolytes, to prevent decomposition [4.52–56]. This concept has lead to remarkable success in some cases which will be discussed in the next section. However, as can be easily shown, even this does not result in full stability against photodecomposition. If the semiconductor is inserted into a saturated solution, we have equilibrium between the solid and the dissolved species in the electrolyte. For example, for a compound semiconductor MX which dissolves in form of its solvated ions as

$$\text{MX} + \text{solv} \rightleftarrows \text{M}^{z+} \cdot \text{solv} + \text{X}^{z-} \cdot \text{solv} \tag{4.39}$$

the free energy difference ΔG_{39} between the solid and the dissolved compound is zero. The decomposition potential for reaction (4.28) is therefore the same as that for the oxidation of the component $\text{X}^{z-} \cdot \text{solv}$ from the solution.

$$\text{M}^{z+} \cdot \text{solv} + \text{X}^{z-} \cdot \text{solv} + zh^+ \rightleftarrows \text{M}^{z+} \cdot \text{solv} + \text{X} + \text{solv} \tag{4.40}$$

with $\Delta G_{40} = \Delta G_{28} + \Delta G_{39} = \Delta G_{28}$, for $\Delta G_{39} = 0$.

Again, it is a question of kinetics whether the components of the crystal or of the electrolyte are oxidized by holes. Usually, it will be the species in solution which reacts faster because the bonds with the solvation shell are less rigid than those to other atoms in the solid. Atoms in the crystal surface, however, have a better chance for hole capturing since they are strongly coupled to the electrons of the crystal lattice and therefore have a large reaction cross section. The difference of the real reaction rates will depend on the activation energy needed to form the intermediates in both reactions. In the case of CdS where this problem has been investigated, it turned out that the crystal slowly decomposes even in a saturated solution containing the redox system $\text{S}^{2-}/\text{S}_2^{2-}$ [4.57]. The result of decomposition in this case is an oversaturation of CdS due to an increased Cd^{2+} concentration at the interface, and consequently microcrystals of CdS are formed as a very thin layer on top of the original single crystal. This thin polycrystalline layer seems to slow down further decomposition after some time, but its presence decreases the photocurrent quantum yield.

No photocorrosion could be detected in the form of weight loss with CdSe in sulfide solution [4.54]. It turned out, however, that the surface of a CdSe crystal is converted into a CdS layer in such a photoelectrochemical cell [4.51–59]. This CdS layer blocks the photocurrent if it becomes too thick for electron transfer by tunneling. This blocking can be prevented if the electrolyte contains a mixture of sulfide and selenide that probably causes the formation of a mixed $\text{CdS}_x\text{Se}_{1-x}$ layer on the electrode [4.58].

A particularly interesting class of materials are the transition metal chalcogenides which form layer crystals like MoS_2, WS_2, and others. Their electrochemical behavior has been studied by *Tributsch* et al. [4.60–65]. These crystals are semiconductors with band gaps between 1.4 and 1.9 eV. They are available as p-type or n-type specimens. Figure 4.17 shows schematically the structure of such a layer crystal and the energy position of the electronic levels of the valence and conduction bands of MoS_2 and MoSe_2 with respect to the electronic levels in the electrolyte [4.64]. The thermodynamic decomposition

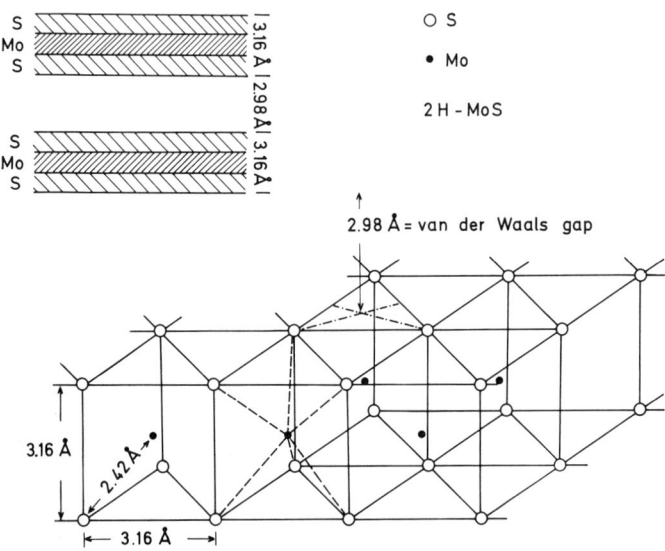

Fig. 4.17a. Structure of an MoS_2 crystal

potential of MoS_2 is shown in Fig. 4.16 for one particular final product. The very negative value indicates that MoS_2 should be very easily decomposed by oxidation. It turns out, that $MoSe_2$ should, thermodynamically, be even less stable. In the presence of redox couples which are easier to oxidize than water, MoS_2 and $MoSe_2$ are, however remarkably stable [4.61, 64].

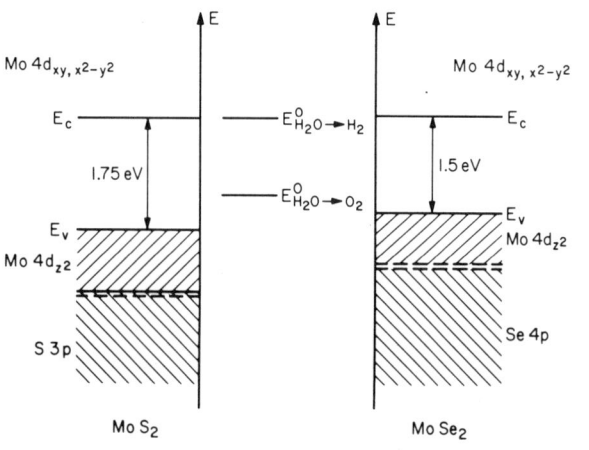

Fig. 4.17b. Position of electron energy bands of MoS_2 and $MoSe_2$ with respect to redox energy levels of aqueous electrolytes [4.63, 64]

Tributsch has attributed this stability to the fact that the electronic states of the highest valence band in which the reacting holes are generated are formed by the $4d_{z^2}$ electron states of the Mo atoms which are nonbonding states in the theoretical description of the crystal electrons. They indeed contribute very little to the bonding between the Mo atoms and the two layers of S atoms between which the metal atoms are positioned. However, these $4d_{z^2}$ states interact strongly enough with each other to be split over an energy range of 0.6 eV due to degeneracy. The term nonbonding states comes from a comparison with the energy of the electrons in the isolated atoms. A transition to higher energies in the d band leads to antibonding states which let the crystal energy increase and make the surface more reactive.

A somewhat more adequate description of the reasons for the high stability of these layer crystals can be found in the very weak interaction between the electronic states forming the highest valence band and possible reactants in the electrolyte. Since the $4d_{z^2}$ states of the Mo atoms do not interact with the p states of the sulfur atoms above and beneath them, they scarcely extend into the electrolyte in great contrast to the surface atoms in the model of Fig. 4.13 where the holes are found in states which extend largely beyond the interface and can easily interact with components of the electrolyte. In spite of this very weak interaction with the electrolyte, the holes in the $4d_{z^2}$ states of the Mo atoms can capture electrons from H_2O molecules and oxidize water to some intermediates which are radicals, and in turn react with the sulfur atoms of the layers, and finally, after the addition of further holes, oxidize them to SO_2 and SO_3, or SO_3^{2-} and SO_4^{2-}, respectively [4.60]. The mechanism of these reactions is not yet known, but the fact that holes at the upper edge of the valence band of MoS_2 are certainly not energetic enough to form free OH^- radicals clearly indicates that the oxidation of MoS_2 must pass an intermediate state in which oxygen and water strongly interact with the sulfur atoms at the surface. Recently, experimental evidence has been obtained which shows that steps and dislocations in these layer crystals play an important role in the kinetics of their decomposition reactions [4.66]. This suggests that for this material, protection against decomposition might be achieved by merely stabilizing the local structural defects on the surface.

We have seen that stabilization of a semiconductor electrode can be obtained by competition with solvent decomposition. Instead of the solvent molecules, redox couples in the electrolyte which have a redox Fermi level above the anodic decomposition Fermi level of the crystal can do this even better. Holes will be captured then by this redox couple as long as its reduced component is present in high enough concentration. Especially efficient for this kind of competition are redox systems with a single electron-transfer step. The rate of such electron-transfer reactions is normally quite high, much higher than of reactions in which two or more electrons are needed for the net reaction. Such reactions are slow because they have to proceed via intermediates like the crystal decomposition process. Such fast redox reactions, characterized by a large exchange-current density, can even compete with the

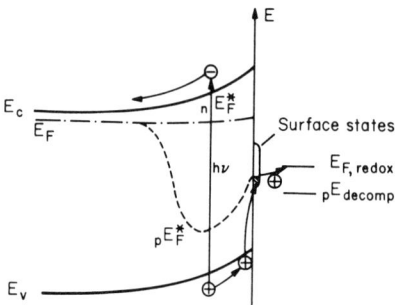

Fig. 4.18. Role of surface states at semiconductors in redox reactions and for stabilization against photodecomposition

decomposition reactions if their redox Fermi level is below the anodic decomposition Fermi level of the crystal. This has been studied at CdS electrodes by some authors using the rotating ring-disc electrode arrangement for analyzing the current efficiency for oxidation of the redox couple [4.67, 68]. It was found that the photodecomposition of CdS can be prevented if the transport of the redox species to the electrode is fast enough and the light intensity does not reach too high values. This competition, however, depends very much on the properties of individual electrodes. The presence of a large number of dislocations on the crystal surface or of grain boundaries favor the decomposition reaction.

The competition of other redox reactions with photodecomposition can be promoted to a large extent by the presence of electronic surface states on the semiconductor surface in an energy range above the anodic decomposition Fermi level. If these surface states have a large cross section for hole capturing they can pin the quasi-Fermi level of holes to the range of their energy position. They also catalyze the electron transfer reaction with redox systems in solution [4.33, 37, 69]. This favorable situation is schematically represented in Fig. 4.18. There are some indications which lead to the conclusion that surface states in the right energy range are present on TiO_2 electrodes [4.70, 71] and help to stabilize this material against photodecomposition.

We have discussed here in detail only the anodic decomposition by holes, because up to now this is the process which has limited the use of photoelectrochemical cells with n-type material. Many of the semiconductor materials used as electrodes are stable against cathodic photodecomposition as Fig. 4.16 shows. It appears therefore that p-type semiconductors seem to be more suitable for electrochemical solar cells than n-type material.

If cathodically stable, a p-type semiconductor having a cathodic decomposition Fermi level above the conduction band edge can be brought into contact with a redox system having a redox Fermi level close to this band edge. Thus an inversion layer would be formed beneath the semiconductor surface at equilibrium. This gives a chance of obtaining a large photovoltage because the initial band bending is large. However, with increasing light intensity and increasing photovoltage, the band bending decreases and the concentration of

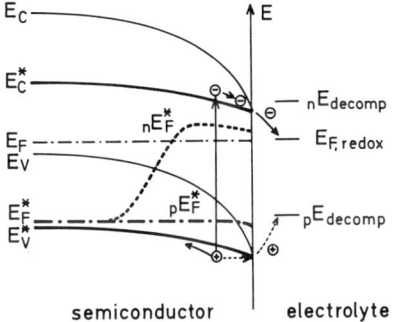

Fig. 4.19. Electron energy correlation for a *p*-type semiconductor in contact with a Schottky barrier forming redox electrolyte, at equilibrium and under illumination

holes at the surface and their free energy increase. If the quasi-Fermi level of holes reaches the anodic decomposition Fermi level, the decomposition of the crystal can proceed, although the net current at the semiconductor electrode will still be cathodic, and might increase further with illumination intensity. Since the kinetic conditions for decomposition by holes are the same as discussed for *n*-type material, one sees that even in this case, the tendency to anodic decomposition limits the stability. If the position of the Fermi level for anodic decomposition is far above the valence band edge, one cannot exploit the large band bending in the initial state to obtain the high photovoltage available from such *p*-type semiconductors without risking decomposition.

Figure 4.19 shows the energy correlations for a system where the Schottky barrier is formed by depletion of holes and the photoredox reaction is due to electrons generated in the space charge layer. The critical photovoltage which just reaches the thermodynamic stability limit is indicated in this picture. One sees that there is no difference in principle with respect to the stability of *n*-type or *p*-type specimen of the same material. However, for some types of photoelectrochemical cells, the use of *p*-type material could be quite interesting.

4.3 Function and Efficiency of Photoelectrochemical Solar Cells

4.3.1 Regenerative Cells

a) Principles of Operation

The photovoltage generated at a semiconductor electrolyte interface can be used for electrical work in just the same way as that of other photovoltaic devices [4.9]. For this purpose, the electrolytic process at the semiconductor electrode has to be exactly reversed at the counterelectrode to prevent any chemical change in the system. Since electrolysis causes local changes in chemical composition of the electrolyte next to the electrode interface, one cannot fully avoid such changes. These concentration changes go in opposite directions at the two electrodes. Equilibration of the electrolyte composition is

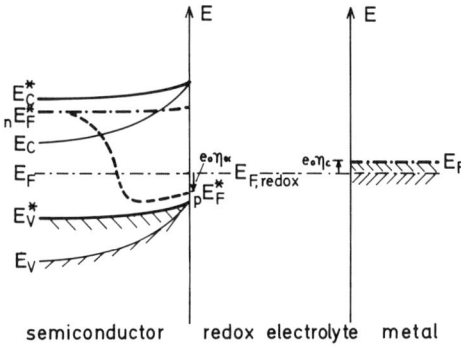

Fig. 4.20. Principle of a regenerative photoelectrochemical cell with an n-type semiconductor electrode operating as a photovoltaic device

achieved by diffusion and convection. To minimize such concentration differences, either the distance between the two electrodes has to be as short as possible, or means have to be provided for achieving efficient transport rates by convection (e.g., stirring).

Figure 4.20 shows in terms of electronic energies, the scheme of such a cell with an n-type semiconductor, in the dark and at illumination. It also gives the energy losses needed as driving forces for the electrode reactions ($|e_0 \cdot \eta|$). Further energy losses are caused by ohmic potential drops in the system, particularly in the electrolyte, since electrolytic conductivity never reaches the high values of good electronic conductors, which is one more reason to keep the electrolyte layer of such a cell thin.

We have seen in Sect. 4.1 that the photovoltage is limited by the band bending at equilibrium [cf. (4.22)]. Maximal band bending is therefore desirable for the situation of darkness. In Sect. 4.2, we discussed the restrictions which one has to observe to prevent photodecomposition. The Fermi level of the redox system in the electrolyte should not exceed the decomposition Fermi level of the semiconductor for the reaction with the minorities. If it does, one is no longer on the thermodynamically safe side, and the kinetics have to be considered. The farther the redox Fermi level of the electrolyte is below the anodic decomposition Fermi level for n-type specimen (or above the cathodic decomposition Fermi level for p-type specimen), the riskier is the use of the device.

On the other hand, the band bending is controlled by the flat-band potential which depends on the concentration of donors or acceptors in the semiconductor. It appears attractive to use high donor concentrations for n-type or high acceptor concentration for p-type specimens to obtain large band bending at dark. However, high concentration of ionized atoms in the space-charge layer has the disadvantage that the extension of the space-charge layer shrinks, with the consequence that the light is only partially absorbed in the space-charge layer. Since many, and often most, of the electron-hole pairs generated outside the depletion layer are lost by recombination, the extension of the depletion layer should correspond to the reciprocal of the absorption coef-

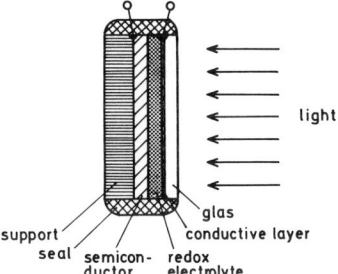

Fig. 4.21. Model of a photoelectrochemical cell for power generation

ficient, averaged over the effective part of the absorbed light, $\langle a \rangle$. Assuming that the necessary band bending for efficient charge separation is in the order of 0.25 V, this condition gives,

$$\delta_{sc} = \left(\frac{2\varepsilon\varepsilon_0}{eN} \times 0.25 \right)^{1/2} \approx \frac{1}{\langle a \rangle}, \tag{4.41}$$

where N is the donor or acceptor concentration.

For $\langle a \rangle = 3 \times 10^4$, this gives a value for N/ε of approximately 10^{15} cm^{-3}. This corresponds to a relatively low donor concentration (if ε is not extraordinarily large) which will usually be insufficient to get a high enough bulk conductivity. Therefore, one will have to compromise and must try to find materials where the lifetime of the minority carriers is long enough so that they can reach the space-charge layer by diffusion also.

There is one more condition for the electrolyte. It must be transparent to the useful wavelengths of the irradiating light, since maximal light intensity must hit the depletion layer from the front. Therefore, redox couples which absorb too much light cannot be used. This condition has serious consequences for the counterelectrode. The best arrangement is to employ transparent counterelectrodes, because this permits the use of a very thin electrolyte layer and the use of a cell configuration which optimizes the current distribution in the cell. The current distribution should be as uniform as possible to avoid local depletion of the redox species which could initiate deterioration processes.

The relatively low conductivity of most transparent electrodes, like glass coated with heavily doped SnO_2 or In_2O_3, can itself cause already a distortion in the current distribution between the ohmic contacts of these electrodes and the free parts needed for the illumination of the semiconductor. Some kind of metallic grid structure on the transparent counterelectrode will therefore be needed to keep this current distribution homogeneous enough.

Figure 4.21 gives the principle of a cell configuration which should fulfil all forseeable conditions for the operation of such a system. It has to be sealed, to avoid losses of the electrolyte, but the seal must be elastic enough to allow thermal expansion and contraction. This will set some limit for the size of single cells. Such problems will have to be solved if a technical device of this kind is to be built.

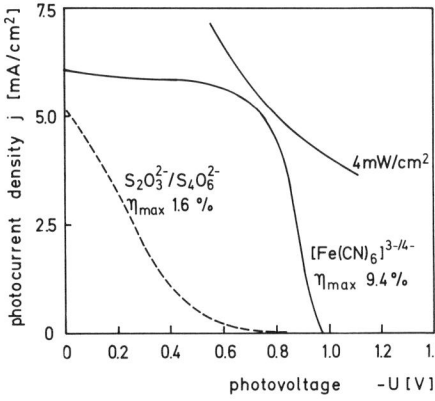

Fig. 4.22. Power characteristics of photocells with a CdS electrode and two different redox electrolytes forming a Schottky barrier. Illumination by a Xenon lamp with 40 mW cm^{-2} irradiation intensity

b) System Analysis

All photoelectrolytic cells which have been studied experimentally up to now were laboratory devices. Emphasis was put on the behavior of the semiconductor electrodes but little concern was given to the overall cell performance. A photovoltaic cell is characterized by its current-voltage output under variable load. This gives the so-called power characteristics of a cell, simply obtained by varying an external resistance between the two electrodes. The first cell studied in this way was a CdS electrode in contact with the redox system $[Fe(CN)_3]^{3-}/[(Fe(CN)_6]^{4-}$ in a concentrated KCl solution [4.9, 72]. The power characteristics of such cells vary with the quality of the semiconductor. Figure 4.22 gives two curves with different redox electrolytes and a single crystalline semiconductor electrode as the energy converter.

The fill factor of this cell is up to 70%. It depends not only on the properties of the semiconductor and the cell geometry but also on the redox couple used. This is clearly demonstrated by the power characteristics of the same cell filled with the redox couple $S_2O_3^{2-}/S_4O_6^{2-}$ in a concentrated Na_2SO_4 electrolyte [4.72]. The open voltage of this cell is somewhat lower than the previous one because $|V_{redox}^0 - V_{fb}|$ is smaller. More important, however, is the much lower fill factor which is caused by the slow rate of electron transfer in this redox couple. The reason is that energy-rich intermediates like $S_2O_3^-$ and $S_4O_6^{3-}$ have to be formed in the single steps of this redox reaction, which causes large overvoltages in anodic and cathodic directions. It is mainly the large polarization of the counterelectrode which causes the high voltage losses in this system.

In principle, one can derive the power characteristics of a photoelectric cell from the two separate current-voltage curves of the semiconductor electrode and the counterelectrode. If both are measured with a potentiostat and plotted on the same scale with respect to the equilibrium potential of the redox system as reference zero, one has only to connect the points of equal anodic and cathodic currents to find the corresponding points of the power characteristics.

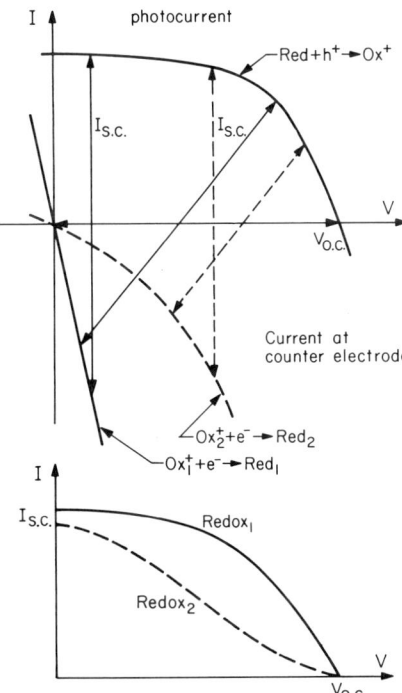

Fig. 4.23. Construction of power characteristics of a solar cell from separate current-voltage curves of the irradiated semiconductor and of the conterelectrode. Each arrow connects points on the current-voltage curves of equal anodic and cathodic currents; its slope, $-1/R$, represents the influence of the external resistance R at which the output voltage V would be obtained

This is shown in Fig. 4.23 for a semiconductor in contact with two different redox systems having the same redox potential at equilibrium, but very different polarization behavior. It is assumed for simplicity that the photocurrent-voltage curve of the semiconductor in this case is not affected by the kinetics of the redox couple.

Semiconductors with a band gap between 1.4 and 1.8 eV are particularly interesting for solar energy conversion, since they reach the highest efficiencies. In order to get a high absorption in the space-charge layer, the absorption coefficient should be above 10^4. This condition is fulfilled by GaAs and some selenides and tellurides. These systems, however, are all susceptible to anodic photodecomposition as we have seen in Sect. 4.2.

Semiconductor electrodes in contact with the redox couples S^{2-}/S_2^{2-}, Se^{2-}/Se_2^{2-}, and Te^{2-}/Te_2^{2-} are particularly stable, and have been intensively studied by *Wrighton* et al. [4.52, 56, 73, 74] and by *Heller* et al. [4.54, 55, 58, 75–77]. Although saturation of the electrolyte up to the solubility product of the semiconductor cannot provide absolute protection against decomposition, as we have explained in the preceding section, it can slow down the deterioration to such an extent that extended use becomes possible. The power characteristics of an n-CdSe cell with a S^{2-}/S_2^{2-} redox electrolyte and of an n-GaAs cell with a Se^{2-}/Se_2^{2-} redox electrolyte is reproduced in Fig. 4.24 from studies of *Heller*'s group [4.75, 76]. The stability of the CdSe electrode in the

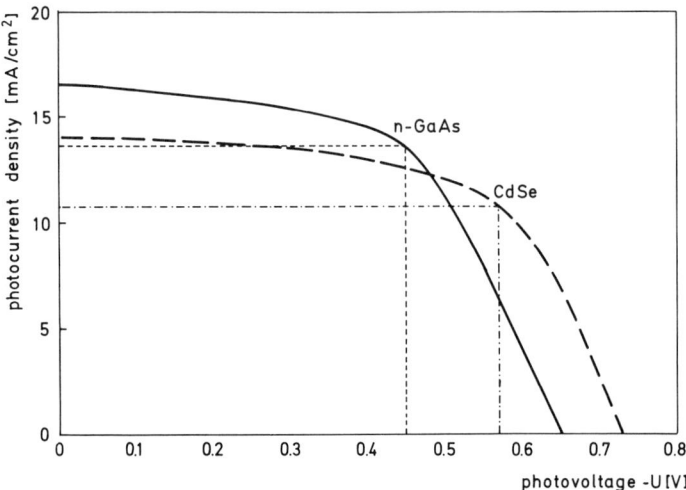

Fig. 4.24. Two examples of power characteristics for regenerative photoelectrochemical solar cells in S^{2-}/S_n^{2-} redox-electrolytes. (After [4.75, 76])

sulfide solution is reached by a transformation of the surface into CdS, as we have mentioned before. This transformation causes some losses in efficiency due to an increase in recombination at this newly formed surface layer. The wider band gap of CdS tends to create a barrier for hole transfer from the CdSe to the surface. This can be avoided by addition of Se to the S^{2-}/S_2^{2-} electrolyte [4.58]. The GaAs electrode is stabilized by the very negative redox potential of the selenide/polyselenide redox system ($V_{NHE}^0 = -0.7$ V) which is above the decomposition potential of GaAs. This, however, reduces the photovoltage to about 0.45 V at the optimum load. In spite of this, a conversion efficiency for sunlight of the order of 9% has been obtained in this cell, the highest value reached up to now with a photoelectrochemical system [4.76].

For practical application of photoelectrochemical solar cells it would be very advantageous for economic reasons to use thin layers of semiconductors and polycrystalline materials. Encouraging experiments with anodically deposited CdSe have been performed by *Hodes* et al. [4.53, 78], with anodically formed CdS on Bi_2S_3 by *Miller* and *Heller* [4.54] and with sintered material of CdSe by *Miller* et al. [4.79].

Regenerative cells with *p*-type semiconductors have not yet been tested in much detail. The reason is that materials which can be made *p*-type (like GaP, GaAs, CdTe) have a rather high position of the valence band edge. Consequently, one needs redox couples with very negative redox potentials to form a depletion layer with large enough band bending at equilibrium in the dark. Such redox systems are not stable in aqueous solution, as they react with water. Experiments with *p*-GaAs and *p*-GaP have been reported by *Memming* [4.67]. The use of non-aqueous electrolytes should be more favorable for such

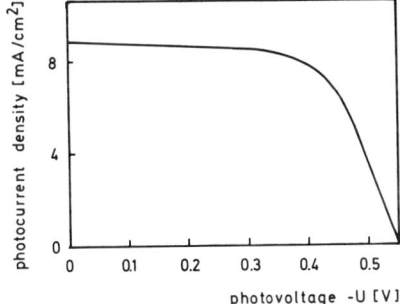

Fig. 4.25. Power characteristics of an n-MoSe$_2$/aqueous KI/KI$_3$ solar cell [4.80]

experiments, if suitable redox systems and solutions with high enough conductivity can be found.

Photocells with semiconducting layer crystals which are supposed to be particularly stable against photodecomposition for kinetic reasons have been studied by *Tributsch* [4.63, 64]. The system consisting of n-type MoSe$_2$ with the redox couple I^-/I_3^-, has proved to be quite promising with regard to stability and constancy of photocurrent output. With selected crystals, a rather high fill factor can be achieved as is shown in Fig. 4.25 [4.80]. The fill factor decreases in these systems with increasing illumination intensity as in most other systems. It is not yet clear whether this is an inherent property of layered materials, or is caused by individual deficiencies of the specimen used.

4.3.2 Storage Cells

a) Principles of Operation

For direct energy storage in photoelectrolysis, two different redox reactions have to proceed in opposite directions at the two electrodes of the cell. The energy stored is the difference in free energy between these two products. The simplest example is the loading of a redox battery consisting of two redox couples in solution separated by an ion selective membrane which is impermeable to the components of the two redox systems. Figure 4.26 shows the scheme of such a cell, in which one electrode generates the photovoltage needed as the driving force of photoelectrolysis. Again, in this figure, an n-type semiconductor is taken as an example. It could equally be a p-type semiconductor. In this case, only the direction of the band bending needs to be reversed, and the redox couple exchanged.

Figure 4.26b shows the electronic energies at open circuit. The membrane is needed to prevent electronic interaction between the two redox couples. Otherwise, electronic equilibrium would be approached by irreversible electron transfer from redox system II to I. This will also occur if the circuit is closed in the dark. Since electrons will flow through the external connection, redox couple II will be oxidized at the metal and redox couple I reduced at the

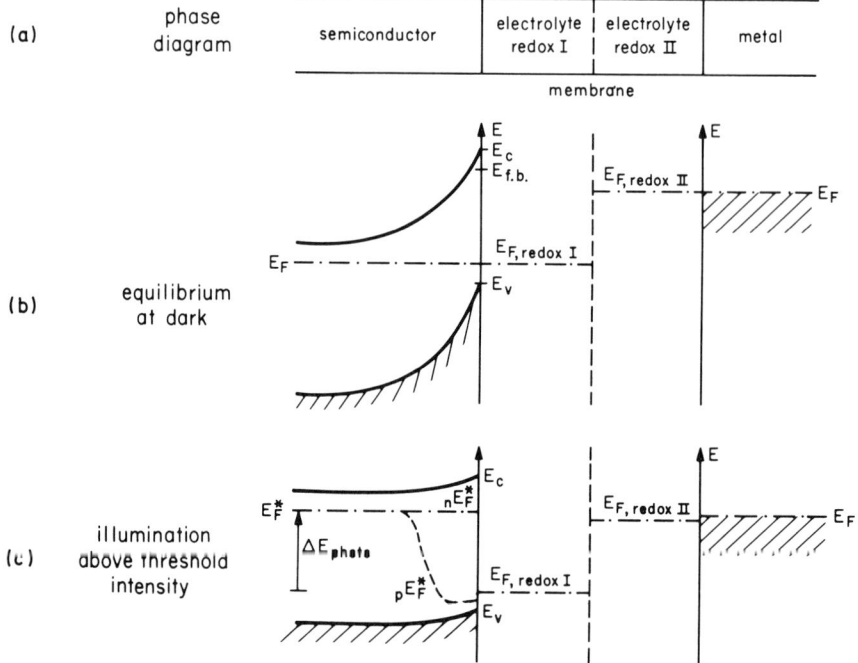

Fig. 4.26a–c. Principle of a photoelectrochemical storage cell. (a) Scheme of cell; (b) electron energy diagram at open circuit; (c) electron energy diagram in operation under illumination

semiconductor electrode. To avoid such a discharge, the cell must be kept open whenever the photovoltage generated in the semiconductor electrode is insufficient to overcompensate the open-cell voltage V_{oc} of this device.

We see that such a system can only store energy above a critical threshold of illumination intensity. The height of this threshold depends on the energy difference between the two redox couples, on the position of the band edges of the semiconductor in relation to the two Fermi levels of the redox systems, and on the recombination losses in the depletion layer of the semiconductor. No theoretical prediction is possible except the obvious one that this threshold will decrease for one particular semiconductor with the distance between the position of the valence band edge and the redox Fermi level of redox system I. Clearly the Fermi level of redox II must be below the flat-band Fermi level, E_{fb}. Otherwise the photovoltage could never reach the critical height for reduction of this redox couple.

Figure 4.26c shows the situation in terms of quasi-Fermi levels of electrons and holes under illumination above the threshold intensity. ΔE_{photo} is the shift of the Fermi level of the majorities in the bulk of the semiconductor relative to the redox Fermi level $E_{F,redoxI}$ in solution. This corresponds to the photovoltage for the regenerative cell discussed in the proceeding section.

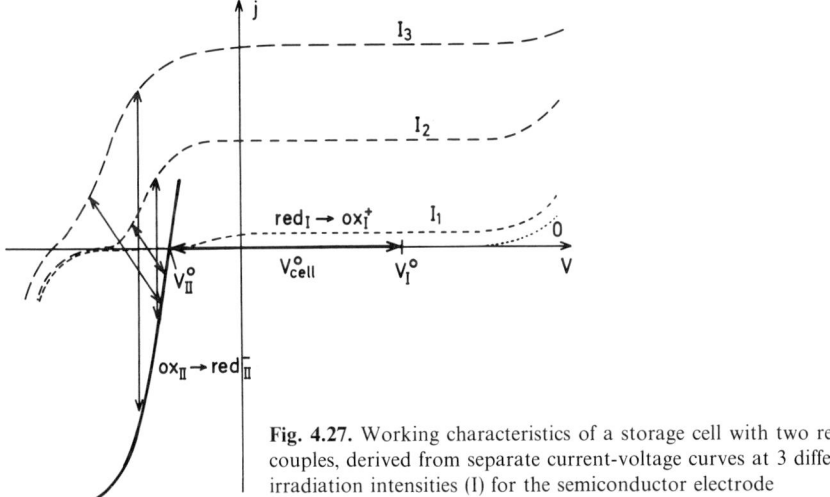

Fig. 4.27. Working characteristics of a storage cell with two redox couples, derived from separate current-voltage curves at 3 different irradiation intensities (I) for the semiconductor electrode

The function and the energy conversion efficiency of a storage cell can be derived from the current-voltage curves of the single electrodes in a similar way as was shown for the power characteristics of regenerative cells. One has, however, to take into account that the two electrode reactions have very different equilibrium potentials. Figure 4.27 represents current-voltage curves for the two redox systems, one at a metal electrode where it behaves quite reversibly (redox system II), and the polarization is therefore mainly concentration polarization; the other at an n-type semiconductor with a depletion layer at equilibrium where the anodic reaction is fully blocked and the cathodic reaction begins only at high cathodic overvoltages when the electron concentration at the surface of the semiconductor has reached a high enough value. If one connects the two electrodes in darkness, their electrode potential will be very close to the redox potential of system II, because the metal electrode is much less polarizable than the semiconductor electrode. A small current will flow in discharge direction controlled by the rate of the reduction of redox system I at the semiconductor.

The figure shows photocurrent-voltage curves for the semiconductor electrode with redox system I at three different light intensities from which a cell analysis can be obtained. The storage cell operates ideally with negligible resistance between the two electrodes. In this case, the cell operates at the voltage where the currents at the semiconductor and at the metal electrode balance each other. This is shown in Fig. 4.27 by the perpendicular arrows. The stored energy is given by the time integral of the current multiplied with the reversible cell voltage V_{cell}^0

$$G_{stored} = \int V_{cell}^0 \cdot I dt. \qquad (4.42)$$

Figure 4.27 also shows the energy loss caused by the overvoltage at the counterelectrode. The steeper the current-voltage curve for redox system II, the smaller is this loss. The ohmic loss has to be added to this energy loss which will depend mainly on the electrolyte resistance. The construction of Fig. 4.27 then has to be modified by finding the compensating currents on the two current-voltage curves being connected by a straight line with the slope $-1/R$ as is also shown in Fig. 4.27 in two examples.

The generation of gaseous products with low solubility in the electrolyte is particularly attractive because a separating membrane is not necessary in this case. This is one of the reasons why photodecomposition of water appears so desirable apart from the fact that hydrogen is an ideal fuel for storage and later use in fuel cells or in combustion. In addition to the low solubility, the products H_2 and O_2 are very inert and will not easily react at the counterelectrode if they should reach it by diffusion or convection (this is not quite true for O_2, which is reduced at all metal electrodes, active for hydrogen evolution catalysis; but the low solubility limits such losses). However, this constitutes no difference in the operating principles of such photocells from the redox cells discussed before. One has only to use the redox Fermi levels of the oxygen and the hydrogen electrode reaction in aqueous solution for the description of this cell to arrive at the same picture as in Fig. 4.26. In contrast to the redox battery described before, however, one has lost the freedom of combining different redox couples with various semiconductors, which would permit an adjustment of the redox Fermi levels to the position of the band edges of the semiconductors. With the redox potentials given by these particular reactions – which depend on the pH of the solution – one is left with the band edges of the semiconductors as the only adjustable parameter. We shall see that this implies serious restrictions.

Before we come to the various attempts to achieve water photoelectrolysis, we shall discuss briefly the energetics of a cell for water photoelectrolysis and some general problems for the design of such cells. Figure 4.28 describes the energetics analogously to Fig. 4.26. Part (a) shows the ideal open-circuit situation if both electrodes were at equilibrium with their respective redox reactions. Since in reality equilibrium is not reached due to the sluggishness of the redox reactions for water decomposition,

$$2H_2O + 4h^+ \rightleftharpoons O_2 + 4H^+ \cdot \text{solv}, \tag{4.43}$$

$$2H_2O + 2e^- \rightleftharpoons H_2 + 2OH^- \cdot \text{solv} \tag{4.44}$$

the open cell situation is rather poorly defined depending on the oxygen and hydrogen concentration in the electrolyte and their reaction rates at the electrodes. If one closes the circuit in the absence of illumination, the Fermi levels of both electrodes will become equal and stay somewhere between the two equilibrium Fermi levels of the redox system, as is shown in Fig. 4.28a. This undefined situation cannot, however, be taken as the reference state for measuring photovoltages, as sometimes has been done. The photovoltage has to be related to the ideal equilibrium situation of the semiconductor electrode

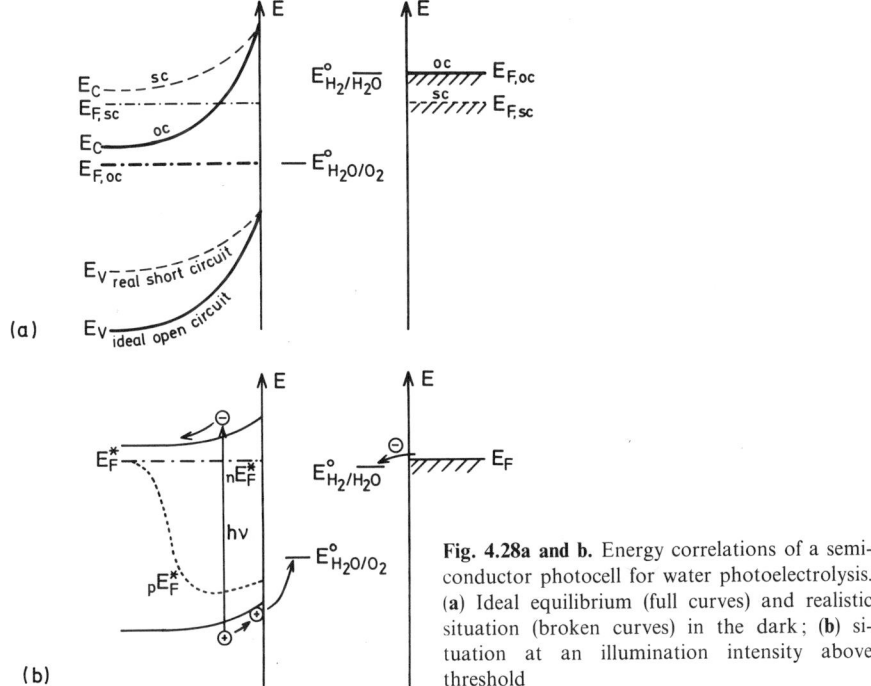

Fig. 4.28a and b. Energy correlations of a semiconductor photocell for water photoelectrolysis. (a) Ideal equilibrium (full curves) and realistic situation (broken curves) in the dark; (b) situation at an illumination intensity above threshold

in contact with that redox reaction which occurs there at illumination. This is shown in Fig. 4.28a.

Consequently, a threshold in illumination intensity must be surpassed to reach the driving force for photoelectrolysis of water. Figure 4.28b represents a working system under illumination with the split quasi-Fermi levels of electrons and holes in the space-charge layer of the semiconductor. The difference between the quasi-Fermi level of holes at the semiconductor surface and the redox Fermi level for water oxidation is needed to overcome the overvoltages of this sluggish reaction. To a lesser extent, the same is necessary for the water reduction. These free energy differences are lost for the energy conversion. They are unfortunately particularly high for the water oxidation reaction. In order for holes to reach sufficiently low quasi-Fermi levels to oxidize water sufficiently fast, the valence band edge of the semiconductor must be located far below the redox Fermi level for water oxidation. Since the Fermi level of electrons in the counterelectrode (which is supposed to be at equilibrium with the bulk of the semiconductor) must be above the Fermi level for water reduction, these conditions can only be met by semiconductors with a much larger band gap than the 1.23 eV thermodynamically needed for the water decomposition. We shall come back to this point in the discussion of the conversion efficiencies.

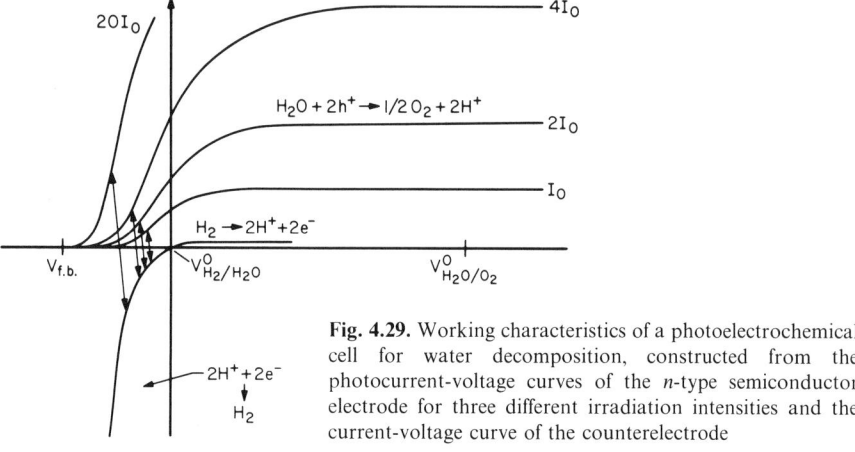

Fig. 4.29. Working characteristics of a photoelectrochemical cell for water decomposition, constructed from the photocurrent-voltage curves of the n-type semiconductor electrode for three different irradiation intensities and the current-voltage curve of the counterelectrode

Figure 4.29 shows the analysis of the water photodecomposition cell in terms of single current-voltage curves for the two redox reactions analogously to Fig. 4.27 for the redox battery. The metal electrode of this system is presumed to be a good catalyst for hydrogen evolution and the flat-band potential of the n-type semiconductor is assumed to be far enough above the redox Fermi level of the water reduction reaction.

In the construction of storage cells, one encounters a particular difficulty if one intends to irradiate the photovoltage-generating depletion layer with as much light intensity as possible. Since one needs a counterelectrode which is a good catalyst for hydrogen evolution, the transparent highly doped semiconducting glasses employed for the regenerative cells can not be used here. At least, no transparent electrode with such catalytic properties has yet been found. One has to use therefore metallic nontransparent counterelectrodes. If they are used in the form of a net in front of the semiconductor electrode, they absorb part of the incoming light. *Nozik* has proposed an arrangement where the counterelectrode forms the reverse side of the semiconductor electrode with a minimum of electric resistance between them. He has called this a photoelectrochemical diode [4.81]. Since the electric circuit has to be closed by an electrolytic connection, such a system must have electrolytic pathways circumventing the diodes. If a separating membrane is needed it has to be in this electrolytic path. Such cells have the disadvantage of giving an inhomogeneous current distribution over the electrode surfaces because of the differences in the resistance of the respective electrolytic paths. A cell construction is shown in Fig. 4.30 where this difference is minimized, and all sunlight can be absorbed by the photoactive semiconductor, if oriented correctly to the position of the sun. The diodes of this cell are separated by ion-selective membranes to prevent the back reaction between the redox couples. One can use such a system to charge the redox battery by sunlight and store the

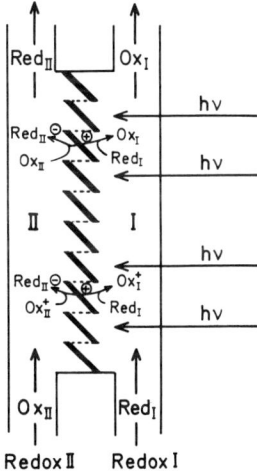

Fig. 4.30. Model of a photoelectrochemical storage cell with diode structures separated by membranes for storage of the two different redox electrolytes in separate tanks

charged solution in separate tanks from which they can be transferred to a redox battery for later use, as is indicated in this figure.

The same arrangement would be useful for water photoelectrolysis if suitable diodes can be made. Since, in this case, only the gases have to be kept apart, a simple diaphragm would be sufficient as a separator. There are certainly many other constructions possible; but the requirement of locating the separator at a place where it does not absorb sunlight excludes having the counterelectrode in front of the semiconductor electrode.

b) Cells for Water Photoelectrolysis

In the first report on water photoelectrolysis by *Fujishima* and *Honda* [4.7], n-type TiO_2 was used as the power generating electrode. It turned out that the photovoltage generated in this cell was not sufficient for the electrolysis. It needed an additional voltage in the order of 0.25–0.5 V to reach simultaneous oxygen evolution at the TiO_2 electrode and hydrogen evolution at a Pt counter electrode [4.10, 82–84]. The band gap of TiO_2 (rutile) is 3.0 eV. The crystal therefore absorbs only the small part of the sun spectrum in the uv. This results in a very low quantum yield [4.11]. Most of the experiments with the TiO_2 system have been performed with a uv light source and the high quantum yields often reported in the literature are obtained in this way. Quantum yields for monochromatic light in the order of 0.5–0.9 are not unusual in semiconductor photoelectrochemistry. This is, however, necessary for the whole spectral range of maximal sun radiation to obtain a high efficiency in solar energy conversion.

Besides its wide band gap, another reason for the deficiency of TiO_2 electrodes for water photoelectrolysis is the unfavorable position of the band edges (compare Fig. 4.16a). The conduction band edge is just slightly above the Fermi level for water reduction. The quasi-Fermi level of the electrons can

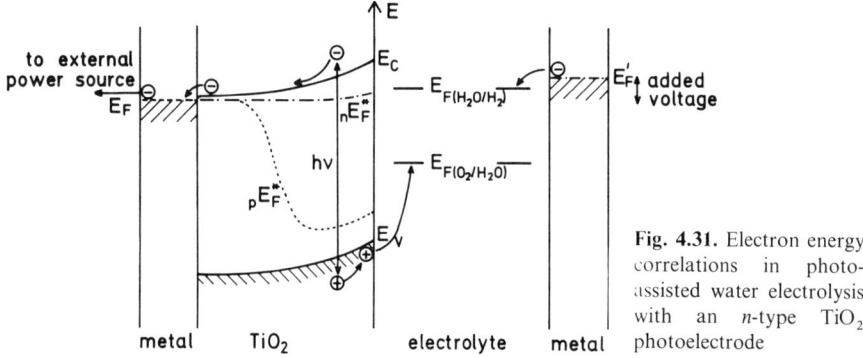

Fig. 4.31. Electron energy correlations in photoassisted water electrolysis with an n-type TiO_2 photoelectrode

therefore not reach a large enough driving force for hydrogen evolution. This requires additional external voltage. On the other hand, the valence band edge of TiO_2 is far below the Fermi level for water oxidation. The quasi-Fermi level of holes will therefore reach and surpass this energy at very weak illumination as long as the energy of the light exceeds the band gap. Therefore even at weak illumination intensity, this system has a much higher driving force for the oxidation of water than necessary and a lot of excess energy is dissipated. Figure 4.31 shows the energy correlations for the TiO_2 electrode in aqueous electrolytes and gives an indication of the origin of the energy losses.

More suitable semiconductors for water photoelectrolysis should have a somewhat higher position of the conduction band edge than TiO_2 and a position of the valence band edge much closer to the redox Fermi level for water oxidation. The first condition can be fulfilled with some other oxides like $SrTiO_3$ [4.85–87], $BaTiO_3$ [4.88], $KTaO_3$ [4.89], ZrO_2 [4.90], Nb_2O_5 [4.90] or SnO_2 [4.91]. However, these oxides have even wider band gaps than TiO_2 and are therefore less appropriate for solar light absorption. An interesting result has been reported by *Augustynski* et al. [4.92] who produced mixed oxides of the composition TiO_2–M_xO_y on Ti metal by spray deposition of the chlorides and consecutive oxide formation by heating in air. After "activation" by heat treatment in oxygen, these electrodes exhibited quite different photocurrent-voltage curves at illumination with a high pressure mercury lamp, although only slight changes in the spectral response were observed compared with a TiO_2 electrode. The electrode with a mixture of TiO_2–Al_2O_3 gave remarkably high photocurrents and appeared particularly stable. Increased photocurrent yields by doping of TiO_2 with Al or Cr have also been reported by *Gosh* and *Maruska* [4.93].

A systematic study of the suitability of various n-type oxidic semiconductors has been executed by *Kung* et al. [4.94], including many oxides with similar or smaller band gaps than TiO_2 which have also been investigated by others (e.g., WO_3 [4.95, 96], Bi_2O_3 [4.97], $YFeO_3$ [4.98] and Fe_2O_3 [4.97, 99, 101]). This was done in the hope of increasing the efficiency by absorbing more sunlight. However, the flat-band potential of all these oxides is below the Fermi level for hydrogen evolution and their use would therefore require external voltage assistance. The most interesting of them with the lowest band gap, Fe_2O_3, has besides this a rather low quantum yield of the photocurrent.

Although the solid-state properties of the oxides used in these experiments varied widely with the particular method of preparation or pretreatment and were often rather ill defined, it seems to be obvious that water photoelectrolysis with a semiconducting oxide as the only power generator has little chance of performing at reasonable efficiency. The valence band in the oxides investigated is located at an energy level that is too deep owing to the high electron affinity of the oxygen atom. The consequence is that the band gap has to be large – too large for efficient absorption of sunlight – to bring the conduction band up to the energy range where the electrons can generate hydrogen. One will have to search for different oxidic compounds with very different crystal and band structures to achieve an efficient photooxidation of water.

The deep position of the valence band in oxides is related to their stability against photodecomposition in aqueous solutions which we have discussed in Sect. 4.2. Nonoxidic semiconductors are more susceptible to photochemical deterioration. In view of the position of their valence band edge, sulfides would be much better adjusted to the energy range of water oxidation. One would further have the advantage of being able, by variation of the pH, to shift the Fermi level of the water oxidation reaction over a range of about 700 mV without affecting the flat-band potential of the semiconductor, provided that no oxide is formed on the sulfide surface. The unfortunate obstacle for the use of sulfide is their instability against photodecomposition. All efforts to prevent this decomposition by covering the surface of unstable semiconductors with coatings of stable oxides have so far either failed to protect the surface, or have blocked the photocurrent of the semiconducting substrate [4.102–105]. The same problems have been observed with coatings by noble metals [4.106, 107].

Two alternatives have been proposed which could help solve this problem. The first alternative studied by *Nozik* [4.108] and *Yoneyama* et al. [4.109], is to upgrade the driving force of the photoelectrolytic cell by combining an *n*-type and a *p*-type semiconductor electrode, both being illuminated simultaneously. One will loose half the sunlight in such a cell, but since two photovoltages act together here in one electrolytic cell, each of them has to generate only part of the photovoltage needed for the cell reaction. This system resembles the power generation in photosynthesis where two photoredox reactions are coupled in series to get the driving force for water oxidation and CO_2 reduction [4.6]. The advantage is that one could use electrodes with smaller band gaps, since the Fermi level of the majority carriers in the illuminated material does not have to reach the energy needed to drive the reaction at the counterelectrode. For instance, *n*-type oxides with a smaller band gap which needed external voltage assistance in some of the cells previously described could be combined with a *p*-type material in which the necessary Fermi level is reached by the additional

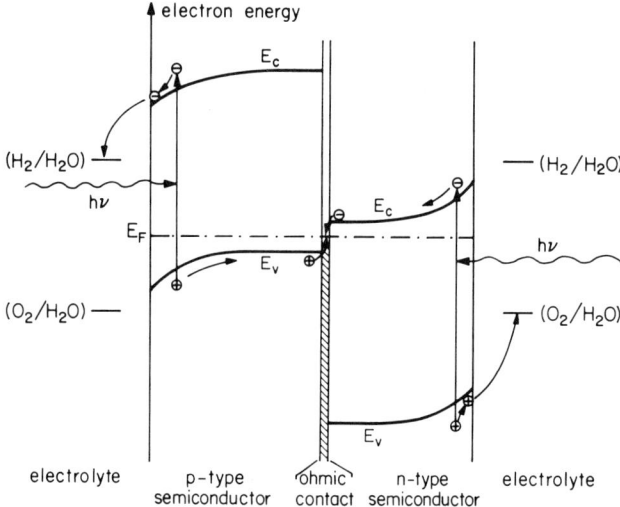

Fig. 4.32. Energy correlations necessary for a water photoelectrolysis cell with two irradiated semiconductor electrodes, one n type, the other p-type

photovoltage generated therein. Figure 4.32 shows the energy correlations which have to be met for such systems.

In order to get the same photocurrent from both electrodes of the $(n+p)$ cells, they should have similar band gaps. If both electrodes are arranged parallel to each other side by side in one plane, balancing different efficiencies can be accomplished by compensation in surface area. However, the stability problem is doubled by such a device since both semiconductors must be stable in the same aqueous electrolyte in the dark and under illumination. Since the redox reactions are fixed to the water decomposition Fermi levels, this imposes a serious limitation on the semiconductors which can be used. The Fermi levels at the flat-band situation should be about in the middle between the redox Fermi levels for water decomposition. The band edges of the minority carriers must be far enough above or below these redox Fermi levels to provide for the overvoltage in the decomposition reactions.

In such a system, a higher voltage output will be needed to compensate for the hydrogen reaction overvoltage which is much larger at semiconductors than at metals. In the experimental studies made with a combination of n-TiO_2 and p-GaP [4.108, 109] the quantum yield was fully controlled by the TiO_2 electrode because of its excessively large band gap. So far, this has shown no improvement in efficiency. Combinations of n-type $SrTiO_3$ with p-type CdTe or GaP can do no better because the even wider band gap of $SrTiO_3$ will decrease the quantum yield [4.110]. No combined systems with smaller band gap have been studied yet, obviously because of the difficulties in adjustment of the band edges between two different semiconductors and of the stability problem.

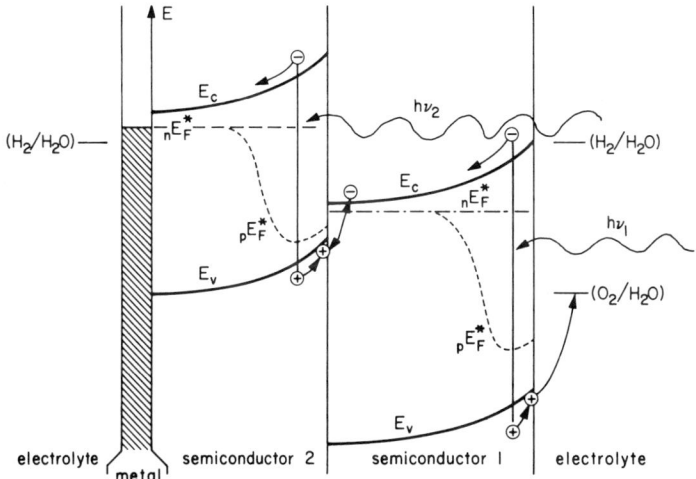

Fig. 4.33. Energy correlations necessary for photoelectrolysis of water with a two-layer n-type semiconductor electrode forming two Schottky barriers in series

Another, possibly more promising alternative, is a combination of two junctions in series. The first junction must be transparent for light of longer wavelength in order to allow this to be absorbed in the second junction. The first semiconductor must therefore have the wider band gap. Such an arrangement has theoretically advantages for solar energy conversion. The loss in current density is more than compensated by the gain in voltage in the second junction. For electrolytic systems where additional overvoltages have to be overcome in the electrode reactions, this appears particularly favorable. We shall come back to this point in the efficiency analysis.

This type of cell was used in a regenerative cell by *Wagner* and *Shay* [4.111] where the first Schottky barrier was formed by an electrolyte–n-type semiconductor interface, the second between two n-type semiconductors. The same principle could be used for water photoelectrolysis. Figure 4.33 shows the necessary energy correlations for such a cell. The condition for the position of the band edge of the valence band is the same as for the normal n-type semiconductor in water photoelectrolysis. The conduction band however can be at a much lower energy level if this difference is compensated by the photovoltage of the second junction. Optimal efficiency is reached if the photocurrents in both junctions are equal. This was far from being the case in the n-CdS/n-GaAs combination studied by *Wagner* and *Shay*. However, their experiment represented in Fig. 4.34, shows that such a device can be built. The advantage over the $(n+p)$ combination is that no area is lost for light absorption by the electrodes and no complicated construction is needed to bring the light to both electrodes.

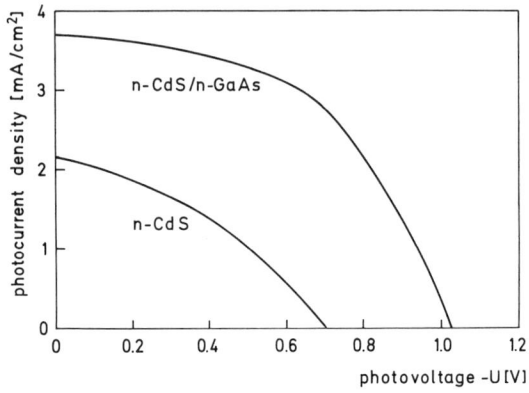

Fig. 4.34. Power characteristics of a regenerative cell with a two-layer photoelectrode (n-CdS/n-GaAs) in the redox electrolyte S^{2-}/S_n^{2-}, and for comparison of a cell with an n-CdS photoelectrode in the same electrolyte. (After [4.111])

4.3.3 Conversion Efficiency in Relation to Materials Properties

We have seen that photoelectrochemical solar cells are based on Schottky barriers as the power generating element. They are therefore subject to the same limitations as photovoltaic cells with solid-state heterojunctions with regard to the free energy conversion efficiency. It has been shown for such systems that the optimum efficiency is obtainable with semiconductors having band gaps between 1.2 and 1.6 eV [4.112–116]. Depending on the particular assumptions which are made with respect to the entropy production in such systems, conversion efficiencies between 25 and 30% have been derived as the theoretical limit.

The efficiencies of photoelectrochemical cells should reach similar values. However, some reduction might be expected due to additional entropy generation at the solid/electrolyte interfaces. This, however, does not mean that photoelectrochemical cells are generally inferior in practical terms, because solid-state devices also do not reach this theoretical limit, since in reality additional losses occur at the solid/solid interface too, and the ideal conditions for the charge separating processes in the junctions are never met. The question of which system will reach better efficiencies at an economic cost will be decided by future research.

In the following we shall try to get a rough estimate of the efficiencies for the conversion of light to free energy in photoelectrochemical cells with similar assumptions as have been made for solid-state photovoltaic devices. We shall make such an estimate for the various types of photoelectrochemical cells, the regenerative cell with a simple redox system, the redox storage cell, and the cells for water decomposition.

For simplicity, we shall assume that the semiconductor has a steep light absorption edge at the energy of the band gap. Consequently, all photons are absorbed in a narrow depth beneath the interface. We assume further that the energy of the generated electron-hole pairs is immediately degraded to the energy of the gap. If this were the only energy loss, the hypothetical conversion efficiency into electronic energy would be given by the following expression

$$Y_{hyp} = \frac{E_{gap} \int_{E_{gap}}^{\infty} f(hv)d(hv)}{\int_0^{\infty} hv f(hv)d(hv)}, \qquad (4.45)$$

where $f(hv)$ means the photon flux density of the solar spectrum. This hypothetical conversion efficiency has a maximum around $E_{gap} = 1.2\,\text{eV}$ of the order of 47% for sunlight at air mass 1.

The real conversion efficiency for electronic energy and particularly into free energy is lower due to unavoidable further energy losses and for the free energy due to entropy production. To get a rough estimate of the losses in terms of free energy, we shall make some simplifying assumptions based on experimental experience. Some energy is lost by the electron-hole pair separation in the space-charge layer. The minimum will be of the order of 0.2 eV corresponding to a minimal band bending needed for a sufficient degree of charge separation. Assuming that efficient electron-hole pair separation occurs mainly in the space-charge layer while light absorbed in the field-free bulk is for the most part lost by recombination, a sufficiently large extension of this layer is necessary, as has been pointed out before. This sets a limit for the doping concentration which can be used in such cells and gives a minimal distance between the Fermi level in the bulk and the respective band edge. We shall assume that this difference must be about 0.2 eV. This estimate shows that the free energy or the quasi-Fermi level of the majority carriers will be at least 0.4 eV below the respective band edge at the optimal working conditions of such a cell, which is however only part of the loss in free energy, since the quasi Fermi level of the minority carriers will also deviate from the position of the respective band edge.

In an optimal semiconductor-electrolyte Schottky barrier, the position of the Fermi level of the redox system in the electrolyte with which the minority carriers react should be as close as possible to the position of the band edge of the minority carriers. This means that, for the regenerative cell, an inversion layer would be formed at equilibrium by injection of minority carriers from the redox system into the semiconductor surface. Although even degeneracy could be reached in principle at the surface in this way, some difference between the Fermi level of the minority carriers and the respective band edge will remain. In fact a large amount of excess surface charge would result in an increasing voltage drop in the Helmholtz double layer and cause a shift of the band-edge position relative to its location under flat-band conditions. Besides this, an excessively high surface concentration of minority carriers will allow increased recombination in the space-charge layer and decrease the efficiency of charge separation. We can assume therefore that a distance of at least 0.10 eV should be left over between the quasi-Fermi level of the minority carriers E_F^* and the

Table 4.1. Data for the estimation of free energy losses

E_{gap} [eV]	i_{max} [mA cm^{-2}]	Y_{hyp} [%]	metal η_{redox} [mV]	semicond. η_{redox} [mV]	semicond. η_{H_2} [mV]	catalyt. metal η_{H_2} [mV]	semicond. η_{O_2} [mV]	catalyt. metal η_{O_2} [mV]
1.0	52	46	12	87	380	23	538	320
1.2	44	47	11	79	370	20	530	315
1.4	37	46	10	70	360	17	520	310
1.6	29	42	8	60	345	13	505	305
1.8	22	35	6	49	330	10	490	300
2.0	15	28	4	37	315	7	475	295
2.2	11	21.5	3	27	300	5	460	290
2.4	7.6	15	2	19	280	3	440	285
2.6	4.7	10.5	1	13	260	2	420	280
2.8	2.6	6	1	7	240	1	400	275
3.0	1.2	2.7	–	3	220	1	375	270

respective band edge under working conditions. This means that the rate of the electron-transfer reaction which consumes the minority carriers must reach the height of the absorbed quantum flux at this surface concentration of the minority carriers. The difference between the equilibrium Fermi level of the redox system and E_F^* is then $e\eta_{min}$, where η_{min} is the overvoltage necessary for the corresponding reaction rate with the minority carriers. The energy loss caused by this part of the entropy production will therefore amount to at least $0.1 + e\eta_{min}$ eV.

Another free energy loss occurs at the counterelectrode of an amount of $e\eta_{maj}$ where η_{maj} means the overvoltage for the electrode reaction supplied by the majority carriers. Finally there will be some loss by ohmic resistances, iR. The whole loss in free energy relative to the band gap is therefore according to this estimate,

$$|\Delta G_{loss}| \geq 0.5 + e(\eta_{min} + \eta_{maj}) + iR. \tag{4.46}$$

The overvoltages individually depend on the particular electrode reactions and on the current density. They are relatively small for one-electron transfer reactions but can be quite large for complicated redox reactions passing through various one-electron transfer steps and connected with the formation of intermediates of sometimes very high energy. The redox reactions for water electrolysis belong to the latter unfavourable class, particularly the oxygen-formation process. A discussion of the efficiencies which might be obtainable has to take into account the differences in the kinetics of the redox reactions. To get some estimates, we have listed typical values of overvoltages for one-electron transfer redox reactions and for the reactions of water electrolysis in Table 4.1 in the range of current densities to be expected for solar cells with different absorption thresholds.

The exchange current densities assumed in these data are: 10^{-1} A cm^{-2} for the redox reaction of the metal electrode, 10^{-2} A cm^{-2} for the same redox reaction at the semiconductor (a somewhat optimistic assumption) and 10^{-1}–10^{-2} A cm^{-2} for the hydrogen electrode. The data for the oxygen electrode, which is in any case irreversible at normal temperatures, are a conservative guess for the semiconductor electrode, and correspond to the catalyst RuO$_2$ on titanium for the metal electrode.

The conversion efficiency into free energy as a function of the band gap is given for the regenerative cell by

$$Y_{regen} = Y_{hyp}(1 - \Delta G_{loss}/E_{gap}). \tag{4.47}$$

For the storage device, one can define a "free energy storage efficiency" Y_{stor} by

$$Y_{stor} = Y_{hyp}(\Delta E_{stor}/E_{gap}) \tag{4.48}$$

with the condition that $\Delta E_{stor} \leq E_{gap} - |\Delta G_{loss}|$. Otherwise, storage would not be possible. ΔE_{stor} is the equilibrium cell voltage of the photoredox battery or the hydrogen-oxygen fuel cell in case of water photoelectrolysis.

By using the data of Table 4.1 together with the relation (4.46), one can derive the energy conversion efficiencies for the different systems discussed in this chapter. Using a constant loss by band bending of 0.2 V for all semiconductors would, however, give too optimistic a picture for semiconductors with a larger band gap, because such small band bending will not be reached in the stationary state at the lower current densities obtained with semiconductor of wider band gap. A better comparison can be made if one assumes that the light would be so much concentrated for each system that the same photocurrent is obtained independent of the band gap. We shall make this assumption, and use a light intensity corresponding to the maximal current for a semiconductor with $E_{gap} = 1.2$ V as the reference for the regenerative cell, and, for the storage cells, a current density of a semiconductor with the smallest possible band gap fulfilling condition (4.48). The latter obviously depends on the open cell voltage and on electrode reaction rates.

The free energy loss in the Schottky barrier is assumed to be 0.5 eV as a minimum, to which the values for the overvoltage and ohmic losses have to be added. The ohmic loss may be kept as low as 0.05 eV for the regenerative cell, but will be higher for all storage cells, because the cell geometry will be less favourable for these. A conservative estimate in the latter case may be 0.15 eV at a current density of 45 mA cm^{-2} or 0.05 eV at 15 mA cm^{-2}. The result of such calculations is shown in Fig. 4.35. This figure represents besides Y_{hyp} the estimates for the optimal conversion efficiencies of a regenerative cell and the estimated storage efficiencies of a photoredox battery with 1 V open-circuit voltage and of various storage cells based on water photoelectrolysis. The efficiency curves of the storage cells have their maximum just at those band gap values where the condition $E_{stor} = E_{gap} - |\Delta G_{loss}|$ is fulfilled for $|\Delta G_{loss}|$ values

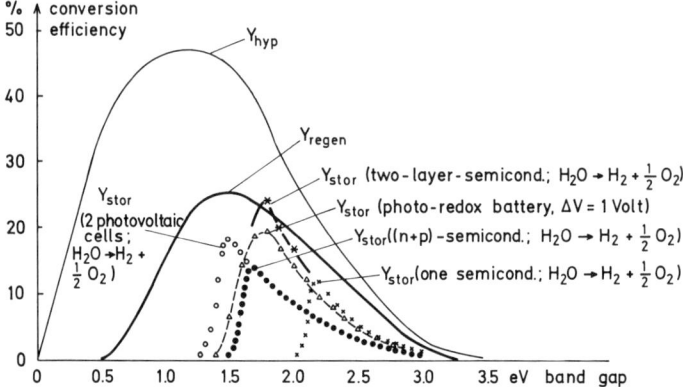

Fig. 4.35. Estimated optimal conversion efficiencies of photoelectrochemical solar cells for power generation (regenerative cells), Y_{regen}, and for storage of free energy, Y_{stor}. Y_{hyp} is the theoretical energy conversion efficiency for solar light (AM 1) if all absorbed light quanta were degraded to the magnitude of the band gap

estimated from our assumptions and the table of overvoltages. The calculation cannot be made in this way below this point, since ΔG_{loss} was calculated for the highest photocurrent which can be reached. But in any case, the efficiency will decrease drastically below this band gap as is indicated in Fig. 4.35.

The result for the regenerative cell is very close to the estimates for solid-state photovoltaic devices. The optimum should be reached at a somewhat higher band gap (around 1.5 eV) than expected for solid-state cells. The simplest case of a storage cell is that with one active, illuminated semiconductor electrode. If such a system is used to charge a redox battery with cell voltage of 1 V, a storage efficiency of 18% could be obtained with a semiconductor of 1.8 eV band gap provided that our assumptions can be met. The situation is much less favourable for water photoelectrolysis in such a device. The calculation made here results in an optimal conversion efficiency of 12% with a semiconductor of 2.2 eV band gap which, however, must also have the optimal position of the band edges in relation to the redox reactions for water decomposition. It is very unlikely that such a semiconductor exists.

A higher efficiency should in principle be reached by a combination of two illuminated semiconductor electrodes as described previously (cf. Fig. 4.32). The efficiency in this case can be calculated by the following equation

$$Y_{stor} = Y_{hyp}(E_{stor}/2E_{gap}) \tag{4.49}$$

with the condition

$$2E_{gap} - [\Delta G_{loss}(\text{cath.}) + \Delta G_{loss}(\text{anode})] \geq E_{stor}.$$

We see in Fig. 4.35 that indeed a somewhat higher yield by water photoelectrolysis could be reached with an $(n+p)$ photoelectrolysis cell if two different semiconductors of equal band gap can be combined. However, to obtain this efficiency optimal adjustment of the band-edge positions to the redox reactions would be necessary, which is even more unlikely than for one semiconductor alone. For comparison, the storage efficiency of a cell for water electrolysis is included which is driven by two simultaneously irradiated photovoltaic cells with equal band gaps, electrically connected in series as voltage generators. In this case no adjustment between band edges and redox Fermi levels is needed, for the electrolysis can be performed in a separate electrolysis cell with optimized catalytic electrodes. Therefore, electrolysis in conventional cells with catalytic metal electrodes, driven by two photovoltaic cells in series, appears to give better efficiencies than direct photoelectrolysis of water by any methods.

The conclusion is that a photoelectrolysis cell with separation of the power generating element and the electrolysis cell seems to have the better chance to reach a high storage efficiency than direct photoelectrolytic generation of a fuel. Only the photoelectrochemical redox battery can reach a similar conversion efficiency according to Fig. 4.35. Regenerative photoelectrolysis cells should be able to reach similar efficiencies as those of solid-state photovoltaics.

An improvement in efficiency can be reached, at least in principle, by the construction of multilayer solar cells with two or more junctions passed consecutively by the sunlight [4.113, 117, 118]. We have seen how this principle has been applied by *Wagner* and *Shay* [4.111] to regenerative photoelectrochemical cells (cf. Fig. 4.34). Even direct photoelectrolysis of water could become much more efficient, as shown by a calculation made with the same assumptions as for the other curves of Fig. 4.35. This result is also shown in this figure. With regenerative cells, efficiencies above 30% could be obtained if the right combinations can be found.

At present, the conversion efficiencies obtained with photoelectrochemical cells are far below the optimal values calculated with the assumptions made above. A great deal more materials research will have to be performed to approach the calculated values.

Acknowledgement. This review has been compiled during a stay at the California Institute of Technology, Pasadena, as a Sherman Fairchild Distinguished Scholar. The hospitality and help of my colleagues at Pasadena, particularly Professors F.C. Anson, H.B. Gray, J.O. McCaldin, T.C. McGill, and Dr. J. Turner is gratefully acknowledged.

References

4.1 H.T. Tien, S.P. Verma: Nature (London) **227**, 1232 (1970)
4.2 D.S. Berns: Photochem. Photobiol. **24**, 117 (1976)
4.3 M. Calvin: Photochem. Photobiol. **24**, 425 (1976)
4.4 J.J. Katz, J.R. Norris: In *Current Topics of Bioenergetics*, Vol. 5, ed. by D.R. Sanadi, L. Packer (Academic Press, New York 1973) p. 41
4.5 H.T. Witt: In *Excited States of Biological Molecules*, ed. by J.B. Birks (Wiley-Interscience, London 1976) p. 245

4.6 Govindjee (ed.): *Bioenergetics of Photosynthesis* (Academic Press, New York 1975)
4.7 A. Fujishima, K. Honda: Nature (London) **238**, 37 (1972)
4.8 A. Fujishima, K. Kohayakawa, K. Honda: J. Electrochem. Soc. **122**, 1487 (1975)
4.9 H. Gerischer: J. Electroanal. Chem. **58**, 263 (1975)
4.10 M.S. Wrighton, D.S. Ginley, P.T. Wolczanski, A.B. Ellis, D.L. Morse, A. Linz: Proc. Nat. Acad. Sci. U.S.A. **72**, 4 (1975)
4.11 T. Onishi, Y. Nakato, M. Tsubomura: Ber. Bunsenges. Phys. Chem. **79**, 523 (1975)
4.12 A. Heller (ed.): *Semiconductor Liquid-Junction Solar Cells* (Proceedings of a conference held at Airlie, May 1977); Proc. Vol. 77-3 (Electrochemical Society Princeton 1977)
4.13 H. Gerischer: In *Solar Power and Fuels*, ed. by J.R. Bolton (Academic Press, New York 1977) pp. 77–118
4.14 A.J. Nozik: Rev. Phys. Chem. **29**, 189 (1978)
4.15 V.A. Myamlin, Yu.V. Pleskov: *Electrochemistry of Semiconductors* (Plenum Press, New York 1967)
4.16 H. Gerischer: In *Physical Chemistry*, Vol. IX A, ed. by H. Eyring, D. Henderson, W. Jost (Academic Press, New York 1970) p. 463
4.17 M.J. Spaarnay: *The Electrical Double Layer* (Pergamon, Oxford 1972)
4.18 Yu.V. Pleskov: In *Progress in Surface and Membrane Science*, Vol. 7, ed. by J.F. Danielli, M.D. Rosenberg, C.H. Cadenhead (Academic Press, New York 1973) p. 57
4.19 V.A. Myamlin, Yu.V. Pleskov: *Electrochemistry of Semiconductors* (Plenum Press, New York 1967)
4.20 F. Lohmann: Ber. Bunsenges Phys. Chem. **70**, 428 (1966)
4.21 F. Dewald: Bell Syst. Tech. J. **39**, 615 (1960)
4.22 F. Lohmann: Z. Naturforsch. **22**a, 843 (1967)
4.23 S. Trasatti: Adv. Electrochem. Electrochem. Eng. **10**, 213 (1977)
4.24 R. Gomer, G. Tryson: J. Chem. Phys. **66**, 4413 (1977)
4.25 W.P. Gomes, F. Cardon: In [Ref. 4.12, p. 120]
4.26 J.O. McCaldin, T.C. McGill: In *Thin Films – Interdiffusion and Reactions*, ed. by J.M. Pate, K.N. Tu, J.W. Maeyer (Electrochemical Society, Princeton 1978)
4.27 M.A. Butler, D.S. Ginley: In *Electrode Materials and Processes for Energy Conversion and Storage*, ed. by J.D.E. McIntyre, S. Shrinivasan, F.G. Will, Proc. Vol. 77-6 (Electrochemical Society, Princeton 1977) p. 54
4.28 M.A. Butler, D.S. Ginley: J. Electrochem. Soc. **125**, 228 (1978)
4.29 L. Blok, P.L. DeBruyn: J. Colloid. Interface Sci. **32**, 518 (1970)
4.30 A.W. Adamson: *Physical Chemistry of Surfaces*, 3rd ed. (Wiley, New York 1976)
4.31 R.A. Marcus: Can. J. Chem. **37**, 155 (1959); Ann. Rev. Phys. Chem. **15**, 155 (1964)
4.32 V.G. Levich: In *Advances Electrochemistry and Electrochemical Engineering*, ed. by P. Delahay, C.W. Tobias, Vol. 4 (Wiley-Interscience, New York 1966) p. 249
4.33 H. Gerischer: Z. Phys. Chem. (Frankfurt am Main) **26**, 223 (1960); **27**, 48 (1961)
4.34 R.R. Dogonadze: In *Reactions of Molecules at Electrodes*, ed. by N.S. Hush (Wiley-Interscience, New York 1971) p. 135
4.35 H. Gerischer: In *Advances in Electrochemistry and Electrochemical Engineering*, ed. by P. Delahay, C.W. Tobias, Vol. 1 (Wiley-Interscience, New York 1961) p. 139
4.36 H. Gerischer: Photochem. Photobiol. **16**, 243 (1972)
4.37 H. Gerischer, I. Mattes: Z. Phys. Chem. (Frankfurt am Main) **52**, 60 (1967)
4.38 W.W. Gärtner: Phys. Rev. **116**, 84 (1959)
4.39 M.A. Butler: J. Appl. Phys. **48**, 1914 (1977)
4.40 R.H. Wilson: In [Ref. 4.12, p. 67]
4.41 D. Laser, A.J. Bard: J. Electrochem. Soc. **123**, 1833, 1837 (1976)
4.42 H. Reiss: J. Electrochem. Soc. **125**, 937 (1978)
4.43 W. Shockley: *Electrons and Holes in Semiconductors* (Van Nostrand, New York 1950)
4.44 A.J. Nozik, F. Williams: Submitted to Nature (London)
4.45 H. Gerischer, W. Mindt: Electrochim. Acta **13**, 1329 (1968)
4.46 H. Gerischer: J. Electroanal. Chem. **82**, 133 (1977)
4.47 A.J. Bard, M.S. Wrighton: In [Ref. 12, p. 195]

4.48 H. Gerischer: J. Vac. Sci. Technol. **8**, 1422 (1978)
4.49 B. Pettinger, R. Schoeppel, H. Gerischer: Ber. Bunsenges. Phys. Chem. **78**, 1024 (1974)
4.50 K. Hauffe: Ber. Bunsenges. Phys. Chem. **76**, 616 (1972)
4.51 L.A. Harris, R.H. Wilson: J. Electrochem. Soc. **123**, 1010 (1976)
4.52 A.B. Ellis, S.W. Kaiser, M.S. Wrighton: J. Am. Chem. Soc. **98**, 1635, 6418, 6855 (1976)
4.53 G. Hodes, J. Manassen, D. Cohen: Nature (London) **261**, 403 (1976)
4.54 B. Miller, A. Heller: Nature (London) **262**, 680 (1976)
4.55 A. Heller, K.C. Chang, B. Miller: J. Electrochem. Soc. **124**, 697 (1977)
4.56 A.B. Ellis, S.W. Kaiser, J.M. Bolts, M.S. Wrighton: J. Am. Chem. Soc. **99**, 2839 (1977)
4.57 H. Gerischer, J. Gobrecht: Ber. Bunsenges. Phys. Chem. **82**, 520 (1978)
4.58 A. Heller, G.P. Schwartz, R.G. Vadimsky, S. Menezes, B. Miller: J. Electrochem. Soc. **125**, 1156 (1978)
4.59 D. Cahen, G. Hodes, J. Manassen: J. Electrochem. Soc. **125**, 1623 (1978)
4.60 H. Tributsch, J.C. Bennett: J. Electroanalyt. Chem. **81**, 97 (1977)
4.61 H. Tributsch: Ber. Bunsenges. Phys. Chem. **81**, 361 (1977)
4.62 H. Tributsch: Z. Naturforsch. **32**a, 972 (1977)
4.63 H. Tributsch: Ber. Bunsenges. Phys. Chem. **82**, 169 (1978)
4.64 H. Tributsch: J. Electrochem. Soc. **125**, 1086 (1978)
4.65 J. Gobrecht, H. Gerischer, H. Tributsch: Ber. Bunsenges. Phys. Chem. **82**, 1331 (1978)
4.66 H. Gerischer, F. Turner: To be published
4.67 R. Memming: J. Electrochem. Soc. **125**, 117 (1978), and in [Ref. 12, p. 38]
4.68 T. Inoue, T. Watanabe, A. Fujishima, K. Honda: J. Electrochem. Soc. **124**, 719 (1977)
4.69 H. Gerischer: In *Electrocatalysis on Non-Metallic Surfaces*, N.B.S. Special Publ. **455**, p. 1 (1976)
4.70 S.N. Frank, A.J. Bard: J. Am. Chem. Soc. **97**, 7427 (1975)
4.71 J.G. Mavroides, V.E. Henrich, H.J. Zeiger, G. Dresselhaus, J.A. Kafalas, D.F. Kolesar: "Electrode Materials and Processes for Energy Conversion and Storage", Proc. Vol. 77-6 (Electrochemical Society, Princeton 1977) p. 45
4.72 H. Gerischer, J. Gobrecht: Ber. Bunsenges. Phys. Chem. **80**, 327 (1976)
4.73 A.B. Ellis, J.M. Bolts, S.W. Kaiser, M.S. Wrighton: J. Am. Chem. Soc. **99**, 2848 (1977)
4.74 A.B. Ellis, J.M. Bolts, M.S. Wrighton: J. Electrochem. Soc. **124**, 1603 (1977)
4.75 A. Heller, K.C. Chang, B. Miller: J. Electrochem. Soc. **124**, 697 (1977)
4.76 K.C. Chang, A. Heller, B. Schwartz, S. Menezes, B. Miller: Science **196**, 1097 (1977), and in [Ref. 12, p. 132]
4.77 A. Heller, K.C. Chang, B. Miller: J. Am. Chem. Soc. **100**, 684 (1978)
4.78 G. Hodes, J. Manassen, D. Cahen: Nature (London) **124**, 532 (1977)
4.79 B. Miller, A. Heller, M. Robbins, S. Menezes, K.C. Chang, J. Thomson, Jr.: J. Electrochem. Soc. **124**, 1019 (1977)
4.80 J. Gobrecht, H. Gerischer, H. Tributsch: J. Electrochem. Soc. **125**, 2085 (1978)
4.81 A.J. Nozik: Appl. Phys. Lett. **30**, 567 (1977)
4.82 J.G. Mavriodes, D.I. Tchernev, J.A. Kafalas, D.F. Kolesar: Mater. Res. Bull. **10**, 1023 (1975); and in "Electrocatalysis on Non-Metallic Surfaces", N.B.S. Special Publ. **455**, 221 (1976)
4.83 K.L. Hardee, A.J. Bard: J. Electrochem. Soc. **122**, 739 (1975)
4.84 A.J. Nozik: Nature (London) **257**, 383 (1975)
4.85 J.G. Mavroides, J.A. Kafalas, D.F. Kolesar: Appl. Phys. Lett. **28**, 241 (1976)
4.86 M.S. Wrighton, A.B. Ellis, P.T. Wolczanski, D.L. Morse, H.B. Abrahamson, D.S. Ginley: J. Am. Chem. Soc. **98**, 2774 (1976)
4.87 T. Watanabe, A. Fujishima, K. Honda: Bull. Chem. Soc. Jpn. **49**, 355 (1976)
4.88 R.D. Nasby, R.K. Quinn: Mater. Res. Bull. **11**, 985 (1976)
4.89 A.B. Ellis, E.W. Kaiser, M.S. Wrighton: J. Phys. Chem. **80**, 1325 (1976)
4.90 P. Clechet, J. Martin, R. Oliver, C. Vallony: C.R. Acad. Sci. Ser. C **282**, 887 (1976)
4.91 M.S. Wrighton, D.L. Morse, A.B. Ellis, D.S. Ginley, H.B. Abrahamson: J. Am. Chem. Soc. **98**, 44 (1976)
4.92 J. Augustynski, J. Hinden, C. Stalder: J. Electrochem. Soc. **124**, 1063 (1977)

4.93 A.K. Gosh, H.P. Maruska: J. Electrochem. Soc. **124**, 1516 (1977)
4.94 H.H. Kung, M.S. Jarrett, A.W. Sleight, A. Ferretti: J. Appl. Phys. **48**, 2463 (1977)
4.95 M.A. Butler, R.D. Nasby, R.K. Quinn: Solid State Commun. **19**, 1011 (1976)
4.96 W. Gissler, R. Memming: In [Ref. 4.12, p. 241]
4.97 K.L. Hardee, A.J. Bard: J. Electrochem. Soc. **124**, 215 (1977)
4.98 M.A. Butler, D.S. Ginley, M. Eibschutz: J. Appl. Phys. **48**, 3070 (1977)
4.99 R.K. Quinn, R.D. Nasby, R.J. Baughman: Mater. Res. Bull. **11**, 1011 (1976)
4.100 L.S.R. Yeh, N. Hackerman: J. Electrochem. Soc. **124**, 833 (1977)
4.101 J.H. Kennedy, K.W. Freese, Jr.: J. Electrochem. Soc. **125**, 709 (1978)
4.102 J.O'M. Bockris, K. Uosaki: Energy **1**, 95 (1976)
4.103 P.A. Kohl, S.W. Frank, A.J. Bard: J. Electrochem. Soc. **124**, 225 (1977)
4.104 S. Gourgaud, D. Elliott: J. Electrochem. Soc. **124**, 102 (1977)
4.105 M. Tomkiewicz, J.M. Woodall: J. Electrochem. Soc. **124**, 1436 (1977)
4.106 Y. Nakato, T. Ohnishi, H. Tsubomura: Ber. Bunsenges. Phys. Chem. **80**, 1002 (1976)
4.107 R.H. Wilson, L.A. Harris, M.E. Gerstner: J. Electrochem. Soc. **124**, 1233, 1511 (1977)
4.108 A.J. Nozik: Appl. Phys. Lett. **28**, 150 (1976)
4.109 H. Yoneyama, H. Sakamoto, H. Tamura: Electrochim. Acta **20**, 341 (1975)
4.110 K. Ohashi, J. McCann, J.O'M. Bockris: Int. J. Energy Res. **1**, 259 (1977)
4.111 S. Wagner, J.L. Shay: Appl. Phys. Lett. **31**, 446 (1977); and in [Ref. 12, p. 231]
4.112 J.J. Loferski: J. Appl. Phys. **27**, 777 (1956)
4.113 M. Wolf: Proc. I.R.E. **48**, 1246 (1960)
4.114 W. Shockley, H.J. Queisser: J. Appl. Phys. **32**, 510 (1961)
4.115 J.J. Loferski: Proc. IEEE **51**, 677 (1963)
4.116 H. Fischer: In *Festkörperprobleme XIV*, ed. by H.J. Queisser (Pergamon-Vieweg, Braunschweig 1974) p. 153
4.117 E.D. Jackson: Trans. Conf. on the Use of Solar Energy, Tucson 1955, Vol. 5 (University of Arizona Press, Tucson 1958) p. 122
4.118 R.T. Ross, Ta-Lee Hsiao: J. Appl. Phys. **48**, 4783 (1977)

5. Carrier Lifetime in Silicon and Its Impact on Solar Cell Characteristics

K. Graff and H. Fischer

With 44 Figures

In most semiconductor devices, because of their small dimensions and the correspondingly short distances the excess charge carriers have to travel by diffusion, the carrier lifetime is not a critical parameter of the material. In some devices, however, the active layer is thicker and the carrier lifetime or, alternatively, diffusion length becomes important. This is the case in silicon solar cells, where the thickness of the base is increased in order to increase the amount of light absorbed, and hence the efficiency of the cell. A silicon layer of about 350 μm is necessary to absorb about 90% of AM 0 and 93% of AM 1 sunlight[1]. In the present study, we are interested in applications to solar cells of experimental results on carrier lifetimes in silicon and so, in general, only p-type silicon was investigated. Only in some cases are experimental results for n-type silicon presented for comparison.

To understand the operation of the solar cell, the carrier lifetime in the finished device has to be considered. Since the silicon crystal undergoes several temperature treatments during the preparation of a cell, emphasis was not only put on the as-grown crystal but also on the crystal after various temperature-annealing procedures. We tried to investigate to what extent process-induced variations of carrier lifetime should be attributed to material parameters or to experimental procedure.

The recombination carrier lifetime is defined as the time required for the concentration of excess charge carriers in a semiconductor material to decay by recombination to $1/e$ of its initial value. In general, extrinsic semiconductors are investigated and the concentration of excess minority carriers is considered. The recombination carrier lifetime depends to some extent on the excess carrier concentration. This dependence saturates for high and for low concentrations of the excess carriers, and correspondingly we define the high-level and low-level carrier lifetimes. Unless otherwise stated, "carrier lifetime" means the low-level recombination minority carrier lifetime.

The diffusion length L of excess carriers is related on the carrier lifetime τ according to the equation:

$$L^2 = D\tau, \tag{5.1}$$

---

[1] AM0 sunlight is sunlight outside the earth's atmosphere, with a spectral distribution of [5.7] and an integrated energy of 135.3 mW cm^{-2}; AM1 sunlight is the sunlight at sea level that has passed through the earth's atmosphere; its spectral distribution is [5.2], and its integrated energy is approximately 100 mW cm^{-2}.

where D is the diffusion coefficient of the relevant charge carriers. The diffusion coefficient can be calculated with the aid of the Einstein relation

$$D = kT\mu/e, \tag{5.2}$$

where μ is the mobility of the charge carriers, and the other symbols have their usual meaning.

Improved results for carrier lifetime measurements cannot be obtained without improving the measurement technique. After an introduction concerning the solar cell photovoltaic parameters determined by the carrier lifetime, various techniques for the measurement of these lifetimes are discussed with particular emphasis on the photoconductive decay method. In Sects. 5.3, 4, experimentally obtained results of carrier lifetime measurements are presented concerning as-grown silicon crystals as well as processed silicon samples. Finally, limitations for solar cell parameters caused by material properties are discussed.

5.1 Photovoltaic Parameters of Solar Cells as Determined by Carrier Lifetime

The conversion of light into electricity in a solar cell is caused by the photovoltaic effect at a boundary layer. When photons with an energy greater than the band gap are incident on a solar cell, the absorption of the photons creates free electron-hole pairs. These excess minority carriers may diffuse to the space-charge region, cross the junction and give rise to photocurrent, photovoltage, and power into a load. These photogenerated excess carriers diffuse towards the *pn* junction; at the same time, these carriers are lost by bulk or surface recombination. The bulk recombination process is described by the lifetime of the excess carriers in the specific material.

The relationship between solar cell parameters and minority carrier lifetime for various types of solar cells has been investigated by many researchers. A comprehensive collection of the most important contributions can be found in [5.1, 2]. To analyze the dependence of photovoltaic parameters on carrier lifetime, a one-dimensional model of a solar cell is used. The cell consists of a *p*-doped base and a very thin *n*-doped surface zone, which is usually prepared by a diffusion process.

The lifetime of minority carriers in the diffused zone depends strongly on the type of doping, the surface concentration, and its associated stress, dislocation density, and band-gap perturbations [5.3]. The parameters which determine the carrier lifetime in the base of a solar cell are treated intensively in the following sections. This section will deal with the quantitative effects of carrier lifetime or base diffusion length on the solar cell parameters.

Table 5.1. Properties of the reference cell

Solar cell structure	n^+p high-efficiency type
Junction depth x_j	0.15 µm
Thickness T	350 µm
Front contact	20-finger grid, 94% active area
AR coating	TiO_x, $\lambda_{min} = 0.6$ µm, $n_{AR} = 2.3$
Diffused region	
Average carrier lifetime τ_p	5×10^{-11} s
Surface concentration N_s	10^{20} cm^{-3}
Surface recombination velocity S_p	10^5 cm s^{-1}
Diffusion profile	erfc distribution
Base region	
Resistivity ϱ_n	10 Ωcm
Diffusion constant D_n	33.6 cm^2 s^{-1}
Rear side contact	ohmic, recombination velocity $S_n \to \infty$
	BFS, recombination velocity $S_n \to 0$
Material data	
Absorption coefficient α	[5.5]
Refractive index n	[5.6]
Spectral distribution of AM0 sunlight	[5.7]

For illustration, a reference cell has been selected the structure of which fits to a typical currently available high-efficiency silicon solar cell. Table 5.1 summarizes the physical properties for the reference cell treated here. In the computations, only the base minority carrier lifetime is changed; diffused region properties will be kept constant. Best available material data and AM 0 illumination have been used.

5.1.1 Collection Efficiency

The ability of a solar cell to generate photocurrent at a given wavelength of the incident light is measured quantitatively by its collection efficiency Q. This parameter represents the ratio of those carriers which contribute to the photocurrent to the total number of carriers generated by the absorption of light.

The analytical expression for Q can be derived by solving the one-dimensional steady-state diffusion equation for excess minority carriers in both the n surface and the p base region with the appropriate boundary conditions [5.2, 4]. For the reference cell with uniformly doped base, ohmic and back-surface-field (BSF) contacts, Q has been calculated as a function of wavelength with minority carrier diffusion length L_n as parameter. The base contribution to

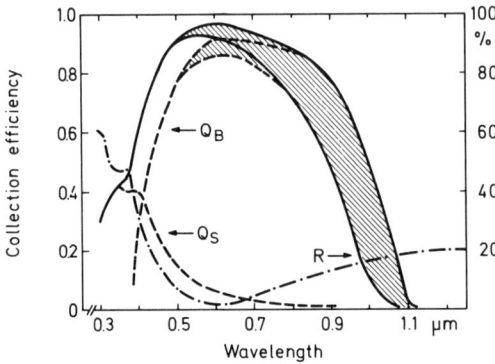

Fig. 5.1. Collection efficiency vs wavelength. Hatched area indicates variation of lifetime from 0.1 to 100 μs

Q, Q_B, is given by

$$Q_B = (1-R)\frac{\alpha L_n \exp(-\alpha x_j)}{\alpha^2 L_n^2 - 1}$$
$$\cdot \left[\frac{\exp(-\alpha T)(\alpha L_n - S_n L_n/D_n) + \sinh(T/L_n) + (S_n L_n/D_n)\cosh(T/L_n)}{\cosh(T/L_n) + (S_n L_n/D_n)\sinh(T/L_n)} - \alpha L_n\right].$$

(5.3)

The surface contribution of Q, Q_S, was calculated for the diffused region. Q_S was found not to depend essentially on the base diffusion length in the range of interest and remained constant through the following considerations.

Figure 5.1 displays Q as a function of wavelength. The contributions of the surface region Q_S and the base region Q_B are plotted separately. As indicated, the blue part of the spectrum is used in the diffused layer and the red part in the base of the cell. The reflectance R of the solar cell surface is monitored separately. The hatched area at long wavelengths is caused by the variation of the base carrier lifetime from 0.1 μs up to the 100 μs range. This indicates that it is mostly the red response of a cell that depends on variation of the carrier lifetime.

5.1.2 Photogenerated Current

The total short-circuit current obtained from a solar cell is the product of collection efficiency and the number of photons N_λ in the incident light, integrated over the useful wavelength range,

$$j_{sc} = e \int_{0.2}^{1.2} QN_\lambda d\lambda.$$

(5.4)

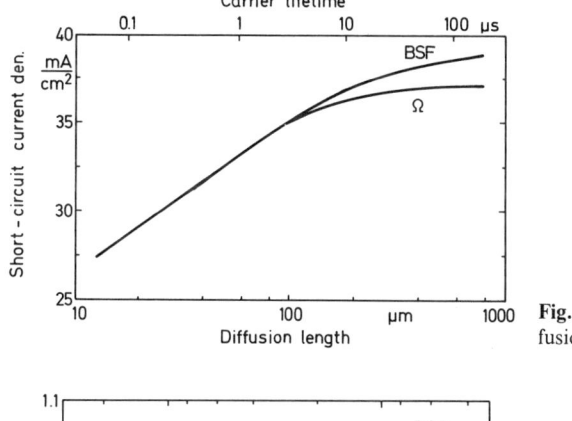

Fig. 5.2. Short-circuit current vs diffusion length

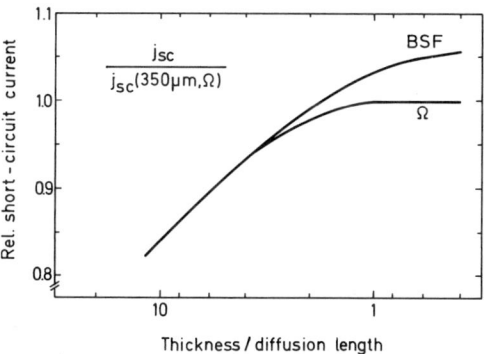

Fig. 5.3. Relative short-circuit current vs ratio of thickness to diffusion length

Figure 5.2 displays the short-circuit current as a function of carrier diffusion length and lifetime. For low values of L_n, j_{sc} increases approximately proportionally to $\log L_n$. It is essential to establish that if the diffusion length approaches the cell thickness, a back-surface-field contact can improve the current by minimizing the number of photocarriers that normally recombine at an ohmic contact. In the case of diffusion length equal to cell thickness, about 68% of all incident photons are collected for an ohmic contact, and 70% for a BSF contact. Figure 5.3 displays the relative short-circuit current as function of the ratio of thickness to diffusion length. As can be seen, ideal collection conditions are achieved for the ohmic contact if the diffusion length is equal to the cell thickness. The back contact conditions are essential if the diffusion length exceeds the cell thickness. This general trend is of growing importance for very thin cells (e.g., less than 200 μm).

5.1.3 Current-Voltage Characteristic

The current-voltage characteristic of a cell in the dark is of equal importance as the photocurrent. This is because it determines the output of the cell, while the junction characteristic determines how much of the energy of the incident light

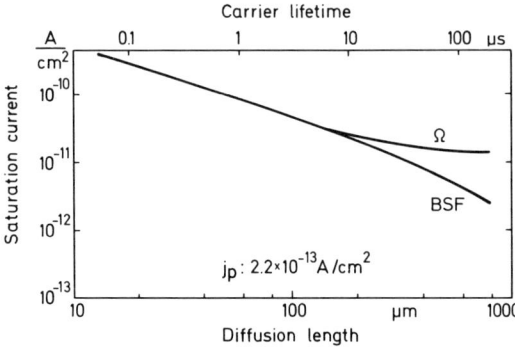

Fig. 5.4. Saturation current vs diffusion length. j_p indicates magnitude of surface-zone contribution

will be available as electrical energy to a load and how much will be lost as heat. When power is taken out of a cell, a voltage exists across the junction and a "dark current" exists which opposes the photocurrent. This is expressed by the standard solar cell equation,

$$J = J_0[\exp(V/V_T) - 1] - j_{sc} \tag{5.5}$$

with $V_T = kT/e = 25.8$ mV at 300 K, and j_0 the diode saturation current. At a real cell, series resistance effects and normal diode losses have to be considered [5.8]. The most important contribution results from the saturation current, which comes from injection of minority carriers from one side of the junction to the other. Solving the diffusion equation for the dark with the same boundary conditions as in Sect. 5.1.1 results in an analytical expression for the saturation current,

$$j_0 = \frac{D_n n_i^2}{L_n P_p} \frac{\sinh(T/L_n) + (S_n L_n/D_n)\cosh(T/L_n)}{\cosh(T/L_n) + (S_n L_n/D_n)\sinh(T/L_n)}. \tag{5.6}$$

This injection current is determined mostly by the conditions and properties of the base and has been calculated for the reference cell as a function of base diffusion length as displayed in Fig. 5.4. For the reference cell, the saturation current j_0 is dominated by the electron flow from the surface to the base region j_n; the opposite contribution j_p is generally more than one order of magnitude lower and does not affect the solar cell characteristics in this doping range. As can be seen from Fig. 5.4, the saturation current will also be influenced by the incorporation of a BSF contact if diffusion length exceeds the thickness of the cell.

From the solar cell equation (5.5), analytical expressions for the open-circuit voltage V_{oc} and the curve fill factor CF can be derived [5.2, 4]. Figure 5.5 displays the open-circuit voltage as a function of diffusion length and carrier lifetime. The voltage of the cell is less sensitive than the photocurrent to variation of carrier lifetime. A change of four orders of magnitude causes only

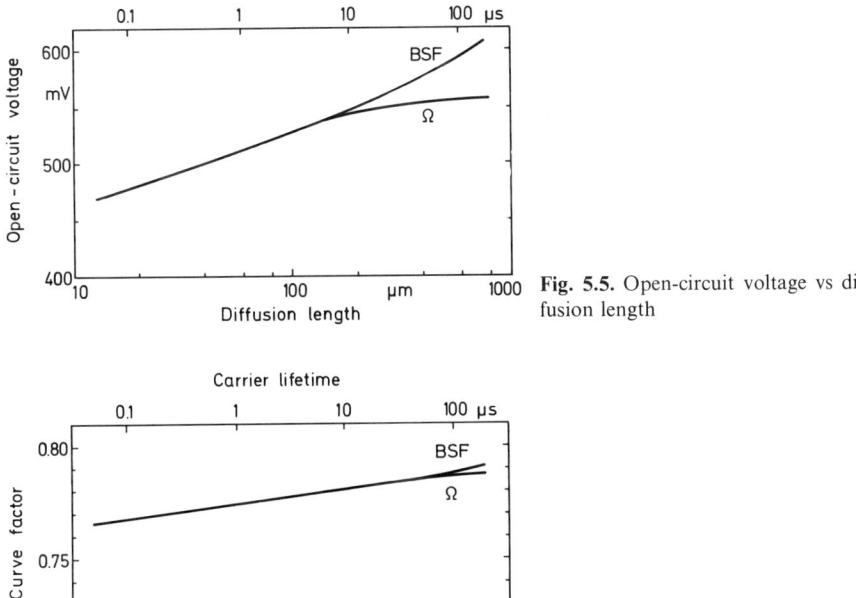

Fig. 5.5. Open-circuit voltage vs diffusion length

Fig. 5.6. Curve fill factor vs diffusion length

variation of about 10% in the open-circuit voltage. The curve factor CF depends even less sensitively on diffusion length, as shown in Fig. 5.6. In practice, this parameter is much more affected by series and shunt resistance losses.

5.1.4 Conversion Efficiency

The efficiency of the conversion of sunlight into electric energy is the most important parameter which characterizes the quality of a cell. It depends on the previously derived quantities according to

$$\eta = \mathrm{CF}\, V_{oc} j_{sc}/P_{in}, \tag{5.7}$$

where P_{in} is the total energy of the incident light (AM 0 for the reference cell [5.7]). The dependence on diffusion length is strongly marked, as indicated in Fig. 5.7. A good approximation to the relative change of efficiency due to a change in the carrier lifetime is

$$\Delta\eta/\eta \sim 0.1 \ln(\Delta\tau/\tau). \tag{5.8}$$

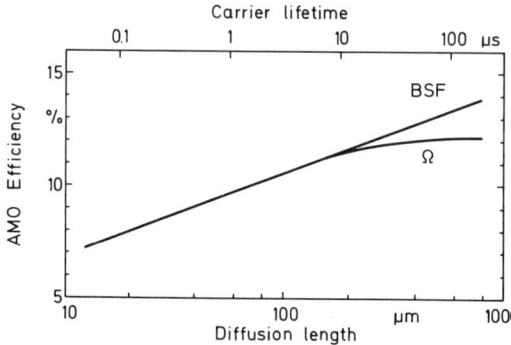

Fig. 5.7. AM 0 efficiency vs diffusion length

As indicated in Fig. 5.7, a considerable increase in efficiency can be obtained if the diffusion length exceeds the cell thickness and a BSF contact is incorporated. This strong increase of efficiency with the value of the carrier lifetime is the reason for the many efforts devoted to the improvement of this parameter by improving the quality of the material and reducing the process-induced deterioration. For the finished silicon solar cell, the lifetime of the base minority carrier should be appreciably greater than 10 μs and can even be in the range of 100 μs for cells with BSF contacts.

5.2 Methods for Measuring Carrier Lifetime

In order to determine the carrier lifetime in a semiconductor crystal, excess minority carriers must be created and detected. They can be created by irradiation by light, X-rays, or electrons. Furthermore, carriers can be injected via rectifying contacts. To detect these excess carriers, photoconductivity, photovoltage or photocurrent, and the photoelectromagnetic effect can be used. Methods published in the literature represent various combinations, some of which will be discussed here. The photoconductive decay method (PCD) is reported in detail since it is the most common technique. Furthermore, the surface photovoltage technique and a method applying the spectral response of solar cells are discussed because they allow the determination of diffusion lengths, and so are very suitable for the present problems.

5.2.1 The Photoconductive Decay Method

The PCD technique is based on the creation of excess carriers by irradiation by near-infrared light. The change in conductivity in the sample is monitored on an oscilloscope as a function of the time elapsed after the light is turned off. In order to reduce the lowest measurable lifetime, which is determined by the fall time of the irradiation, flash light sources were applied frequently. The

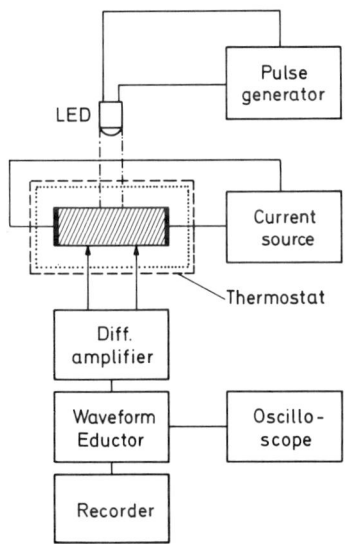

Fig. 5.8. Experimental arrangement of improved PCD method

precision of this method was estimated to $\pm 135\%$ for measurements in silicon [5.9]. In the following, improved equipment is described, which yields results of higher precision.

The experimental arrangement is shown in Fig. 5.8. The size of the silicon samples equal those proposed in the ASTM Standards [5.9]. Nonrectifying contacts cover the entire ends of the specimens. They are prepared by rubbing gallium on the freshly lapped surfaces of the sample. The sample holder is electrically shielded and connected to a thermostat. One advantage compared to usual arrangements is the use of infrared-emitting diodes for creating excess carriers. Special GaAs:Si diodes were used with the maximum of the emitted radiation at a wavelength near 1 μm. Their fall time is about 100 ns. On the other hand, GaInAs diodes are commercially available which emit radiation at a wavelength of 1.06 μm. Their spectral half-width is 550 Å and the fall time is 5 ns or less.

The diode is connected to a pulse generator which allows the repetition rate, the pulse duration, and the intensity of the radiation to be adjusted. With the aid of the pulse duration, a steady state of excess carriers in the irradiated sample can be arranged before the diode is turned off. This increases the reproducibility of the measurement [5.10] and facilitates theoretical treatment using the Shockley-Read model. In order to measure the photoconductivity, the sample is supplied with a constant current. The photovoltage resulting from the photo-induced variation of the conductivity is taken from the sample by the same contacts. The signal is amplified by a factor of 10^4 and low-frequency as well as high-frequency noise can be suppressed separately. As a further improvement, a Waveform Eductor (Princeton Applied Research) is applied which increases considerably the signal-to-noise ratio. The photovoltage decay

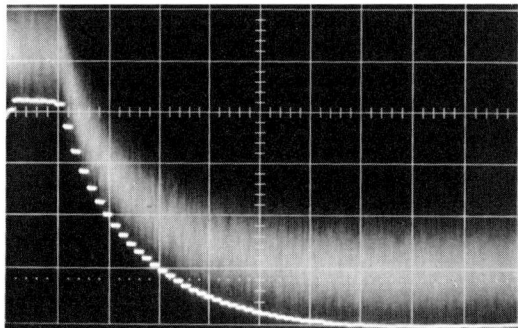

Fig. 5.9. PCD signal with and without application of a Waveform Eductor

curve is stored in 100 capacitors for an adjusted time. Signals obtained are shown in Fig. 5.9 showing a photograph taken from the screen of the oscilloscope with and without application of the Waveform Eductor. The steps in the decay curve represent an accurate time base of one-hundredth of the sweep duration. They facilitate the evaluation of the decay curve which can be read out from the Waveform Eductor by a recorder. The reduction of the high-frequency noise is clearly seen in Fig. 5.9, whereas the reduction of low-frequency noise in this case is carried out by taking the photograph with short exposure times.

Using this apparatus, far smaller signals can be detected. Carrier lifetime can be determined in crystals of low specific resistivity such as 0.1 Ωcm. The injection level of excess carriers can be drastically reduced; thus increasing the accuracy of the results. The carrier lifetime in n-type silicon as a function of injection level is shown in Fig. 5.10 for gold acting as recombination center. For the calculation, gold acceptors and also donors were taken into account in a concentration of 2.9×10^{10} cm^{-3} using recombination parameters from the literature [5.11]. Injection levels of 10^{-2}, as allowed in ASTM Standards [5.9], result in a systematic deviation from the low-level value of 25%. Injection levels of 10^{-3}, however, reduce this deviation to about 3% which is an acceptable measurement error. This injection level may be sufficiently low, since gold in n-type silicon causes one of the largest increases of carrier lifetime with injection of excess carriers known so far.

In order to increase the reproducibility of the carrier lifetime measurement, the sample must be kept at a definite temperature during measurement. In the present arrangement, this is done by mounting it in a thermostat. The worst case is again found to be with gold as recombination center. Its temperature dependence can be reconstructed from the Arrhenius plot of the carrier lifetime. For an acceptable measurement error, the temperature of the sample during measurement must be kept within $\pm 0.5\,°C$ of a fixed value, for example, 25 °C, since this can be easily maintained.

For the accurate measurement of the recorder plot, it is best transferred to semilogarithmic axes. In this way, the exponential decay of the signal with time

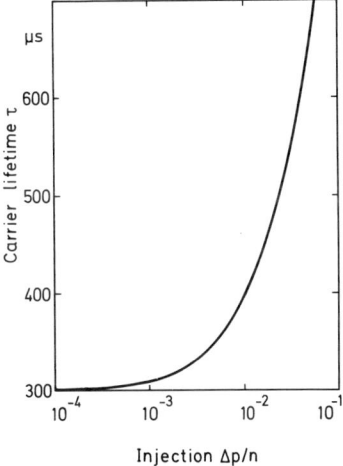

Fig. 5.10. Carrier lifetime in n-Si:Au as function of injection level

can be easily controlled. Deviations from exponential decay usually were not observed, except

I) for small samples with high carrier lifetimes where the carriers mainly recombine at the surfaces;

II) for crystals with highly inhomogeneous carrier lifetimes;

III) for sweep-out of carriers at one of the contacts because of high carrier lifetimes and high electric fields; and

IV) for samples containing traps.

For moderate trap concentrations, the decay curve transferred to semilogarithmic axes can be replaced by a straight line of gradient equal to the highest gradient observed. The influence of traps can be eliminated by additional illumination of the sample with cw light or by raising its temperature. Both procedures, however, increase the injection level of excess carriers.

While sweep-out of carriers at one contact can be avoided by reducing the electric field in the sample, the recombination of carriers at the surfaces can not be eliminated. Carrier lifetimes determined in filament-shaped specimens, τ_f, must be corrected for specimen size to obtain bulk values, τ_B. If the crystal perfection in the surface region of the sample is disturbed by sawing or lapping, an infinite surface recombination velocity can be assumed. The correction is then carried out following the equation

$$1/\tau_B = 1/\tau_f - \pi^2 D(1/a^2 + 1/b^2 + 1/c^2), \tag{5.9}$$

where D is the diffusion coefficient of the minority carriers, and the dimensions of the sample are a, b, c. For corrections exceeding a factor of 2, the errors increase considerably and another measurement is recommended using specimens of larger dimensions. Correction curves for three specimen sizes [5.9] and

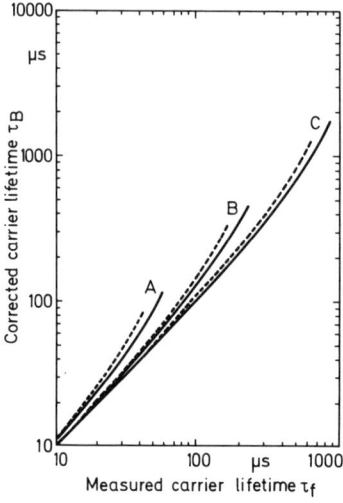

Fig. 5.11. Correction curves for specimen size in 1 (solid) and 10 Ωcm p-Si (dashed). (A) 2.5 × 2.5 × 15; (B) 5 × 5 × 25; (C) 10 × 10 × 25 mm

for 1 and 10 Ωcm p-type silicon samples are shown in Fig. 5.11, which assumes infinite surface recombination velocity.

Equations to calculate the decay of excess carriers with time as well as the carrier lifetime as a function of injection level were derived from Shockley-Read model by *Schlangenotto* [5.12]. A steady state of excess carriers before turning off the radiation was assumed, as well as small concentrations of recombination centers with respect to the majority carrier concentration in the crystal. Only minority carriers are considered. For a single recombination center in p-type silicon, the following equation holds:

$$t_n = \left(\ln \frac{x(0)}{x} + \frac{\alpha_n}{\alpha_p} \ln \frac{1+x(0)}{1+x} \right) \Big/ \alpha_n N, \tag{5.10}$$

where $x = \Delta p/p_0$; $x(0)$ is the initial injection of excess carriers $\Delta p/p_0$ at the time $t = 0$; α_p and α_n are the recombination probabilities of holes and electrons, respectively, which are proportional to carrier cross sections, with the diffusion velocity as proportionality factor; and N is the concentration of recombination centers. Equation (5.10) represents the temporal decay of the excess carrier concentration. Substituting $x(0)/e$ for x gives the carrier lifetime $t = \tau$ as a function of injection. To apply (5.10) to n-type silicon, the symbols n and p must be interchanged.

As one boundary condition, the high-level carrier lifetime in n-type and p-type silicon turns out to be:

$$\tau_\infty = 1/(\alpha_p N) + 1/(\alpha_n N). \tag{5.11}$$

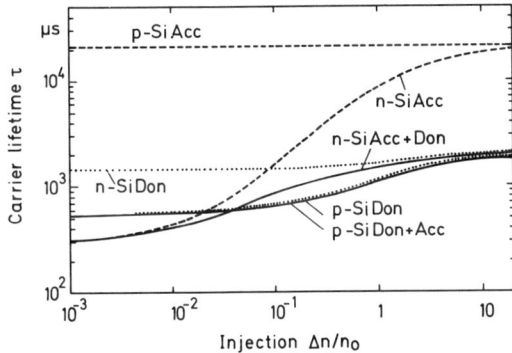

Fig. 5.12. Carrier lifetime in gold-doped n- and p-type silicon as function of injection level

The low-level carrier lifetime in p-type silicon is

$$\tau_0 = 1/(\alpha_n N). \tag{5.12}$$

The carrier lifetime and its dependence on injection differs considerably in n- and p-type silicon of the same impurity content. This is demonstrated in Fig. 5.12 for gold as recombination center in a concentration of 2.9×10^{10} cm^{-3}, using recombination parameters from the literature [5.11]. Since gold in silicon acts both as acceptor and donor, (5.10) was extended for two centers in p-type silicon,

$$t_n = \frac{1}{\beta_n N} \ln \frac{x(0)}{x} + \frac{1}{\alpha_p N} \ln \frac{1+x(0)}{1+x}$$
$$+ \frac{1-\beta_p/\alpha_p - \alpha_n/\beta_n}{N(\beta_p + \alpha_n)} \ln \frac{1+x(0)+x(0)\alpha_n/\beta_p}{1+x+x\alpha_n/\beta_p}. \tag{5.13}$$

To apply this equation to n-type silicon, n and p, as well as α and β, must be interchanged. The results obtained for gold as acceptor and donor are indicated in Fig. 5.12.

Modifications of the measurement equipment presented here can be found in the literature. Since the lower limit of the electronic equipment reported above is about 4 μs, a combination of the Boxcar Integrator and the Waveform Eductor was applied [5.13]. It extends the lower limit to several nanoseconds. Instead of using infrared-emitting diodes, some authors use lasers that emit wavelengths of 1.06 μm [5.14, 15]. A sampling oscilloscope can also be used to detection excess carriers [5.15].

5.2.2 The Surface Photovoltage Technique

In contrast to the PCD method, the surface photovoltage technique is suitable for determining diffusion lengths of minority carriers. It is based on the creation of excess carriers by illuminating the crystal with monochromatic light of various wavelengths. Using the variation in the absorption coefficient of silicon with wavelength, the penetration depth of the light can be changed, and carriers can be created at different distances from the surface. In the electric field of the depletion layer which is usually found at silicon surfaces, electrons and holes diffuse in opposite directions. Thus, a surface photovoltage is built up. It can be taken from the sample by a rectifying contact at the illuminated surface and an ohmic contact on the reverse side [5.16]. On the other hand, the signal can be measured by means of a lock-in amplifier, which requires the use of chopped light [5.17].

For large diffusion lengths compared with the penetration depth of the light, the signal is almost independent of the wavelength. All excess carriers reach the surface of the sample and contribute to the photovoltage. Increasing the wavelength causes increasing penetration depth of the light, and excess carriers partly recombine before arriving at the surface of the sample. The photon intensity that is required to build up a constant photovoltage must be increased. Evaluation of the experimental data is performed by plotting this photon intensity, corrected for reflection, as a function of the penetration depth of the light, which is the reciprocal absorption coefficient for each wavelength. The linear plot obtained is extrapolated to the intercept with the axis of penetration depth. It reveals a negative value of the effective diffusion length.

Experimental arrangements and theoretical treatments can be found in the literature [5.17, 18]. The main difficulty in applying this technique is the accuracy with which the absorption coefficient is known as a function of wavelength. It may differ from crystal to crystal because of lattice stress, carrier concentration, and temperature. Experimentally obtained absorption coefficients of silicon in μm^{-1} [5.19] are given mathematically by the following equation for stress-relieved crystals [5.17]:

$$\alpha = 0.526367 - 1.14425\lambda^{-1} + 0.585368\lambda^{-2} + 0.039958\lambda^{-3}. \tag{5.14}$$

Another equation for non-stress-relieved silicon is usually appropriate for as-grown crystals [5.17, 20]:

$$\alpha = -1.06964 + 3.34982\lambda^{-1} - 3.61649\lambda^{-2} + 1.34831\lambda^{-3}. \tag{5.15}$$

The validity of this equation is limited to crystals with specific resistivities exceeding 0.1 Ωcm. Calculated absorption coefficients as functions of wavelength are shown in Fig. 5.13. Evidently, there are considerable differences between the two functions. It is recommended [5.17] to choose the equation which better fits the data obtained from surface photovoltage measurements.

Applying this method, results usually disagree more or less from those obtained by PCD measurements in the same sample, since corrections for surface recombination can not be performed. In contrast to the PCD method this technique is, however, suitable for measurements on wafers, provided the wafer thickness exceeds the diffusion length by at least a factor of four [5.17].

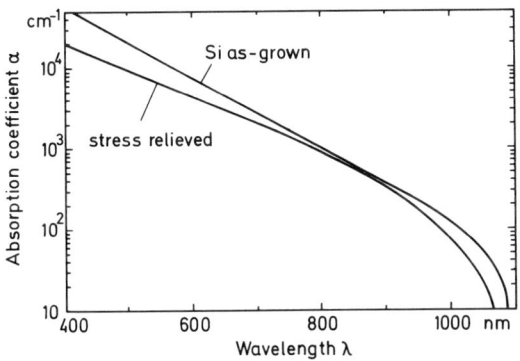

Fig. 5.13. Absorption coefficients of as-grown and stress-relieved silicon as functions of wavelength

5.2.3 Measurements Applying the Spectral Response of Solar Cells

The method which uses the spectral response of solar cells is similar to the surface photovoltage technique and has proven excellent in most solar cell development laboratories. The diffusion length of base minority carriers can be determined from measured spectral response data in a wavelength region around 0.9–1 µm. At this wavelength interval, the response of the diffused surface layer is negligible and the influence of the rear side contact area does not yet dominate (Fig. 5.1). In this case, and with the condition $L_n < D$, all of measured response is base-generated and (5.3) reduces to

$$Q(\lambda) \approx \alpha[1 - R(\lambda)]/(\alpha + 1/L_n). \tag{5.16}$$

This equation is normally written as

$$[1 - R(\lambda)]/Q(\lambda) \approx 1 + 1/(\alpha L_n). \tag{5.17}$$

The reflectivity R and collection efficiency Q are measured for various wavelengths around 0.9–1 µm and plotted against the penetration depth $1/\alpha$ of the incident monochromatic light. For rough approximations, the reflectivity is about 0.3 for a bare cell, and 0.15 for an antireflection coated cell at a wavelength of 1 µm. The gradient of the straight line obtained gives the reciprocal diffusion length. A problem with this method arises again from the absorption coefficient of silicon. Values taken from [5.5] have proved to be the most reliable data. Figure 5.14 gives a plot of experimental data for the reference cell as described in Sect. 5.1.1.

The accuracy of this method can be controlled using the exact formula for $Q(\lambda)$, (5.3), and comparing (5.17) for various diffusion lengths. This is demonstrated in Fig. 5.15, where the exact values of the diffusion length were plotted against values derived via (5.17) for the reference cell. In the same figure, the "relative failure" $\Delta L_n/L_n$ is plotted against L_n. As can be seen, the failure is less than 20% if $L_n < 0.5 D$.

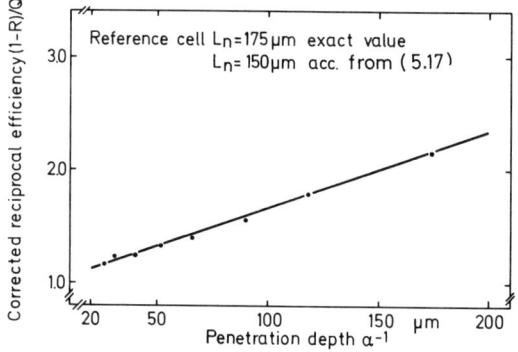

Fig. 5.14. $(1-R)/Q$ vs penetration depth of light. Slope gives diffusion length L_n

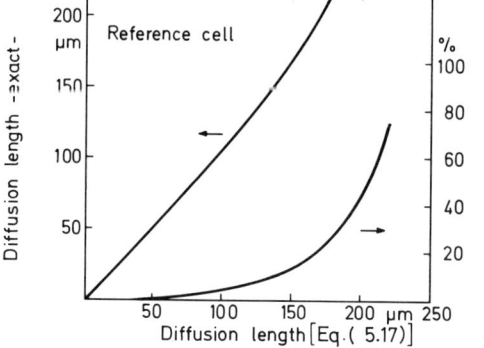

Fig. 5.15. Exact value of L_n vs value determined according to (5.17); and relative failure of this method

5.3 Carrier Lifetime in as-Grown Silicon Crystals

Carrier lifetimes were determined in a large number of p-type silicon crystals with the aid of the PCD method. It was found that the crystals usually exhibit both a radial and an axial profile in the carrier lifetime. For an exact characterization, measurements in several samples from different positions in one crystal rod must be performed. Therefore samples were selected with regard to their position with respect to the distance from the seed end of the rod as well as their distance from the crystal axis. Their usual dimensions were $5 \times 5 \times 25$ mm for pulled crystals, and $10 \times 10 \times 25$ mm for floating-zone silicon.

Results obtained in Czochralski-grown p-type silicon with resistivities of 1 and 10 Ωcm are compared with those in floating-zone crystals. The influence of high doping concentrations upon the carrier lifetime is discussed, and typical examples for the axial and radial profile of the carrier lifetime in pulled and floating-zone p-type silicon are presented.

5.3.1 Comparison of Results in Czochralski-Grown and Floating-Zone p-Type Silicon

Measurements were mainly performed with raw material for solar cell production. Therefore, crystals with the following specifications were investigated:

  1 Ωcm dislocation-free Czochralski   p-Si 35 mm in diameter
10 Ωcm dislocation-free Czochralski   p-Si 65 mm in diameter
  1 Ωcm dislocation-free floating-zone p-Si 35 mm in diameter
10 Ωcm dislocation-free floating-zone p-Si 35 mm in diameter.

The measurements revealed specific properties of these groups. Differences between 1 and 10 Ωcm floating-zone silicon were not considered because of the low number of measurements.

Results obtained in 1 Ωcm pulled crystals are shown in the upper part of Fig. 5.16 in the form of a histogram. The number of samples showing a given carrier lifetime within an interval of 1 µs is plotted as a function of the carrier lifetime. The total number of samples was 1160, taken from various positions in a large number of crystals. Separating results obtained in samples from the seed end of the crystals and those from their tang end revealed distributions of the carrier lifetime as shown in the lower part of the figure. The number of samples was 400 in each case. The crystals show smaller carrier lifetimes at their seed ends compared with their tang ends. Differences in carrier lifetime between samples originating from the central region and from the rim of the crystal rod are not statistically significant. The histograms shown in Fig. 5.16 resemble logarithmic normal distributions. In order to examine this, the results were plotted in probability nomographs. The percentage of results up to a certain carrier lifetime on a scale corresponding to a Gaussian distribution was plotted as function of this carrier lifetime on a logarithmic scale. As demonstrated in Fig. 5.17 the measurement points fit straight lines. The best approximation to linearity is achieved with the total number of results because of the large number of measurements. The following mean carrier lifetimes were determined:

Total	17.5 µs
Seed end	14.0 µs
Tang end	22.0 µs
Center	15.5 µs
Rim	15.5 µs.

Deviations between the mean values of the total number and those from the central and the rim region as well can be explained by the different number of results. There are series of crystals showing higher, and others with lower, mean values of the carrier lifetime. The deviations will reduce with increasing number of measurements. It is to be noted that these results do not indicate a definite carrier lifetime in this material. By reducing the number of impurities during crystal growth, the mean value of the carrier lifetime can be increased.

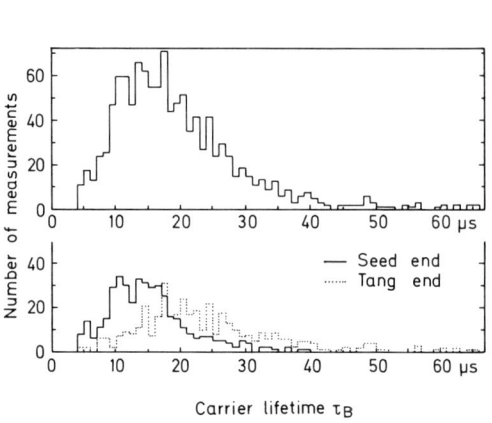

Fig. 5.16. Probability density of carrier lifetime measurements in 1 Ωcm pulled p-Si in total and separated for samples from the seed and the tang end of the crystal

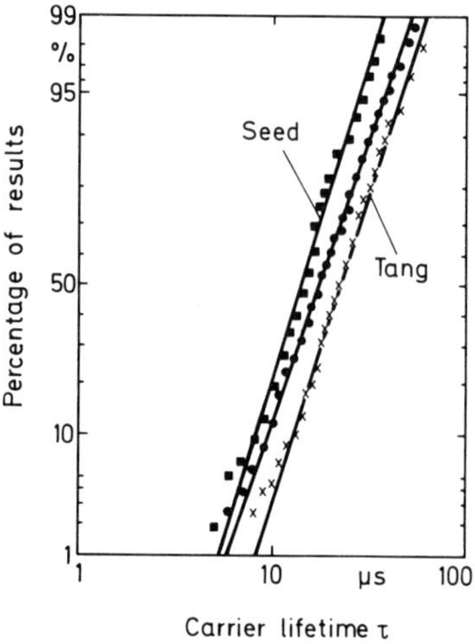

Fig. 5.17. Probability nomographs of carrier lifetime in pulled 1 Ωcm p-Si

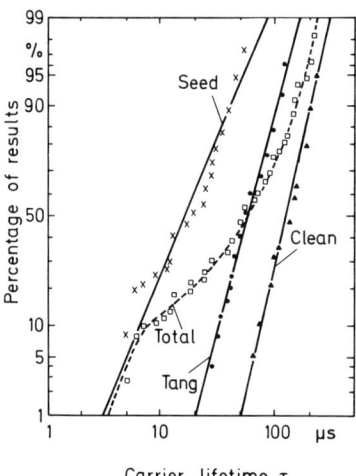

Fig. 5.18. Probability nomographs of carrier lifetime in pulled 10 Ωcm p-Si

Results obtained in 10 Ωcm pulled p-type silicon are demonstrated in Fig. 5.18, also in the form of probability nomographs. The total number of measurements (112) taken together do not fit a straight line. Specific groups of measurements, despite their smaller quantity (20–40), exhibit a linear plot. Smallest carrier lifetimes are observed again in samples originating from the seed end of the crystals. They almost agree with values in 1 Ωcm crystals, in contrast to those from the tang end, which are considerably higher. In

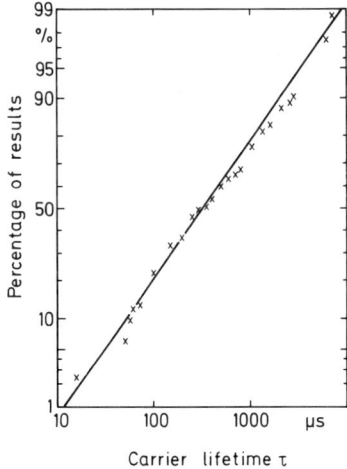

Fig. 5.19. Probability nomographs of carrier lifetime in floating-zone 1 and 10 Ωcm p-Si

additional, we give results for crystals which were grown in particularly clean crucible material. They showed the highest carrier lifetimes.

The following mean carrier lifetimes were deduced:

Total	60 μs
Seed end	16 μs
Tang end	60 μs
Clean conditions	160 μs.

They are much higher than those for the 1 Ωcm crystals grown by the same technique. There are three differences between these two specifications: specific resistivity, crystal diameter, and time of crystal growing – the 10 Ωcm crystals were pulled several years after the others. A definite reason for the increase in carrier lifetime can not be given.

The third group consisted of 1 and 10 Ωcm p-type silicon crystals grown dislocation-free by the floating-zone technique under various conditions. Results are demonstrated in Fig. 5.19. Although the number of measurements was rather small (84), the spread of values is largest, showing carrier lifetimes between 17 and 7000 μs. The mean value works out as about 350 μs, which is considerably higher than in pulled crystals. More detailed investigations could not be performed because of the lack of a sufficient quantity of data concerning sample position in the crystal rod.

The statistical evaluation of carrier lifetimes measured in as-grown crystals reveals mean values for crystals of different specifications. For the production of efficient solar cells, a lower limit in carrier lifetime must be exceeded in the as-grown crystals. This lower limit of course depends on the desired efficiency, the dimensions of the cell, and also the treatment of the material during device production.

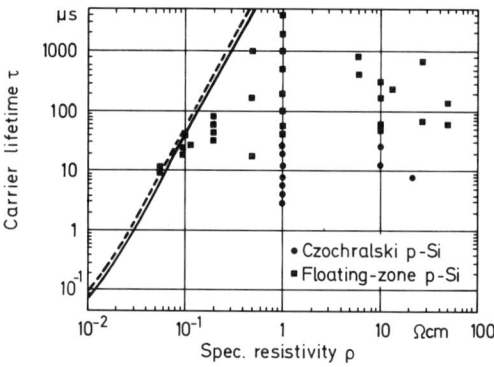

Fig. 5.20. Carrier lifetime in pulled and floating-zone p-Si as function of specific resistivity and limitations by Auger recombination [5.21, 22]

5.3.2 Influence of Doping Concentration

The differences in carrier lifetime reported above are caused by different impurity contents of the crystals. A dependence of the carrier lifetime on the carrier concentration in p-type silicon is not observed before the hole concentration exceeds about 4×10^{16} cm^{-3}, corresponding to 0.5 Ωcm. In this region a further mechanism, Auger recombination, becomes effective. As demonstrated by *Beck* and *Conrad* [5.21], it increases with the squared majority carrier concentration resulting in a steep decrease of the upper limit for the carrier lifetime. This upper limit can only be achieved with negligible impurity recombination compared with Auger recombination. Whereas Auger recombination does not affect carrier lifetimes in 1 Ωcm p-type silicon, the values achievable in 0.1 Ωcm silicon are limited to about 40 μs. This is demonstrated in Fig. 5.20 showing measured carrier lifetimes as functions of specific resistivity in p-type silicon. Measurement points are marked by symbols corresponding to the respective growing technique of the crystals investigated. Two upper limits of carrier lifetime are indicated, differing slightly in Auger coefficients [5.21, 22].

5.3.3 Local Variations of Carrier Lifetime in Silicon Crystals

With the aid of the PCD method, local variations of the carrier lifetime in crystals can be determined in two ways: by cutting several samples in succession from the crystal region of interest, and by sawing one sample from this region and controlling the carrier lifetime at several points. The illuminated area can be displaced successively from one end of the sample to the other. The local resolution of measurements in several samples is limited by their quantity and size. The resolution of measurements on one sample depends on the displacement and the size of the illuminated area which is limited by the scattering of the light in the sample. Sweep-out of carriers can reduce it if the drift length of the minority carriers is of the same order of magnitude as the displacement.

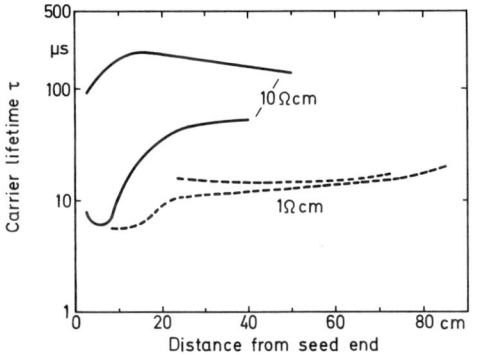

Fig. 5.21. Longitudinal carrier lifetime profiles in 1 and 10 Ωcm pulled p-Si

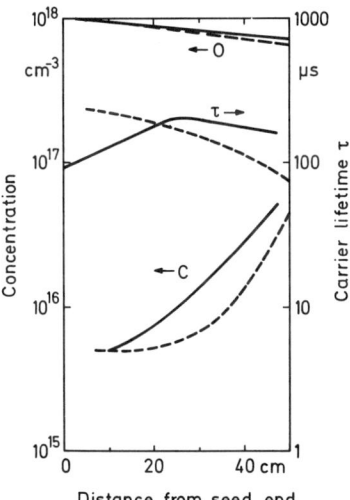

Fig. 5.22. Longitudinal profiles of carrier lifetime, and of oxygen, and carbon concentrations in 10 Ωcm pulled p-Si

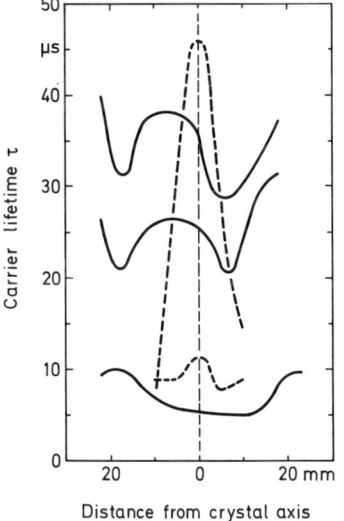

Fig. 5.23. Radial carrier lifetime profile in pulled p-type silicon

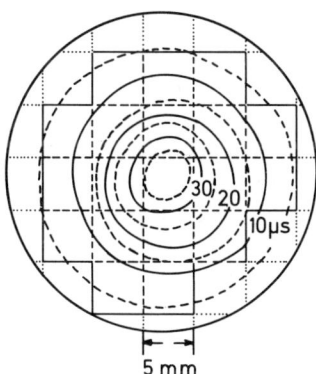

Fig. 5.24. Radial carrier lifetime profile in a pulled 10 Ωcm p-Si crystal near seed end

a) Longitudinal and Radial Profiles of Carrier Lifetime in Czochralski-Grown Silicon Crystals

Differences in carrier lifetime between samples cut from the seed end and the tang end of pulled crystals have already been pointed out. Measurements in greater detail revealed longitudinal profiles of the carrier lifetime in 1 and 10 Ωcm p-type silicon as demonstrated in Fig. 5.21. Usually the values increase

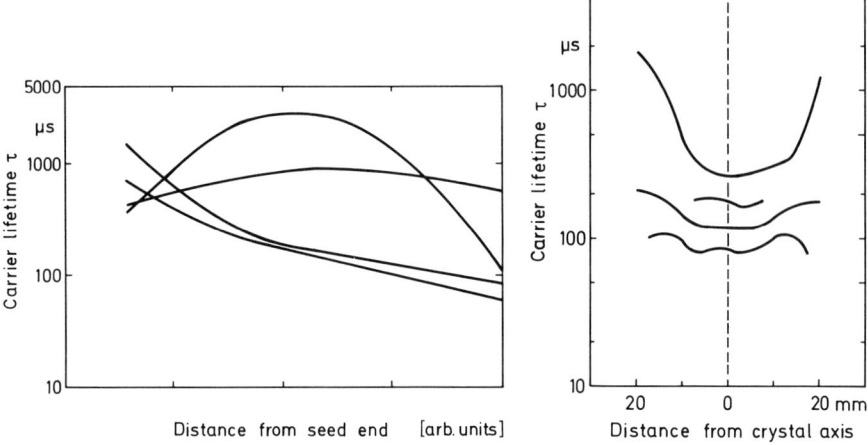

Fig. 5.25. Longitudinal carrier lifetime profile in floating-zone p-type silicon

Fig. 5.26. Radial carrier lifetime profile in floating-zone p-type silicon

towards the tang end. But there are also profiles showing a maximum or a minimum in the middle region of the crystal rod. 10 Ωcm silicon crystals near the seed end exhibit similar carrier lifetimes as 1 Ωcm silicon. The increase with the distance from the seed end was more pronounced, resulting in considerably higher values at the tang end of the crystal. But there are also crystals starting at high values. Two further axial profiles of the carrier lifetime in 10 Ωcm p-type silicon crystals are shown in Fig. 5.22. Their oxygen and carbon concentration profiles are also indicated. It seems that they are independent of one another.

Radial profiles of the carrier lifetime in pulled crystals are strongly marked although no statistically significant differences were observed between central and rim regions of crystal rods. The profiles, more or less symmetrical, can be preserved or changed over the length of the crystal. Some examples are given in Fig. 5.23. Only the variety of results can be demonstrated. A rather simple but pronounced radial profile is shown in Fig. 5.24. In this case, lines of equal carrier lifetimes are reconstructed from results obtained in 19 samples cut in seccession from a cross section of a 10 Ωcm silicon crystal near the seed end. A steep decrease towards the rim of the crystal can be observed.

b) Longitudinal and Radial Profiles of Carrier Lifetime in Floating-Zone Silicon Crystals

Longitudinal profiles of the carrier lifetime determined in floating-zone crystals again show considerable variations. Primarily, a decrease of carrier lifetime towards the tang of the crystal was observed. But there are also others showing maxima in the middle region of the crystal as demonstrated in Fig. 5.25 for four different crystals. The distances from the seed end in these cases were not known quantitatively. The pronounced variations in carrier lifetime over the length of the crystal do not necessarily signify considerable variation in impurity concentration. Because of the high carrier lifetime, the magnitude of the impurity content and its variation is rather small compared with pulled crystals.

Some examples for the radial profile of the carrier lifetime in 1 and 10 Ωcm floating-zone silicon are given in Fig. 5.26. In general, no significant differences to pulled crystals were found.

5.4 Carrier Lifetime in Processed Silicon Crystals

The concentration of recombination centers which determines the carrier lifetime in a crystal can be changed by processing the material. Additional impurities can diffuse from the surfaces of the sample into its volume and impurity reactions can take place changing the recombination parameters of the centers. There is a variety of possible impurity reactions in silicon, many of which may be still unknown. On the other hand, a redistribution of present centers can be caused by gettering and precipitation of impurities. This way, the carrier lifetime of the as-grown crystal can be increased as well as decreased; both effects are observed.

Processes are distinguished in the following according to the temperature range in which treatments were performed. This classification into room temperature and high temperatures, however, is rather arbitrary. The general opinion that diffusion in silicon takes place only at high temperatures, although correct for most impurities, must be modified to take into account recent experimental results. They will be reported in Sect. 5.4.1. In Sect. 5.4.2, variations in carrier lifetime due to annealing are discussed.

5.4.1 Processing Near Room Temperature

a) Effect of Sawing Silicon Samples

To measure carrier lifetimes in materials, the samples must be sawn. This process unfortunately was found to affect the carrier lifetime [5.23]. Deviations between two measurements in the same sample made certain periods of time after sawing indicated reactions so far unknown in silicon. This is demonstrated in Fig. 5.27 showing the carrier lifetime of a 1 Ωcm pulled p-type silicon crystal as a function of distance from the crystal axis. First measurements were carried out 3 days after sawing the samples, and the next measurements 10 days later. The typical structure of the radial profile is preserved but the individual results are more or less reduced. More systematic investigations resulted in the following facts. Only decreases in the carrier lifetime were found. The decay is a function of the time elapsed after sawing the samples. After about 2–3 weeks, saturation takes place. The effect is limited to Czochralski-grown silicon. The

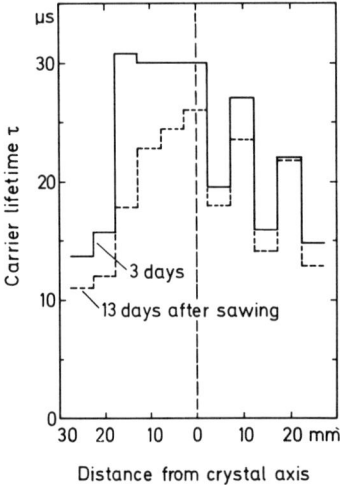

Fig. 5.27. Radial carrier lifetime profile determined in samples 3 and 13 days after sawing them from the crystal rod

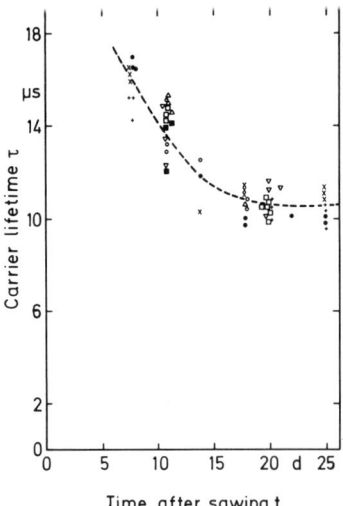

Fig. 5.28. Decay of carrier lifetime in samples of 8 pulled crystals as a function of the time elapsed after sawing

magnitude of the decay varies from crystal to crystal and even from sample to sample, and is also affected by the sawing procedure. Repeated sawing of the same sample does not show the same effect. The decay is independent of the cleanness of the surfaces. In some crystals a remarkable disappearance of traps was observed.

The decay of the carrier lifetime in eight different crystals as function of time is shown in Fig. 5.28. The samples were sawed in succession. Although not uniform for all results, an averaged decay can be plotted, as indicated in the figure. In general, the decay is less pronounced than the one shown.

Since the effect is limited to pulled crystals, it may be caused by a certain impurity content. An explanation, however, was not found before a similar decay of carrier lifetime was observed in samples of floating-zone crystals after quenching them from high temperatures [5.23]. This could be reasonably explained by diffusion of vacancies frozen-in during quenching which subsequently react with impurity atoms. In this way, electrically inactive impurities may become active when combined with a vacancy. The mechanism after sawing pulled samples might be explained in the same manner. The creation of vacancies during the sawing process must be assumed as well as the presence of certain impurities in pulled crystals. Due to their combination, the carrier cross section must be increased. Considerable difficulties are to be expected in characterizing pulled crystals since carrier lifetimes depend on the time elapsed after cutting the samples. The magnitude of the decay is not known, so measurements can be performed at two defined times: at the beginning of the decay, and after it has reached saturation. Most of the results reported in the present study were obtained as soon as possible after cutting the samples, since the decay requires a long period of time.

b) Influence of Etching Silicon Crystals

Etching silicon samples primarily in alkaline media was found to decrease the carrier lifetime. If a silicon sample is treated in NaOH or KOH, the formation of hydrogen begins at a temperature of about 60 °C. At the same temperature the carrier lifetime decreases, as shown in Fig. 5.29. A floating-zone silicon sample was treated repeatedly in NaOH at various temperatures. The carrier lifetime was measured after each etching. The results are plotted as function of the temperature during etching. A final value of 20 μs or less was observed after etching the sample at 100 °C for 5–10 minutes. Similar, although less pronounced, effects were found previously [5.24], assuming the diffusion of hydrogen. The decreased carrier lifetime can be regenerated by storing the sample at room temperature. At high temperatures the restoration is faster, but even annealing the sample at 400 °C will not fully restore the initial carrier lifetime if it originally exceeded 100 μs.

Taking into account the sample size and the duration of etching required to yield saturation of the carrier lifetime, a diffusion coefficient of the order of magnitude of 10^{-5} cm^2 s^{-1} can be estimated at a temperature of 100 °C. This, however, is much higher than all diffusion coefficients known so far in silicon at this temperature including atomic hydrogen and vacancies. On the other hand, it may be expected that ionized hydrogen consisting only of protons may have a diffusion coefficient which is several orders of magnitude higher than that of atomic hydrogen. An explanation for the restoration of the carrier lifetime is easily found in assuming that the protons diffuse to the surfaces of the sample. If they become discharged, however, the diffusion coefficient decreases drastically. So a small amount of the hydrogen may remain in the crystal.

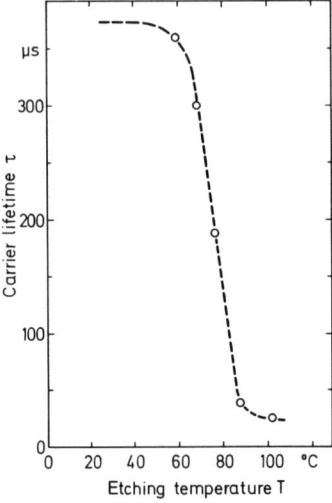

Fig. 5.29. Carrier lifetime in a sample of a floating-zone silicon crystal after repeated etching in NaOH at various temperatures

Since KOH treatment is a common process in solar cell production, this effect is of particular interest. In general, however, the carrier lifetime will be restored by subsequent temperature processes.

c) Illumination of Silicon Samples

Besides by sawing and etching, the carrier lifetime can also be changed by illumination of silicon samples [5.25]. The wavelength of the effective radiation is near 1 μm, so the radiation can penetrate the volume of the sample. The energy of irradiation is still sufficient to initiate impurity reactions. Longer wavelengths do not change the carrier lifetimes.

In Czochralski-grown silicon, the carrier lifetime is generally decreased due to irradiation, whereas it can be increased in swirl-free floating-zone silicon. The amount of change depends on the radiation intensity, its duration, and some unknown properties of the crystal which can be impaired by annealing the sample. So an increase of the carrier lifetime due to irradiation can be transformed to a decrease in the same sample after annealing above 600 °C.

The carrier lifetime in a 1 Ωcm pulled p-type silicon sample as function of the exposure time to illumination is shown in Fig. 5.30 for three different intensities. In order to start with the same initial value, the sample was annealed at 450 °C which increases the carrier lifetime as will be discussed subsequently. Exposure to daylight results in the decay of the carrier lifetime indicated by the dashed line. A further annealing and illumination with a tungsten lamp results in a more pronounced decay, shown by the dotted line. Annealing and focusing a halogen lamp on the sample, which was kept at room temperature, results in a strongly marked decay. It saturates after about 5 h, showing a final lifetime equal to that of the as-grown crystal.

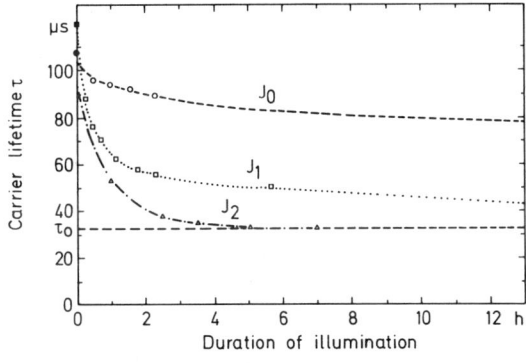

Fig. 5.30. Carrier lifetime in a sample of a 1 Ωcm pulled p-type silicon crystal as a function of exposure time to illumination of three different intensities

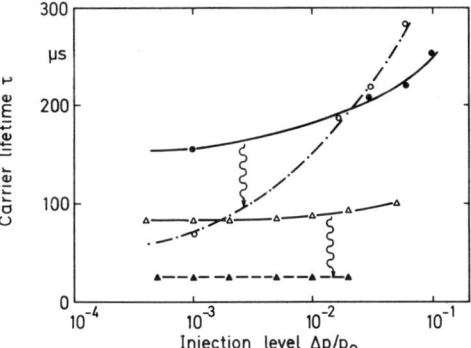

Fig. 5.31. Carrier lifetime of 2 different 10 Ωcm floating-zone p-Si crystals vs injection level before and after photon irradiation

As shown, the carrier lifetime after irradiation is restored by a subsequent heat treatment at a temperature exceeding 170 °C. Sometimes regeneration is observed even at room temperature. Further investigations are shown in Fig. 5.31, which plots the carrier lifetime of two different 10 Ωcm floating-zone silicon crystals as a function of injection level before and after photon irradiation. Both low-level carrier lifetimes decrease due to irradiation. The injection dependence in one crystal is unchanged. In the other crystal, a sharp increase is found, resulting in a crossover of the two functions before and after illumination. Reasonable explanations are found by assuming preservation of carrier cross sections of the effective recombination centers in one crystal and a change of these parameters in the other crystal, because increase of carrier lifetime with injection level is determined only by these parameters.

Although the mechanism is not known in detail, it seems to be caused by impurity reactions in the volume of the crystal. Impurity diffusion coefficients in silicon extrapolated to room temperature would allow reactions if the density of the impurities were sufficient high. Complex generation of two impurities initiated by irradiation can increase or decrease the carrier lifetime according to their inital and final carrier cross sections and concentrations. The effective total carrier cross section can be preserved or changed depending on the contribution of the separate impurities. A subsequent annealing of the sample breaks up the binding and the initial state is restored. The mechanism can hardly be understood without the aid of further techniques since probably more than one recombination center takes part.

The irradiation-induced decrease of carrier lifetime can reduce the efficiency of solar cells during cw illumination. Although the initial state can be restored by a short annealing, this is of no practical use.

d) Electron Bombardment

High-energy particle irradiation, such as electron, proton, or neutron bombardment, decreases the carrier lifetime in silicon, as has been known for a long period of time. Vacancies and silicon self-interstitials are created which can react with doping elements, with impurities, or with themselves because of their high diffusion coefficients at room temperature [5.26]. The resulting variety of centers depends essentially on the purity of the material. Most of the complexes are not stable at high temperatures, and anneal, or form other centers. Carrier lifetime measurements are not suitable for investigating this subject because of crowding of different centers.

Progress in understanding is expected from recent investigations applying Deep-Level Transient Spectroscopy (DLTS) [5.27] and related techniques [5.28–30]. Since high-energy electron bombardment has been used recently for carrier lifetime doping, investigations on this subject are still in progress. Hence a summary of results should be delayed, although there is a large number of special publications (e.g., [5.15, 31–34]). Neutron bombardment, on the other hand, is applied to produce striation-free silicon crystals [5.35]. Results on neutron radiation damage concerning the carrier lifetime, however, are rare so far (see, for example, [5.36, 37]).

One special problem in applying solar cells in space is their degradation due to particle bombardment and additional illumination [5.38]. It has been pointed out that unirradiated solar cells also degrade, due to photon-induced carrier lifetime decay [5.39]. Furthermore, it has been shown that as-grown silicon samples after electron irradiation decrease in carrier lifetime due to strong illumination [5.13]. So the problem is reduced to that of the photon-induced change in carrier lifetime which was discussed above.

5.4.2 Processing at High Temperatures

a) Sample Preparation and Additional Precautions

The final carrier lifetime of silicon crystals after heat treatment is influenced by the following parameters:

    I) atmosphere during heat treatment;
    II) cleanness of the furnace and substrate;

III) rate of cooling the sample to room temperature;
IV) type of crystal conductivity;
V) purity of the crystal; and
VI) cleanness of the sample surfaces.

In order to control these, experiments were performed using n-type dislocation-free floating-zone silicon since reactions of impurities present in the as-grown crystal can be neglected in contrast to p-type silicon.

I) Heat treatments with constant exposure times at various temperatures were carried out in argon, nitrogen, dry oxygen, oxygen with an additional small amount of HCl, and in hydrogen. The results are shown in Fig. 5.32 showing the carrier lifetime as a function of the temperature of the preceding heat treatment. Annealing at low temperatures does not affect the carrier lifetime. Starting at an annealing temperature of about 600 °C, however, it decreases strongly except after annealing in dry oxygen. While the carrier lifetime in samples treated in inert gas atmospheres decreases continually with higher annealing temperatures, the results obtained in oxygen with additional HCl pass a minimum value at about 800 °C and increase again up to the initial value. When annealing the sample in a dry oxygen atmosphere, the constant carrier lifetime can only be preserved by keeping the furnace at a higher temperature during idle time, i.e., without the sample. It was cooled down to the desired temperature only to perform the annealing process. Results obtained by treating samples in a hydrogen atmosphere resemble those in inert gas atmospheres. The values, however, are still smaller, demonstrating the presence of additional deep level impurities.

Pure oxygen and also oxygen with HCl are suitable for cleaning furnace tubes. HCl reacts with silicon at high temperatures, and hence its concentration must be reduced as far as possible to preserve the high quality of polished silicon surfaces. A concentration of 1.5% HCl was found sufficient to preserve the initial carrier lifetime. Higher concentrations do not yield better results.

Samples annealed in a hydrogen atmosphere were found to have considerably higher iron content than those annealed in other ambients. This was observed by means of EPR spectroscopy, which is suitable for detecting neutral interstitial iron in n-type silicon. The iron content of samples after annealing in a mixture of hydrogen and nitrogen increases with hydrogen content. A similar increase was observed for the reciprocal carrier lifetime as a function of hydrogen content. A reasonable explanation was found by assuming a reduction of iron combinations already present in the furnace tube and on the surfaces of the sample. This allows iron to diffuse into the sample whereas the combinations can not diffuse in the same way.

II) The influence of the cleanness of the furnace tube upon the final carrier lifetime of samples after heat treatment is evident and was already mentioned above. It should be noted additionally that the use of silicon tubes instead of silica tubes does not yield better results, provided both are cleaned with the same care.

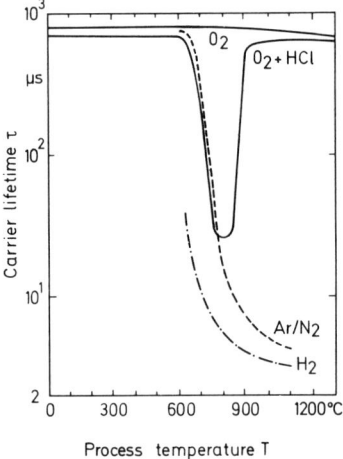

Fig. 5.32. Carrier lifetime in floating-zone n-Si as a function of annealing temperature in various atmospheres

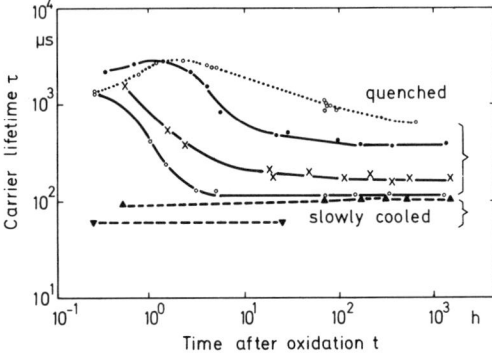

Fig. 5.33. Carrier lifetime in floating-zone n-Si as function of time elapsed after quenching and slowly cooling samples from 1000 °C

III) The rate of cooling the sample to room temperature after annealing processes may affect carrier lifetimes. The formation of deep impurities during quenching of samples from high temperatures [5.40] was not observed. Due to the large volume of the samples, the cooling rates in the present experiments are smaller. After quenching, the resulting carrier lifetime, however, was found to be unstable with the time elapsed [5.23]. The values measured immediately after quenching were comparable to those before heat treatment. Within 100–1000 hours, the carrier lifetime decreased in most cases with the time elapsed as shown in Fig. 5.33. In some cases a moderate increase of carrier lifetime was found before decrease took place. Slowly cooled samples in contrast exhibited a lower but constant carrier lifetime. A possible explanation of this effect was already presented in Sect. 5.4.1, where the effects of sawing silicon samples was discussed, assuming impurity reactions with participation of vacancies which were frozen-in during the quenching of the sample.

IV) As mentioned in the beginning of this section, a general difference in carrier lifetimes after heat treatments was observed in n- and p-type silicon crystals. Whereas the carrier lifetime could be preserved during annealing of n-type silicon, similar results in p-type silicon have not been achieved so far. Details observed after annealing p-type silicon will be discussed below.

V) The purity as well as the perfection of crystals play a significant role in yielding good results after heat treatment. Information concerning the impurity content of commercially available crystals does not exist at present. It is evident, however, that some crystals preserve high carrier lifetimes during heat treatment, while others of the same specification do not. For some elements, differences in impurity concentration can be detected by means of activation analysis. The sensitivity of this method, however, is not sufficient for all elements of interest.

VI) The cleanness of sample surfaces before heat treatment is most important for the resulting carrier lifetime. During and after the cleaning procedure, any contact with heavy metals must be avoided. The carrier lifetimes in samples handled with metal tweezers usually do not exceed several µs after heat treatment. Contaminants such as copper or gold are hard to remove from silicon surfaces, as has been known for a long time [5.41]. The usual preparation of samples consisted of several etching procedures in a mixture of HNO_3, HF, and acetic acid, rinsing in deionized water and drying with the aid of compressed air. Heat treatments were carried out soon after preparation. Plastic tweezers were used to handle the specimens.

b) Annealing of Czochralski-Grown p-Type Silicon

After annealing dislocation-free Czochralski-grown p-type silicon crystals at various temperatures, a characteristic increase of the carrier lifetime can be observed showing a maximum value at a temperature of about 450 °C [5.25, 41–43]. The carrier lifetime in some samples of different 1 Ωcm pulled silicon crystals is plotted in Fig. 5.34 as function of the temperature of the preceding heat treatment. The surfaces of the samples were etched or lapped. A considerable increase in carrier lifetime was found in every case. The enhancement differs from crystal to crystal and can amount to a factor of up to 10. The subsequent decay of the carrier lifetime is strongest for lapped surfaces. Best results were obtained with especially clean conditions in the furnace tube, demonstrating that the decay is partly caused by impurities diffusing into the sample during heat treatment.

A large number of annealing processes at a temperature of 450 °C were carried out using 1 Ωcm pulled crystals. The samples were quenched to room temperature. Carrier lifetimes were determined before and after annealing and shown in Fig. 5.35 showing probability nomographs for 440 measurements. Values in unannealed samples again show logarithmic normal distribution. Results after annealing the samples, however, deviate conderably from a straight line. As can be seen, the increment in carrier lifetime increases

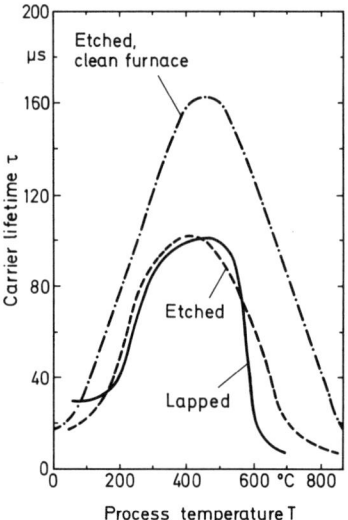
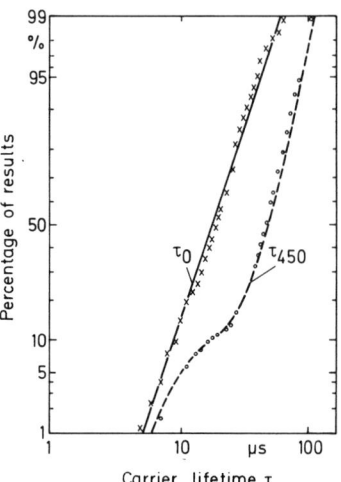

Fig. 5.34. Carrier lifetime in Czochralski-grown p-type silicon as function of annealing temperature

Fig. 5.35. Probability nomograph of carrier lifetime in 1 Ωcm pulled p-Si before and after annealing the sample at 450 °C

continually with the original value. Its gradient has a maximum where the initial carrier lifetime is 10 μs. The getter effect is to some extent reversible. The carrier lifetime, once reduced at high temperatures, can be restored at least partially by reannealing the sample at 450 °C. This plays a significant role in solar cell production, as we shall see. Finally, the getter effect can be increased by slow cooling rates.

Since the influence of the surface upon the getter effect can be eliminated, reactions in the volume of the crystal must be assumed to explain the changes in carrier lifetime. Certain impurities present in the as-grown crystal (or which have diffused from the surface into the bulk of the sample) can be prevented from decreasing the carrier lifetime, by binding them to any kind of complex and thus reducing their carrier cross sections. At higher temperatures, these bindings can break up. Oxygen may be involved in forming these complexes since the getter effect is strongly marked only in pulled crystals. Furthermore the temperature of 450 °C is well known from the formation of oxygen-donor complexes. The reaction mechanism in detail, however, is not known so far.

c) Annealing of Floating-Zone p-Type Silicon

The same treatment carried out on floating-zone silicon as in Czochralski-grown crystals results in rather complicated dependences of the carrier lifetime upon annealing temperature [5.43]. Some examples are given in Figs. 5.36, 37. Carrier lifetimes after annealing of four samples from two different dislocation-free and A-type swirl-defect-free floating-zone p-type silicon crystals are

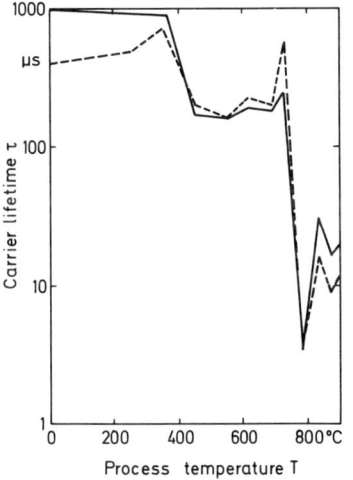

Fig. 5.36. Carrier lifetime at the seed and tang end of an A-type swirl-defect-free floating-zone p-Si crystal as a function of annealing temperature

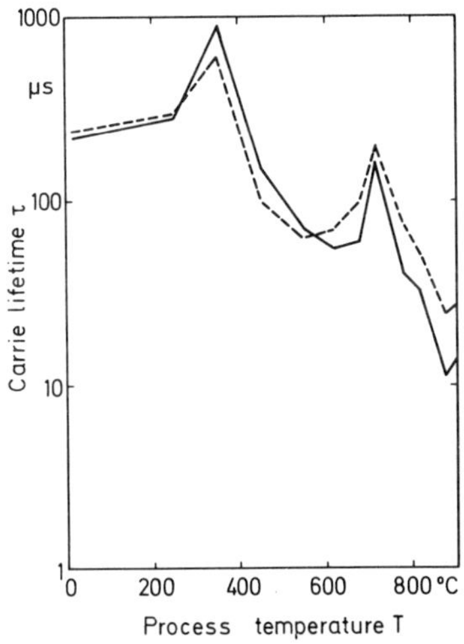

Fig. 5.37. As Fig. 5.36 with different crystal

represented as functions of annealing temperature. One sample was cut from the seed end and one from the tang end of each of the crystals. Both samples from the seed ends were annealed simultaneously, as were those from the tang ends. Nevertheless agreement of results is observed only in the samples of the same crystal although these samples are taken from seed and tang end of the crystal respectively (compare Fig. 5.36). The results obtained after annealing the other crystal (Fig. 5.37) also agree with one another, but show significant differences compared with the first crystal.

The various maxima and minima observed in carrier lifetime after annealing are typical for individual crystals as deduced from numerous investigations although they appear at definite temperatures. So maxima were found at temperatures of about 350 and 700 °C and a narrow minimum sometimes near 800 °C. The characteristic function of the carrier lifetime remained unchanged for different atmospheres during heat treatment. In contrast to pulled crystals, these results are irreversible. Only treatments at higher temperatures can increase the lifetime again.

From measurements in numerous crystals of different specifications it was deduced that the carrier lifetime after annealing depends on residual lattice defects present in the as-grown crystal. In highly dislocated crystals, the carrier lifetime decreases more or less after annealing, depending on the cooling rate [5.44, 45]. In crystals with moderate or low dislocation densities, it can be increased by annealing the sample. The results resemble those of pulled crystals as shown in Fig. 5.38. The etch-pit density of the crystal was 1.5×10^4 cm^{-2}. In contrast to pulled crystals, the increase of carrier lifetime starts at temperatures exceeding about 250 °C. Dislocation-free crystals with high content of A-type

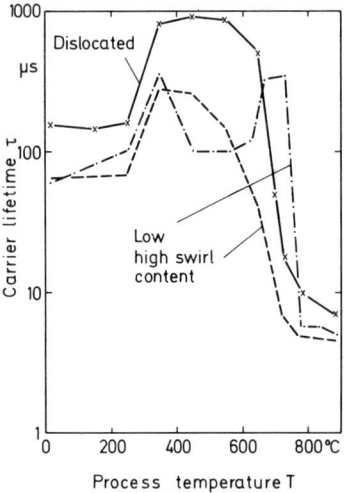
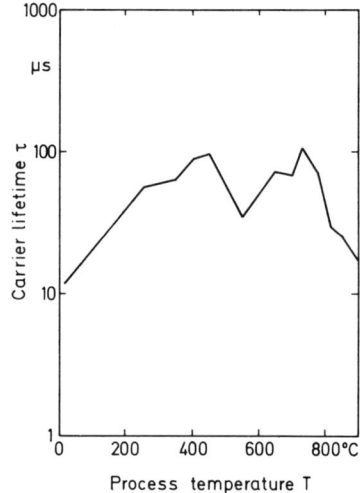

Fig. 5.38. Carrier lifetime in floating-zone p-Si as function of annealing temperature for crystals with moderate dislocation density, and with high and low A-type swirl defect density

Fig. 5.39. Carrier lifetime in a pulled p-Si crystal with reduced swirl content as function of annealing temperature

swirl defects yield similar results as dislocated crystals, which is shown in Fig. 5.38. This agreement is explained reasonably by the nature of the A-type swirl defects which were found to consist of small dislocation loops [5.46]. The occurance of two or more maxima in the carrier lifetime after annealing is limited to A-type swirl-defect-free crystals. It is known from X-ray topography that B-type swirl defects can be decorated with heavy metals only in regions with low densities of dislocations and A-type swirls. Their decoration requires temperatures exceeding 700 °C. The order of getter efficiency was found to be [5.26]: A-type swirls, dislocations, B-type swirls. So it may be deduced that the maximum in the carrier lifetime near 700 °C is produced by a getter mechanism of B-type swirls. In order to investigate this assumption, a Czochralski-grown crystal with reduced swirl content was treated in the same manner. The results are demonstrated in Fig. 5.39. Besides the maximum in the carrier lifetime at a temperature of 450 °C, a second one was observed at 700 °C for the first time in pulled crystals, in agreement with results in floating-zone material.

For solar cell production the knowledge of annealing temperatures responsible for minima in carrier lifetimes is important since diffusion is performed in this temperature range. A small change in diffusion temperature can influence considerably the achievable yield of efficient solar cells.

d) Intentional Getter Processes

The most common getter mechanism usually applied in semiconductor device production is phosphorus diffusion. Heavy metal impurities are accumulated in the diffused layer and bound to phosphorus [5.47]. Because of the high doping

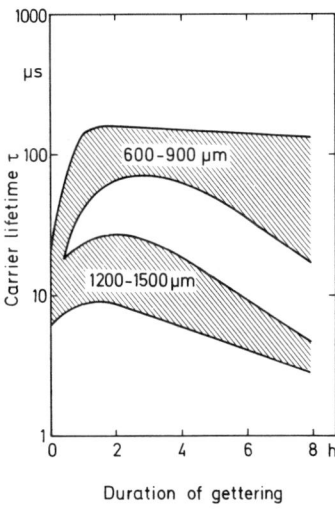

Fig. 5.40. Carrier lifetime in samples of different thicknesses of a pulled p-Si crystal as function of exposure time to phosphorus diffusion

concentration, the carrier lifetime in this layer is determined by Auger recombination. The quantity of impurity atoms dissolvable in the diffused layer, however, is limited. To control the effectiveness of gettering, carrier lifetime measurements were carried out in samples of various thickness of a 10 Ωcm pulled crystal after phosphorus diffusion at 800 °C as a function of exposure time. Results are shown in Fig. 5.40. The carrier lifetime increases by a factor of 10 within 1–2 h in samples which were 600–900 µm thick. The gettering of impurities in thicker samples (1200–1500 µm) is considerably reduced. Results in 400 µm thick samples were not recorded because they exceeded the upper limit where correction for specimen size could be performed. Other getter mechanisms [5.47] are not discussed in this paper because they have not so far been applied to solar cell production.

e) Multiple Annealing Processes

Several heat treatments must be performed in succession to produce devices, so the quantity of impurities diffusing into the crystal will increase; this can be compensated by getter mechanisms. The results depend on the cleanness of the sample surfaces and of the furnace tubes as demonstrated in Fig. 5.41. The carrier lifetimes in two samples of the same floating-zone crystal were measured after heat treatments in normal and specially cleaned furnace tubes. The resulting difference in carrier lifetime amounts almost to a factor of 10. If the same procedures are performed on different crystals the final carrier lifetimes in general correspond to those of the as-grown crystals. This is demonstrated in Fig. 5.42 showing the carrier lifetimes in pulled and floating-zone crystals respectively.

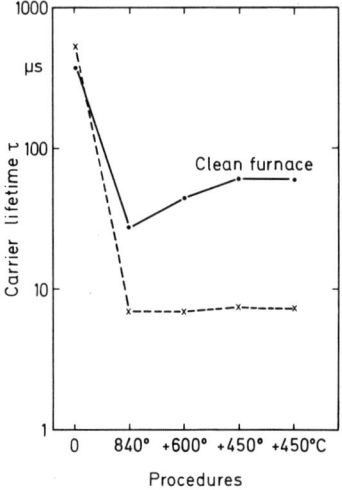
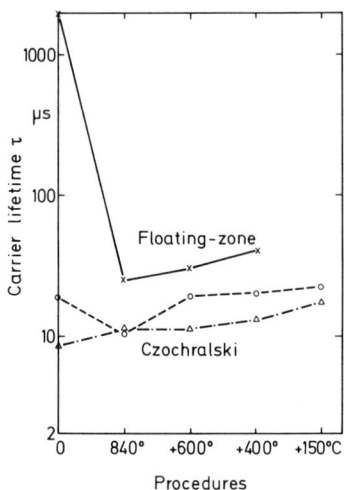

Fig. 5.41. Carrier lifetime in two samples of a floating-zone p-Si crystal after various heat treatments in normal and specially cleaned furnace tubes

Fig. 5.42. Carrier lifetimes in different pulled and floating-zone p-Si crystals after simultaneous heat treatments

After phosphorus diffusion followed by annealing processes, the differences between samples of various thicknesses mentioned above remain recognizable. The final carrier lifetimes, however, are considerably increased. The smallest values, observed in the thickest samples, now exceeded 100 µs after exposure times to phosphorus diffusion of 2 h or more.

In conclusion, the various getter mechanisms discussed above allow the production of solar cells with high carrier lifetimes. Gettering is more effective for low impurity content, and sometimes crystals are found where impurities cannot be reduced sufficiently to achieve good cells.

5.5 Limitation of Solar Cell Parameters by Material Properties

It has been demonstrated in Sects. 5.1.1, 2 that for optimum collection of photogenerated carriers, their lifetime should be long enough to yield diffusion lengths of the order of the cell thickness. This general statement holds independent of doping level. In increasing the efficiency of a solar cell, it is very important to minimize the dark current, as it increases the open-circuit voltage. This can be achieved by increasing the doping level in the base region of a cell, while maintaining carrier lifetime high to keep the light-generated current as high as possible. The influence of doping concentration upon carrier lifetime has been treated in Sect. 5.3.2. As can be seen from Fig. 5.20 for p-type silicon with resistivity below 0.5 Ωcm, the maximum available value of carrier lifetime

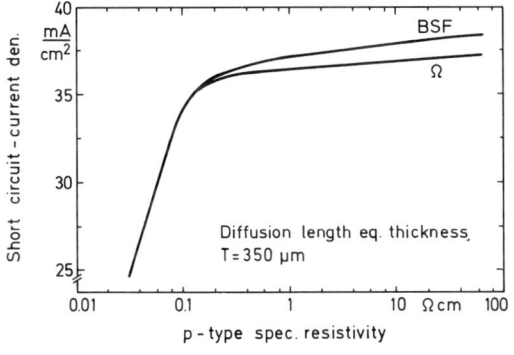

Fig. 5.43. Optimum short-circuit current vs specific resistivity. Decay below 0.5 Ωcm indicates Auger recombination

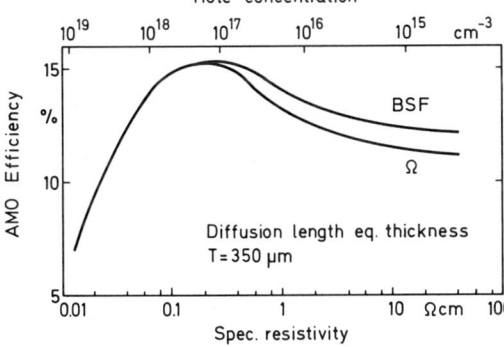

Fig. 5.44. Maximum achievable AM 0 efficiency vs specific resistivity

is limited by the existence of a fundamental recombination mechanism, Auger recombination [5.8, 48]. On the other hand, it can be established for lower doping concentrations, that carrier lifetime in the currently available silicon crystals is high enough to ensure the production of optimum solar cells.

To demonstrate the strong influence of Auger recombination, the short-circuit current was calculated for the reference cell as a function of base resistivity. To ensure optimum collection conditions, a diffusion length equal to the cell's thickness was considered. Figure 5.43 demonstrates how the short-circuit current drops strongly below 0.5 Ωcm.

At high doping levels of the base material, a number of other limiting factors have to be taken into account [5.48, 49], for example, contribution to the dark saturation current from injection of holes from the base into the surface zone, and band-gap shrinkage due to high doping level effects. Figure 5.44 displays AM 0-efficiency for an optimum cell as influenced by the drop of short-circuit current caused by Auger recombination. Only about 15% of the maximum achievable efficiency is possible in a doping range of 0.2 Ωcm. It can be seen that the BSF contact exerts an influence only in the higher resistivity range where diffusion length is unaffected by Auger recombination.

5.6 Conclusions

The most important material parameter for an efficient solar cell is the lifetime of the photogenerated carriers. This paper summarizes the correlation of solar cell parameters with the magnitude of the carrier lifetime. We have reported the present state of knowledge of the dependence of carrier lifetime on the type of silicon crystal used, and we have discussed process-induced variations of this parameter, and present technological and theoretical limitations.

The example of solar cell production shows how problems can be solved which have arisen from insufficient carrier lifetimes. For applications to other devices, the following proposals might be useful since carrier lifetime is a sophisticated material parameter which summarizes properties of impurities of various origins in a crystal. First of all, one must see whether device parameters depend on carrier lifetime, and which kind of carrier lifetime is responsible for their limitations. Differences must be taken into account between recombination- and generation-carrier lifetime, as well as the dependence on the level of injection during device operation. The method of measuring the corresponding value is determined by these results. The sensitivity and accuracy of the measurement technique is to be considered and compared with the expected carrier lifetime and its variations. If as-grown crystals are to be investigated, a sufficient quantity of samples originating from different positions in numerous crystals must be available to yield representative results. Longitudinal as well as radial profiles of the carrier lifetime may impair the efficiency obtained on a wafer or in a certain period of time. Finally processes necessary for device production, cleaning procedures and handling of the material influence the results significantly. The total number of recombination centers in the final device is composed of impurities diffused into the crystal from wafer surfaces and from the furnace tube, as well as those generated by reactions of impurities in the as-grown crystal during heat treatment. The latter can often be avoided by changing the specifications for growing silicon crystals, whereas the former are reduced by changing the processes and cleaning procedures. A getter process performed in additional considerably increases material quality. Consequently, to evaluate the performance of the finished device, the properties of the starting material have to be considered together with their variation during the processing of the material.

References

5.1 M. Wolf: Energy Convers. **11**, 63–73 (1971)
5.2 H.J. Hovel: "Solar Cells", in *Semiconductor and Semimetals*, Vol. 11, ed. by R.K. Willardson, A.C. Beer (Academic Press, New York 1975)
5.3 P.M. Dunbar, J.R. Hauser: Solid-State Electron **20**, 697–701 (1971)
5.4 H. Fischer, W. Pschunder: Final Report: "Experimentalstudie Si-Solarzelle". BMFT Forschungsbericht RVI 1-TO-4/71 (1973)

5.5 W.R. Runyan: Final Rept., NASA Grant NGR 44-007-016. Southern Methodist University (1968)
5.6 H.R. Taft: Phys. Rev. Lett. **8**, 43 (1954)
5.7 M.P. Thekeakara: Appl. Opt. **8**, 1713 (1968)
5.8 H. Fischer: In *Advances in Solid State Physics*, Vol. XIV, ed. by H.J. Queisser (Pergamon, Oxford 1974)
5.9 *1976 Annual Book of ASTM Standards* (American Society of Testing and Materials, Philadelphia 1976) Part 43 F 28
5.10 R.L. Mattis, A.J. Baroody: NBS Technical Note 736 (Washington 1972)
5.11 J.M. Fairfield, B.V. Gokhale: Solid-State Electron. **8**, 685 (1965)
5.12 H. Schlangenotto: Techn. Bericht AEG-Telefunken 10.021/66 (1966)
5.13 K. Graff, H. Pieper: Phys. Status Solidi a **30**, 593 (1975)
5.14 G. Schwab, H. Bernt, H. Reichel: Solid-State Electron. **20**, 91 (1977)
5.15 D. Bielle-Daspet: Solid-State Electron. **16**, 1103 (1973)
5.16 K. Graff, H. Pieper: Solid-State Electron. **15**, 831 (1972)
5.17 *1976 Annual Book of ASTM Standards* (American Society of Testing and Materials, Philadelphia 1976) Part 43 F 391
5.18 W.E. Phillips: Solid-State Electron. **15**, 1097 (1972)
5.19 W.R. Runyan: NASA CR 93154 National Technical Information Service N 68-16510 (1968)
5.20 C. Dash, R. Newman: Phys. Rev. **99**, 1151 (1955)
5.21 J.D. Beck, R. Conrad: Solid State Commun. **13**, 93 (1973)
5.22 J. Dziewior, W. Schmid: Appl. Phys. Lett. **31**, 346 (1977)
5.23 K. Graff, H. Pieper: Phys. Status Solidi a **49**, 137 (1978)
5.24 H. Benda: Z. Naturforsch. **13a**, 354 (1958)
5.25 K. Graff, H. Pieper, G. Goldbach: In *Semiconductor Silicon*, ed. by H.R. Huff, R.R. Burgess (Electrochemical Society, Princeton 1973) p. 170
5.26 A.J.R. DeKock: Philips Res. Rep. Suppl. **1**, 9 (1973)
5.27 D.V. Lang: J. Appl. Phys. **45**, 3023 (1974)
5.28 P. Rai-Choudhury, J. Bartko, J.E. Johnson: IEEE Trans. Electron Devices **23**, 814 (1976)
5.29 A.O. Evwaraye, B.J. Baliga: J. Electrochem. Soc. **124**, 913 (1977)
5.30 H. Lefèvre, M. Schulz: Appl. Phys. **12**, 45 (1977)
5.31 F.A. Abou-el-Fotouh, R.C. Newman: Solid State Commun. **15**, 1409 (1974)
5.32 A.R. Bean, R.C. Newman, R.S. Smith: J. Phys. Chem. Solids **31**, 739 (1970)
5.33 O.L. Curtis, J.R. Srour, R.B. Rauck: J. Appl. Phys. **43**, 4638 (1972)
5.34 Mi. Hirata, Ma. Hirata: Jpn. J. Appl. Phys. **12**, 460 (1973)
5.35 M. Schnöller: IEEE Trans. Electron Devices **21**, 313 (1974)
5.36 H.J. Stein: J. Appl. Phys. **37**, 3382 (1966)
5.37 F.A. Selim, P.D. Blais, P. Rai-Choudhury, R.F. Yut: In *Semiconductor Silicon*, ed. by H.R. Huff, E. Sirtl (Electrochemical Society, Princeton 1977) p. 126
5.38 R.L. Crabb: In *9th Conf. Rec. 9th Photovoltaic Specialist Conf.* (IEEE, New York 1972)
5.39 H. Fischer, W. Pschunder: In *10th Conf. Rec. 10th Photovoltaic Specialist Conf.* (IEEE, New York 1978)
5.40 L.D. Yau, C.T. Sah: Solid State Electron. **17**, 193 (1974)
5.41 W. Kern, D.A. Puotinen: RCA Review **31**, 187 (1970)
5.42 L.M. Nijland, L.J. van der Pauw: J. Electron Control **3**, 391 (1957)
5.43 K. Graff, H. Pieper: J. Electron. Mater. **4**, 281 (1975)
5.44 G. Bemski: Phys. Rev. **103**, 567 (1956)
5.45 V.A. Atsarkin, E.Z. Mazel: Sov. Phys. Solid State **2**, 1874 (1961)
5.46 L.I. Bernewitz, B.O. Kolbesen, K.R. Mayer, G.E. Schuh: Appl. Phys. Lett. **25**, 277 (1975)
5.47 T.E. Seidel, R.L. Meek: *Ion Implantation in Semiconductors and Other Materials*, ed. by B.L. Crowder (Plenum, New York 1973) p. 305
5.48 H. Fischer, W. Pschunder: In *11th Conf. Rec. 11th Photovoltaic Specialist Conf.* (IEEE, New York 1975)
5.49 J.R. Hauser, P.M. Dunbar: IEEE Trans. Electron. Devices **24**, 4 (1977)

6. Problems of the Cu_2S/CdS Cell

M. Savelli and J. Bougnot
1 With the collaboration of F. Guastavino, J. Marucchi, and H. Luquet.

With 30 Figures

Over the last two decades, $pCu_2S/nCdS$ photocells have been studied mainly for their use in space. It is only recently that work has been carried out with terrestrial applications in mind.

In 1956, *Carlson* et al. [6.1, 2] made the first cell from a layer of polycrystalline CdS. Since that time much technological progress has been accomplished paving the way to the development of cells which resist degradation, whose characteristics remain stable during use, and whose efficiency can reach 5%, or even 7–8% in some cases. Most work done on the subject today uses either thermal evaporation or spray reaction to make CdS. A dipping process follows, the purpose of which is to form a thin layer of Cu_2S in order to make a heterojunction. The aim of work done in this field is to lower the price of the cell and to increase its efficiency by improving fabrication parameters. For example, the spray method used for CdS/Cu_2S has been described [6.3, 4] as capable of at least 4% efficiency and lower costs.

In this article we describe briefly thin layer CdS and Cu_2S elaboration techniques. A complete article on the subject was written by *Stanley* [6.5]. After having described the physical properties of CdS, Cu_2S, and of the heterojunction, we review the various conduction mechanism of the cell in light and in darkness. Throughout this article, the authors have emphasized the essential points concerning the most important problems of solid-state physics raised by these studies.

6.1 Cu_2S/CdS Heterojunction Technology

6.1.1 CdS Thin-Film Technology

We shall only deal with CdS fabrication processes applicable to photovoltaic cells.

a) Vapor Deposition

In most cases, thermal evaporation under vacuum has been carried out in open systems with a crucible and the substrate in the same enclosed vacuum chamber. The experiments [6.6, 7] have shown, however, that this process leads to contamination of the CdS film; this is due to the presence of impurities in the evaporation. The first evaporated polycrystalline CdS photovoltaic cells were made by *Carlson* et al. [6.1, 2]. The temperature of the substrate during evaporation has a determinant influence on the properties of the deposition. The CdS molecules dissociate and should recombine onto the substrate if their temperature is less than 150 °C, the sulfur vapor pressure is slightly higher than cadmium vapor pressure, and the deposits have an excess of cadmium. On the other hand, if the temperature of the substrate is higher than 200 °C, cadmium atoms may reevaporate before recombining with the sulfur [6.8]. Thus the stoichiometry of the CdS film depends not only on the material in the crucible but also on the substrate temperature. The temperature range most frequently used is 180–200 °C.

The crucible is an element which tends to limit the uniformity and reproducibility of the CdS film by acting on the distribution of heat in the crucible. Poor heat distribution causes different evaporation rates, and leads to heterogeneity of the deposited film. In addition, it is difficult to heat the CdS to its sublimation temperature without a certain amount of particles being projected onto the substrate. This "splattering" can be avoided by putting a quartz wool plug on the open end of the crucible.

The evaporation crucible is generally made of quartz; tantalum can be used, but after a certain length of time, it is attacked by the CdS and gives off, with the sulfur vapor, Ta_2S_4 which makes the crucible breakable [6.9–11]. In general, evaporation with pure CdS is carried out with the temperature of the crucible at approximately 1000 °C; the material in the crucible can either be in powder form or in pellets. Other evaporation sources, such as flash evaporation, may be used to obtain CdS deposition. However, this method is not suitable for the deposition of thick films, nor, especially, if substrates are used which are destroyed under intense heat, for instance, Kapton [6.12, 13].

Generally, the evaporated CdS films are doped by introducing donors or acceptors such as sulfur or chloride into the charge. Indium is the most common dopant for photovoltaic cells [6.14]; $CdCl_2$ and CuCl have also been used as dopants [6.15]. After evaporation, the films can be annealed in a mixture of CdS and the dopant [6.16, 17]. This decreases the sulfur vacancies, and recrystallizes and dopes the film.

b) Sputtering

Yefremenkova [6.18] made Cu_2S/CdS structures from CdS deposits which were obtained by cathodic sputtering of a cathode of compact CdS doped with $InCl_3$ or CdI_2 in an argon atmosphere. The advantage of this method is that the chemical composition of the pulverized film is, in general, the same as that of the cathode. The films are also more adherent than those obtained by thermal evaporation. In CdS sputtering, it is possible to use a cadmium cathode with gases such as H_2S/Ar or S/Ar as did *Albrand* et al. [6.19, 20]. The sulfur ions which dissociated during the discharge react with the cadmium atoms on the support surface. This makes the film stoichiometry highly dependent on the H_2S partial pressure and pulverization conditions.

c) Chemical Spray Deposition

In recent work, *Jordan* [6.21] has shown that the reactive pulverization method known as "spray" (used by *Chamberlin* and *Skarman* [6.22, 23]) is well adapted to the industrial fabrication of CdS films. This is due to its low cost and its simplicity of operation. The spray method also avoids the considerable loss of CdS powder during preparation of films 40 μm thick by thermal evaporation [6.24]. The spray method is carried out by pulverizing onto the heated substrate a spraying solution which contains the elements necessary for CdS fabrication. If cadmium chloride and thiourea are used as starting materials, the following chemical reaction takes place at $T_s > 250\,°C$:

$$CdCl_2 + SC(NH_2)_2 + 2H_2O \rightarrow CdS\downarrow + 2NH_4Cl\uparrow + CO_2\uparrow.$$

The degree of orientation and crystallinity obtained from a spray deposited film depends on the following parameters:

– starting material
– type of support
– substrate temperature
– anion/cation ratio.

On amorphous supports, the crystalline properties of the spray deposited CdS prepared with solutions whose Cd^{++}/S^{--} ratio equals 1, and whose substrate temperature is higher than 380 °C, are similar to those of evaporated films [6.25–28]. However, as *Micheletti* et al. [6.29] and *Bube* et al. [6.30] have shown, they are very sensitive to oxygen adsorption. CdS doping is carried out by directly adding the dopant ($AlCl_3$, CuCl, ...) to the solution.

d) Sintered Layers

In general, sintered CdS films are prepared from a mixture of CdS powder and $CdCl_2$. The slurry is applied to the substrate and then heated to a temperature ranging from 500 to 600 °C. During the heating process, at 568 °C, the CdS dissolves into the melted $CdCl_2$ which begins to volatilize at 400 °C; thus we observe recrystallization of the CdS. The sintering process promotes particle fusion and recrystallization at a relatively low temperature, and the films are similar to single crystals. It is possible to obtain disks of compact CdS from compressed CdS powder which is then sintered at between 700 and 850 °C in an inert gas. *Nakayama* [6.31] prepared photovoltaic cells from this type of CdS. The size of the CdS grains was anywhere between 5 and 10 μm; their porosity was approximately 0.2 %.

6.1.2 Formation of the Cuprous Sulfide Layer for the Cu_2S/CdS Structure

The cuprous layer is obtained by many alternative processes. Some preliminary treatments of the CdS surface are often beneficial before the formation of this layer. Etching by acids (for instance, HCl) removes surface impurities and increases grain boundaries.

a) The Dipping Process

The dipping process is a topotaxial reaction consisting in the displacement of one cadmium ion by two copper ions, according to the reaction:

$$CdS + 2CuX \rightarrow Cu_2S + CdX_2,$$

where X can be for example a Cl, Br, or I atom.

This reaction is generally performed in aqueous solution at 90–100 °C. To improve the CuX solubility, salts such as ClNa or $ClNH_4$ are added. To avoid oxidation of Cu^+ ions, hydrazine or hydroxylamine is used as a reducing agent.

As an example, at the Institute of Energy Conversion, Delaware, USA, a dipping solution which is used with good results on CdS evaporated layers is as follows:

6 g/l CuCl, 2 g/l NaCl, 1 ml/l hydrazine
pH = 3.5 (adjusted with HCl) [6.32].

Some authors prefer an organic solution instead of an aqueous one. In Fig. 6.1 the influence of the pH and of the nature of the element X on the efficiency of the cell are shown after [6.33]. On the other hand, previous boiling and argon bubbling seem necessary to improve the Cu_xS layer [6.32].

The dipping process is the most often used method for the formation of the Cu_xS layers for CdS–Cu_xS solar cells. One obtains compositions close to Cu_2S. However, as shown in [6.4], the best composition for the cells is $Cu_{1.995}S$, and electrochemical methods are necessary to reach this precise value (see Fig. 6.2).

b) CuCl Evaporation

Instead of dipping, called "the wet method", some authors have tried the "dry method" of CuCl evaporation. After the evaporation of a thin layer of CuCl on the CdS, some thermal annealing enables the ion exchange $Cd^{++} \rightleftarrows 2Cu^+$ to take place.

The purpose of this method used by *Clark* et al. [6.34] and *Mickelsen* et al. [6.35] was to avoid the main problem of the dipping process: too deep Cu_2S migration into the boundary grains of the CdS layer. These authors obtained an efficiency of 3% at AM 0. Recently this method was used by the Philips group on CdS single crystals. The cell obtained was very efficient ($\eta = 8\%$) [6.36].

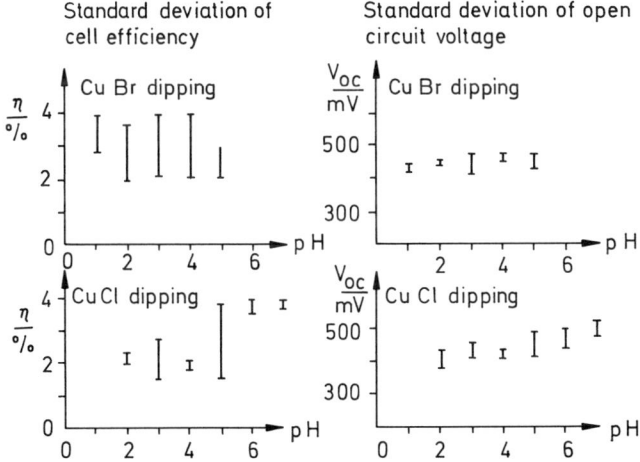

Fig. 6.1. Influence of dipping parameters on cell efficiency after [6.33]

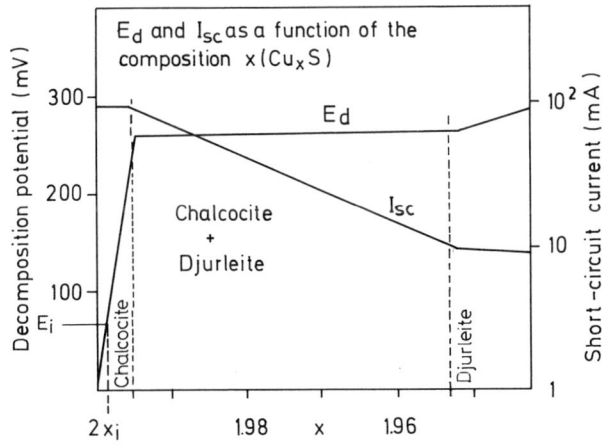

Fig. 6.2. Decomposition emf E_d and short-circuit current (I_{sc}) as functions of x in Cu_xS; after [6.4]

c) Cu_xS Evaporation

Direct evaporation of Cu_2S, or Cu evaporation followed by sulfuration on benzenic solution, or in H_2S gas, has been tried. However, good efficiencies were not obtained [6.37].

d) Electrodeposition

Copper electrodeposition in a sulfate solution has been used by several authors. The copper deposited is transformed into Cu_2S by heat treatment. In [6.38], a Cu cyanide solution is used instead of $CuSO_4$. The results of this method to date are not very encouraging.

Table 6.1. Influence of copper treatment on cell parameters after [6.41]

Cell no.	Short-circuit current [mA/cm^2]		Open-circuit voltage [mV]		AM0 efficiency [%]	
	a	b	a	b	a	b
991	6	29	460	520	<2	6.9
984	12.5	33	475	510	3.3	8.0
986	15	32	475	500	3.7	7.5
961	20	32	505	515	5.4	7.7
963	22	34	507	520	6.0	8.1
933	28	32	520	520	6.7	7.7

[a] Before Cu deposition; [b] after Cu deposition.

e) The Spray Method

This method, used previously, gave interesting results [6.22]. However, the use of water in the spray solution seems to be a factor in oxidizing the Cu_xS layer. In order to avoid this problem, organic solutions are envisaged to day. Recently *Vedel* et al. proposed the use of acetronitrile, CH_3CN. In this solution, copper salts are stable and particularly CuCl. The authors demonstrated the formation of a CuCl-thiourea compound [6.39].

6.1.3 Heterojunction Formation

After the formation of the Cu_xS layer, the next step is the formation of the junction. It is well known that the conditions for the heat treatment depend on the CdS and Cu_xS fabrication method. Generally, a heat treatment at a temperature in the range 150–200 °C for a few minutes is used. The role of this heat treatment is to form the junction itself. The results show a decrease in the shunt resistance and an increase of the open circuit voltage V_{oc} [6.40]. But a heat treatment which lasts too long results in a decrease of the short circuit current I_{sc} [6.40]. Several authors note also an improvement of the fill factor FF following the heat treatment. Also note that some copper treatments by evaporation or by electrochemical methods to obtain the desired composition $Cu_{1.995}S$ for the copper sulfide layer improve the efficiency and the stability of the cell (Table 6.1) [6.41].

6.1.4 Fabrication of Front and Back Electrodes

There are two types of CdS–Cu_xS cells. In the "backwall" type, the light first reaches the CdS layer which presents the higher energy gap (2.4 eV). In the "frontwall" cells, however, the light is absorbed directly by the Cu_xS layer. Depending on the type of cell, the front and back electrodes are of a different nature.

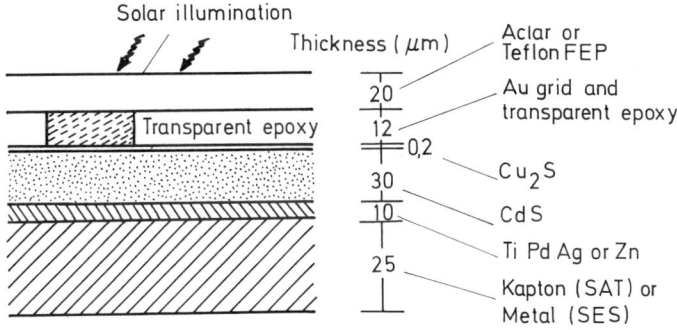

Fig. 6.3. Descriptive drawing of a typical Cu_2S–CdS frontwall cell

a) Frontwall CdS–Cu_xS Cells

Cells made by the Société Anonyme des Telecommunications (SAT) in France, and by Solar Energy Systems (SES) in Delaware, USA, are examples of frontwall cells (see Fig. 6.3). The substrate is usually a plastic film covered by an electrolytically deposited Zn contact or by an evaporated layer of TiPdAg which acts as an ohmic electrode in relation to CdS. On the Cu_xS side, gold, which is an ohmic contact, is used as grid. The gold grids are stuck with gold-loaded epoxy. They are also obtained by electrodeposition. In either case, the cells are stable. Direct evaporation of gold seems to give poor cells with high series resistance and poor adherent contact.

When the grid is attached, the cell has to be protected against atmospheric agents such as uv radiation and oxygen. The film used has to be waterproof and airproof, with good optical transmission and effective mechanical and thermal resistance. Aclar seems to be a good material, and is used by SAT. The lamination operation used to complete the cell also improves the cell parameters V_{oc}, I_{sc}, and R_s [6.32].

b) Backwall CdS–Cu_xS Cells

In the backwall type of cell, shown in Fig. 6.4, tin oxide (SnO_2) or indium oxide (In_2O_3) deposited on the glass before the CdS are used as transparent electrodes. These electrodes are deposited by a spray process using an alcohol solution of $SnCl_4(5H_2O)$ or $InCl_3$.

Doping elements such as F, in the case of SnO_2, or Sn, in the case of In_2O_3, decrease the sheet resistance $R_\square$, and thus decrease the series resistance R_s of the cell. Nowadays, one can obtain an $R_\square$ of 5 to 10 Ω with a high transparency (75% in the solar spectral range). These values allow a larger spacing between the grid which is in any case necessary to reduce the series resistance.

On the copper sulfide side, a copper electrode is used, ensuring good stoichiometry. This electrode should also be protected against elements in the atmosphere by another metal, or by a plastic film as in the frontwall cells.

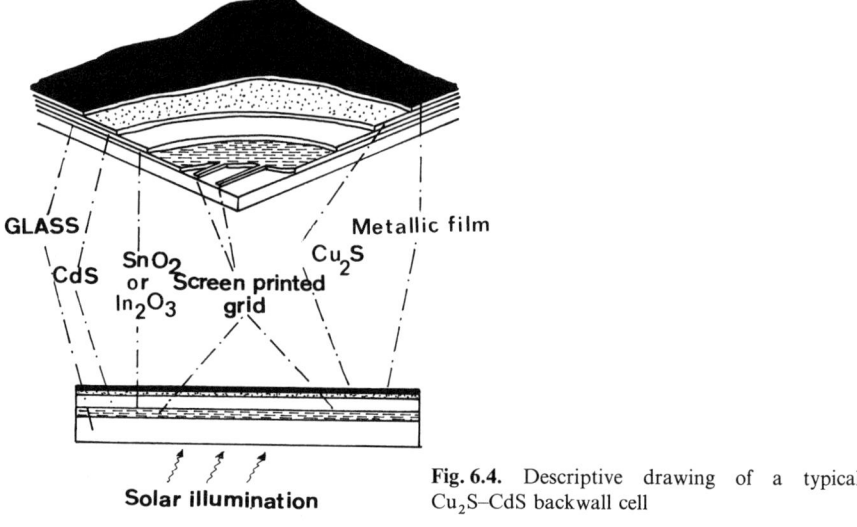

Fig. 6.4. Descriptive drawing of a typical Cu$_2$S–CdS backwall cell

6.1.5 Thin-Film Photovoltaic Structures

Table 6.2 lists the principal thin-film photovoltaic structures that have an efficiency of better than three percent.

6.2 Properties of CdS Films

6.2.1 Review of the Fundamental Properties of Bulk CdS

Cadmium sulfide is a binary II–VI compound which crystallizes in two allotropic forms with the structure of zinc blende and wurtzite. The blende structure can be considered as the result of two interpenetrating cubic close-packed lattices with lattice parameter $a=5.832$ Å. Each cation has four nearest anionic neighbors at a distance of $a\sqrt{3}/4$ at the corners of a regular tetrahedron (Fig. 6.5). The wurtzite structure is composed of two interpenetrating close-packed hexagonal lattices displaced with respect to each other by a distance $3c/8$ along the hexagonal c axis. The nearest neighbor distance from tetrahedral sites is $3c/8$ or $(3/8)^{1/2}a$ (Fig. 6.6). The space group is $C_{6v}^4 - P6_3mc$ and the lattice parameters are: $a=4.613$ Å, $c=6.716$ Å [6.61].

The first Brillouin zone of the blende structure is the same as that of diamond (a truncated octahedron) whereas that of the wurtzite structure is a hexagonal prism (Figs. 6.7, 8). The band structure has been thoroughly studied [6.62–67]. The minimum of the conduction band and the maximum of the

Table 6.2. Thin-film photovoltaic structures with efficiency >3% [a]

Type	Barrier formation process	Ref.
Frontwall CdS evaporated	CuCl dipping, heat at 250 °C CuCl dipping, electroplated Cu Chemiplating CuCl at 90 °C and heat Chemiplating Chemiplating CuCl Evaporation CuCl heat in N_2 or CuCl dipping	[6.42–46] [6.47] [6.48–53, 215] [6.54] [6.41–55] [6.56]
Backwall CdS evaporated	Electroplated Cu, air heat at 275 °C	[6.57]
Backwall or Frontwall CdS sprayed	Spray deposited 0.1 μm heat at 280–320 °C CuCl dipping and heat	[6.22, 23] [6.3, 4]
CdS sintered	CuCl dipping and heat $CuSO_4$ dipping $BiNO_3$ + Cu salt electroplated	[6.58] [6.31] [6.59]

[a] See [6.5].

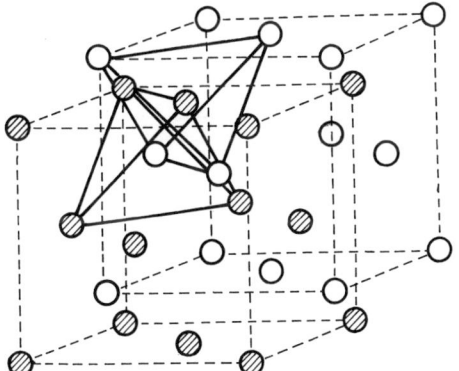

Fig. 6.5. Zinc blende structure [6.60]

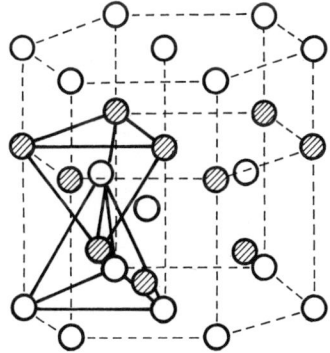

Fig. 6.6. Wurtzite structure [6.60]

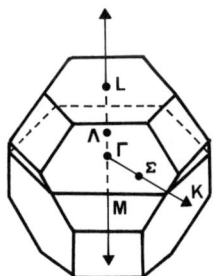

Fig. 6.7. Brillouin zone for zinc blende [6.60]

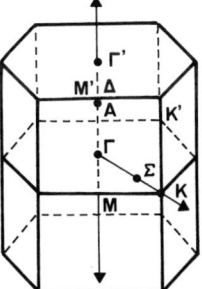

Fig. 6.8. Double Brillouin zone for wurtzite after *Cardona* and *Harbeke* [6.63]

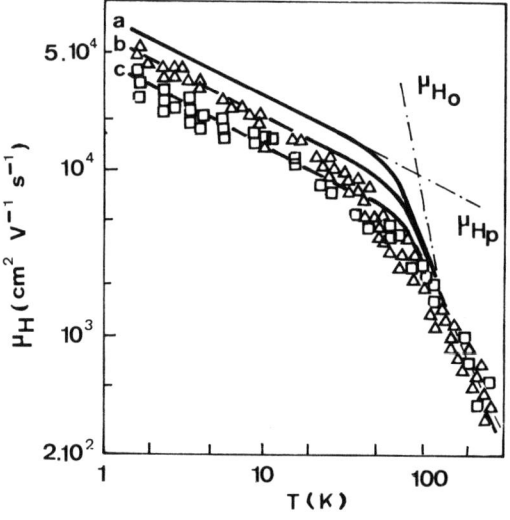

Fig. 6.9. Variation of Hall mobility with temperature in CdS. Triangles and squares indicate measurements taken with current perpendicular and parallel to the c axis. Curve a represents a hypothetical curve combining optical mode and piezoelectric data. Curve b is the best fit of experimental data for $\mu \perp c$ and curve c is the best fit of data $\mu \| c$ undergoing piezoelectric scattering [6.69]

valence band are situated in the center of the Brillouin zone. The lowest conduction band comes from the cadmium 5s levels while the valence band must be associated with the sulfur 3p levels [6.66].

CdS is a direct-gap semiconductor. Experiments on optical reflection and transmission give the following expression for the value of the gap as a function of temperature:

$$E_G = 2.58 - (5.2 \times 10^{-4})T \quad [\text{eV}],$$

where T is measured in K. For $T = 300$ K, i.e., at room temperature, this turns out to be 2.42 eV. The density of intrinsic carriers is very low and the conductivity is controlled by the presence of natural defects and impurities. The resistivity of CdS is generally very high; it is more often thought of as a semi-insulator rather than a semiconductor. Between 100 and 300 K, the electronic mobility follows the behavior predicted for polar scattering; while for temperatures ranging from 25 K down to absolute zero, piezoelectric scattering is dominant [6.68–71] (Fig. 6.9). Substituting chlorine, bromine or iodine for the sulfur, and aluminium, gallium or indium for cadmium, creates donor levels at 0.03 eV from the conduction band [6.72]. Copper and silver give acceptor levels of 0.6 eV and 1.0 eV respectively from the valence band. Sodium, potassium and lithium are acceptor impurities as well. In Table 6.3 we have collected together some important physical properties of bulk CdS.

Copper acting as an acceptor in CdS can, near the Cu_2S–CdS junction, create by diffusion a weakly compensated zone. Studies concerning copper diffusion in this compound have shown that it follows two mechanisms: one which is very quick, but with weak solubility; and another, which is slower, but with high solubility. In Table 6.4 we have grouped the various results found in the literature on the subject.

Table 6.3. Electrical properties of CdS. After [6.60]

Parameters	CdS (Wurtzite)
Relative density	4.92
Molecular weight	144.46
Lattice parameters	$a = 4.136$ Å
	$c = 6.713$ Å
Direct band gap	2.42 eV (300 K)
Effective mass of electrons m_n^*/m_e	0.153–0.171
Effective mass of holes m_p^*/m_e	0.7 light holes
	5 heavy holes
Thermal conductibility	0.20 W K^{-1} cm^{-1} ($\|c$ axis)
Dielectric constant	$\varepsilon \| c = 8.64$
	$\varepsilon \perp c = 8.28$
Refractive index	2.3 ($\lambda = 2$ μm)
	2.26 ($\lambda = 14$ μm)
Electron mobility	~ 400 cm^2 V^{-1} s^{-1}
Hole mobility	15 cm^2 V^{-1} s^{-1}

Table 6.4. Diffusion of copper in CdS

Authors	Method	$D = D_0 \exp(-Q_0/kT)$		Remarks
		D_0 [cm^2 s]	Q_0 [eV]	
Zmija et al. [6.73]	Radioactive tracer ~ 400 °C	8.4×10^{-4}	0.73	Substitutional
	deposited film	2×10^{-4}	0.56	Interstitial
Szeto et al. [6.74]	Optical transmission 500–700 °C		0.58	
Sullivan [6.75]	Capacity 146–300 °C	2.1×10^{-3}	0.96	Copper solubility is $6.6 \times 10^{22} \exp(-0.505/kT)$ in cm^{-3}
Clarke [6.76]	Radioactive tracer	1.6×10^{-3}	0.77	
Woodbury [6.77]	Radioactive tracer		1.20	
Purohit et al. [6.78]	Radioactive tracer	1.6×10^{-9}	0.09	

Mott and *Spear* [6.79] gave the following parameters for minority carriers in single crystal of CdS:

$\mu_p = 15$ cm^2 V^{-1} s^{-1} ; $D_p = 0.36$ cm^2 s^{-1} ;
$\tau_p = 2 \times 10^{-7}$ s ; and $L_p = 4$ μm .

The value of the diffusion length agrees with that determined by *Gill* and *Bube* [6.80]: 3.0 μm $< L_p <$ 7.0 μm.

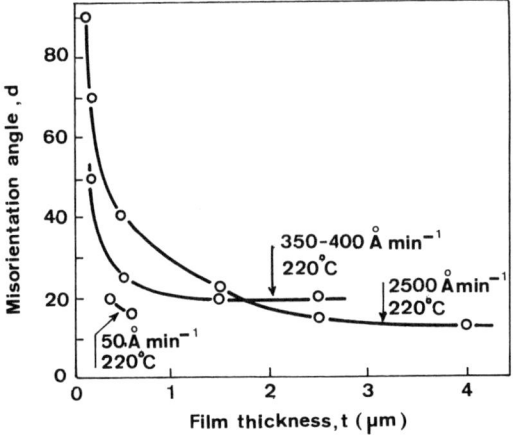

Fig. 6.10. Misorientation angle as a function of thickness for films grown at three different rates [6.82]

6.2.2 Properties of Polycrystalline CdS Thin Films

Thermal evaporation, sputtering, chemical spray reactions, and the methods described in Sect. 6.1 all produce CdS films whose physical properties are dependent on their fabrication parameters.

a) Structures

I) Evaporated CdS films intended for the fabrication of solar cells on amorphous substrates such as glass, Kapton or Aclar are polycrystalline, and have generally a wurtzite hexagonal structure with a preferred orientation: the *c* axis is perpendicular to the substrate. *Shalcross* [6.81] has shown that the degree of preferred orientation increases with the thickness of the deposition. In addition, *Wilson* and *Woods* [6.82] emphasized the influence of the evaporation rate (Fig. 6.10). Other authors [6.16, 30, 83, 84] have shown that the column structure of CdS can be modified by impurities. Chlorine, in particular, destroys orientation, whereas indium and gallium have no effect. The size of the CdS grains is approximately 1 μm.

II) The crystallinity of sprayed CdS layers depends on the type of substrate and on the choice of starting materials. The CdS films fabricated on amorphous substrates are highly polycrystalline if we begin with a solution composed of cadmium chloride and thiourea. However, if the chloride is replaced by cadmium acetate, the CdS films become amorphous. The degree of crystallinity also depends on the temperature of the substrate, whereas the preferred orientation depends on the sulfur to cadmium ion ratio. This fact has been shown by various authors [6.22, 26–28]. CdS films 4 μm thick having crystalline properties similar to that of evaporated CdS [6.27] are obtained at a substrate temperature higher than 380 °C, and with the Cd^{++}/S^{--} ratio of 1 (Fig. 6.11). As in the case of the evaporated CdS, authors have reported that aluminium destroys the preferred orientation of the sprayed CdS film.

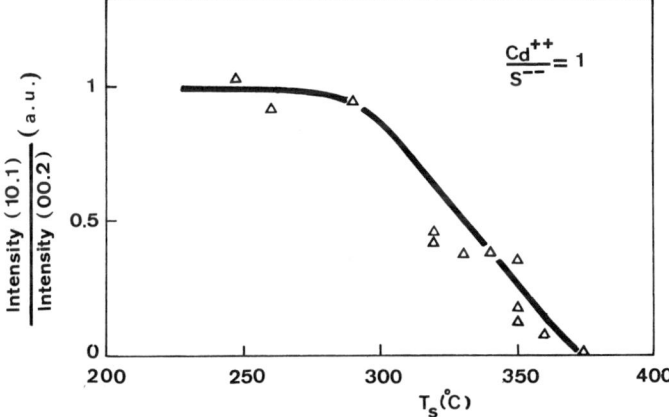

Fig. 6.11. Ratio of X-ray peak intensity (10.1), (00.2) as a function of fabrication temperature [6.27]

III) Strongly oriented films with crystallite sizes between 0.5 and 1 µm can be obtained by using sputtering [6.85].

b) Electrical Properties

I) Electrical properties depend mostly on fabrication parameters. *Wilson* and *Woods* [6.82] have shown that the resistivity of CdS films obtained by thermal evaporation is at a given substrate temperature a function of the thickness of the film and of the evaporation rate (Figs. 6.12, 13). In the case of thick CdS films (10–40 µm) to be used for photovoltaic cells, the necessary resistivity (between 1 and 100 Ωcm) is obtained by doping the CdS. Hall effect measurements have shown that the carrier concentration is between 10^{17} and 10^{18} cm^{-3} and the mobility is between 1 and 20 cm^2 V^{-1} s^{-1}. In general the films are not very photoconductive and above room temperature, the mobility increases exponentially with the temperature. The mobility law $\mu_n = \mu_0 \exp(-e\Phi/kT)$ can be explained by the role of the grain boundaries present in the polycrystal layers [6.86, 87]. *Partain* et al. [6.88] measured the induced current of CdS–Cu$_2$S solar cells with a scanning electron microscope, and were able to determine the diffusion length of minority carriers L_p in the polycrystalline evaporated CdS. They obtained $0.1 < L_p < 0.3$ µm. When we compare this L_p value to that given by *Gill* and *Bube* (~4 µm) for single crystal CdS, we can explain the disagreement only by additional recombination of grain boundaries in the polycrystal.

II) The electrical properties of sprayed CdS are similar to those of evaporated CdS. In particular, we see the same resistivity variation with the thickness of the film [6.25]. To produce CdS$_{spray}$/Cu$_2$S solar cells, the desired CdS resistivity can be obtained by doping the starting solution or by annealing the films in different controlled atmospheres [6.25–28]. Annealing in air, in

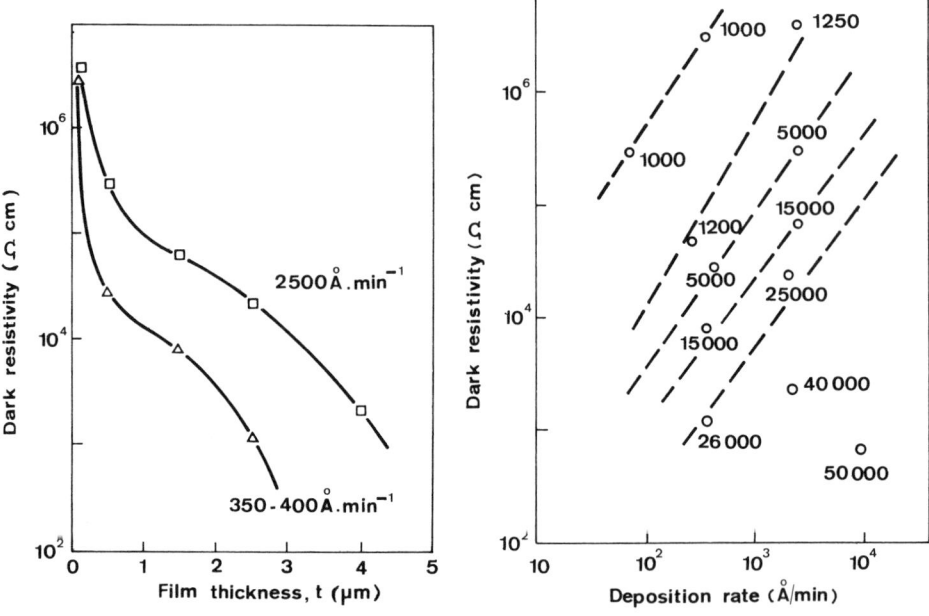

Fig. 6.12. Resistivity at room temperature of two series of CdS films as a function of thickness [6.82]

Fig. 6.13. Resistivity at room temperature of CdS films deposited on glass at 220 °C as a function of deposition rate [6.82]

nitrogen, in hydrogen and in a vacuum results in a sudden increase in the carrier concentration n and the electron mobility μ_n, and thus in a decrease in resistivity ϱ (Figs. 6.14, 15). With annealing in hydrogen, the increase in mobility μ_n to 40–60 cm^2 V^{-1} s^{-1} can be explained by a high oxygen desorption rate [6.27]. The chemisorbed oxygen acts as an acceptor impurity and decreases the free electronic density, and also it is able to trap the free carriers and to increase barrier height of the grain barriers [6.29, 30]. Above room temperature, μ_n increase exponentially with the temperature according to $\mu_n = \mu_0 \exp(-e\Phi/kT)$. This special sensitivity in relation to oxygen even at low temperatures clearly differentiates sprayed CdS films from evaporated films.

c) Optical Transmission: Photoconduction

During fabrication, if the substrate temperature is higher than approximately 380 °C, then the spectral variation of the optical transmission of sprayed CdS is similar to that of evaporated CdS (Fig. 6.16). Sprayed films made at temperatures lower than 330 °C are rough and show strong light diffusion [6.25, 28].

All the undesorbed layers are photoconductive but the photoconduction sensitivity is connected to the oxygen adsorption. The photoconduction spectral variation agrees with that of evaporated layers, except for the intrinsic peak wavelength which varies between 0.49 and 0.52 µm according as the layers are smooth or rough [6.28] (Fig. 6.17).

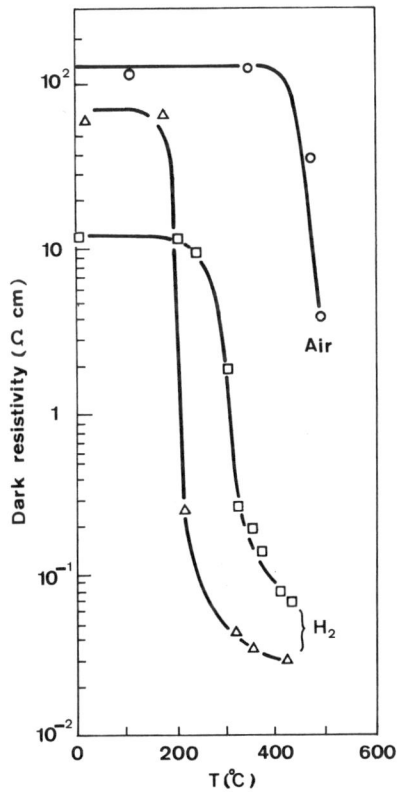

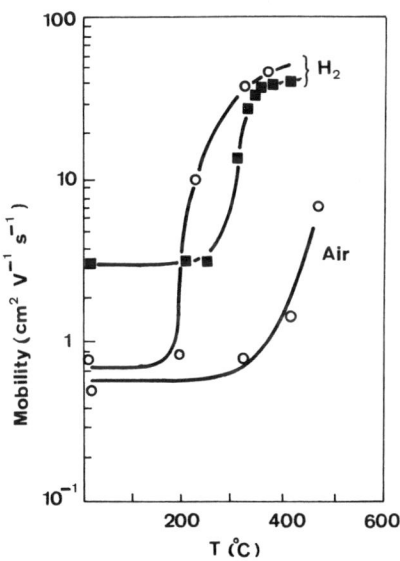

Fig. 6.15. Mobility at room temperature of CdS films sprayed on glass at 340 °C as a function of annealing temperature [6.26]

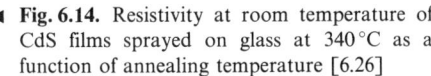

Fig. 6.14. Resistivity at room temperature of CdS films sprayed on glass at 340 °C as a function of annealing temperature [6.26]

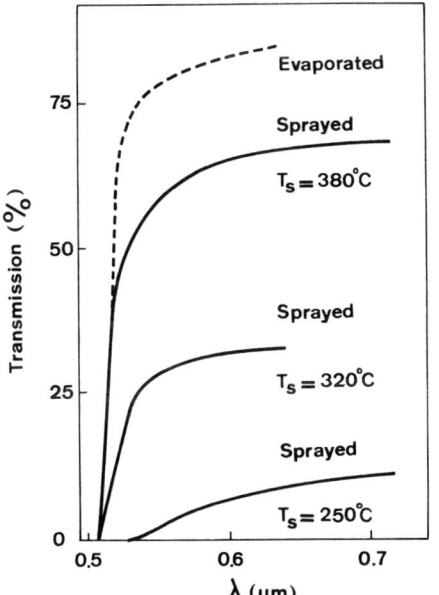

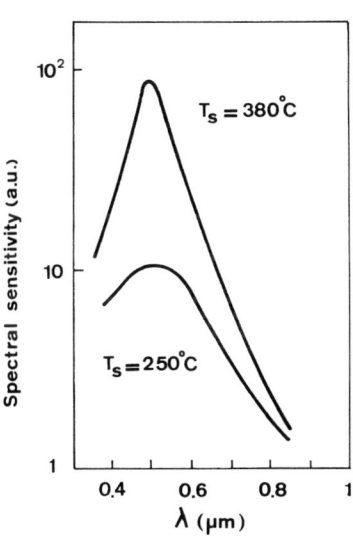

Fig. 6.16. Spectral variation of optical transmission of sprayed CdS [6.25, 28]

Fig. 6.17. Spectral variation of the conduction of sprayed CdS [6.28]

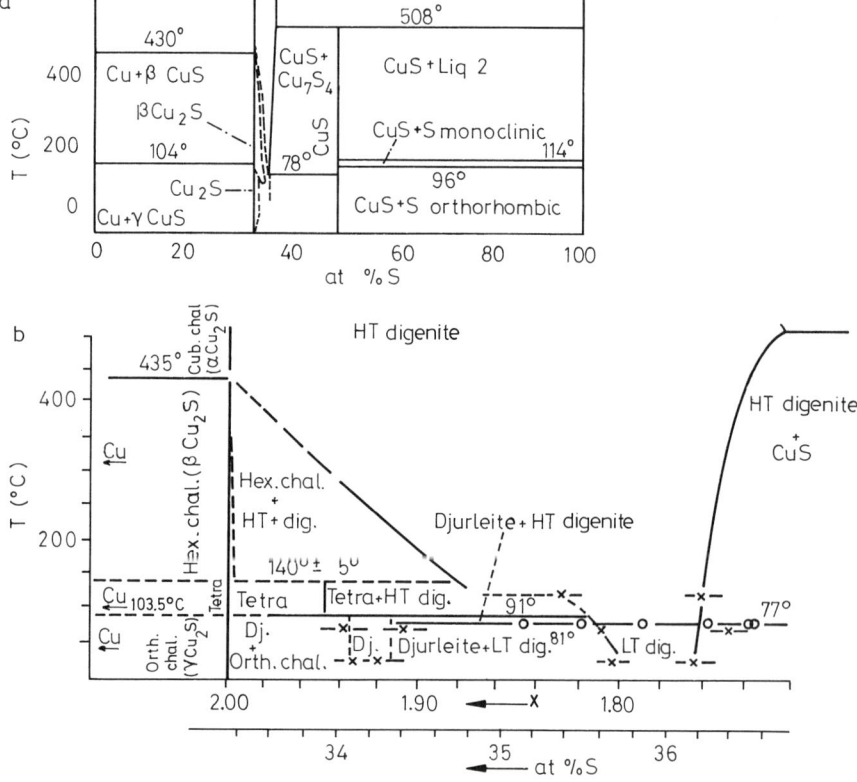

Fig. 6.18a and b. Phase diagram of Cu–S system: **(a)** for all compositions between 0 and 600 °C, **(b)** for compositions involved in Cu_xS–CdS solar cells

6.3 Properties of Cu_2S Films

6.3.1 Phase Diagram of the Cu–S System and Structural Properties of Stable Phases

The copper-sulfur phase diagram is relatively complex (Fig. 6.18). Moreover, especially in the composition range relevant to solar cells, i.e., from Cu_2S to $Cu_{1.78}S$, some limits are still under study [6.89–99].

At room temperature there are five stable phases. On the copper-rich side, we observe orthorhombic *chalcocite* (γCu_2S), and on the sulfur-rich side, *covellite* (CuS). Between these two phases we find: *djurleite* ($x = 1.96$–1.94 or 1.93 depending on the authors), the low-temperature *digenite* ($1.765 < x < 1.79$ at 25 °C after [6.94]), and *anilite* $Cu_{1.75}S$. According [6.99] another phase for $x = 1.90$ would exist at temperatures below 30 °C and similarly for $x = 1.91$ after [6.100].

Table 6.5. Structure and lattice constants of the principal phases of the Cu-S system

Phase	Structure	a [Å]	b [Å]	c [Å]	Ref.
Cu_2S	Orthorhombic	11.881	27.323	13.491	[6.102]
Cu_2S	Hexagonal	3.961		6.722	[6.102]
Djurleite	Orthorhombic	15.71	13.56	26.92	[6.103]
Digenite (Low temp.)	Pseudocubic	5.56			[6.101, 102]
Tetragonal phase	Tetragonal	3.996		11.28	[6.104]
Anilite	Orthorhombic	7.89	7.84	11.01	[6.105]

When the temperature increases the new phases which exist are the mid-temperature form of chalcocite ($\beta\,Cu_2S$) between 103.5 and 435 °C, and the high-temperature form ($\alpha\,Cu_2S$) which is really the copper-rich limit of a large range of solid solutions extending until $Cu_{1.73}S$ on the sulfur-rich side. This is the domain of the high-temperature form of digenite.

Another phase, the tetragonal, exists from $x=1.80$ to $x=2.00$. It was supposed to be metastable until *Cook* showed that it is in fact stable and forms a solid solution between Cu_2S and $Cu_{1.95}S$, at a temperature from 94 °C to approximately 140 °C [6.95]. Note that the domain between djurleite and anilite is still under discussion with regard to the stability and the limits of the phases involved. This has been reviewed in the recent work of *Potter* [6.98].

In Table 6.5 we give the structure and crystallographic parameters of the different phases defined above. It is interesting to note that in the crystallographic structure of these different phases of the Cu-S system, we can distinguish, on the one hand, the sublattice of the chalcogen element, which is the rigid armature of the crystal; and on the other hand, the sublattice of the copper cations which are very mobile and can position themselves in a large number of equivalent sites. This is also found in other compounds of type $A_2^I B^{VI}$, where A can be Ag, Cu, ...; and B stands for S, Se, Te, ...

The following relations concerning the diffusion of the Cu^+ cations have been established by *Etienne* [6.106]:

$D_{Cu^+} = 8.1 \times 10^{-3} \exp(-5.870/T)$ [cm² s⁻¹] with

$D_{Cu^+} = 1.1 \times 10^{-10}$ cm² s⁻¹ at 50 °C for Cu_2S; and

$D_{Cu^+} = 3.6 \times 10^{-2} \exp(-6.100/T)$ [cm² s⁻¹] with

$D_{Cu^+} = 2.4 \times 10^{-10}$ cm² s⁻¹ at 50 °C for digenite.

6.3.2 Electrical Properties of Bulk Copper Sulfides in the Range of Compositions Near the Stoichiometry Cu_2S

Chalcocite Cu_2S is a *p*-type semiconductor, governed by copper vacancies, whatever its form (orthorhombic, hexagonal or cubic).

Table 6.6. Ionic and hole conductivities in β Cu$_2$S

Authors	T [°C]	Hole conductivity σ_p [Ω^{-1}cm^{-1}]	Ionic conductivity σ_i [Ω^{-1}cm^{-1}]	σ_i/σ_p	D_i [cm^{2}s^{-1}]	μ_i [cm^{2}V^{-1}s^{-1}]	μ_p [cm^{2}V^{-1}s^{-1}]	Remarks
Hirahara [6.107]	150	0.0031		0.003 %				Cu excess
	200	0.0144		0.11 %				
	400	0.1994		1.27 %				
Yokota [6.112]	200	3.11	0.55		3.21×10^{-3}	0.077	0.43	
Okamoto et al. [6.115]	200	11.0	0.40		4.1×10^{-4}	0.01	0.28	
	200	1.20	0.41		4.5×10^{-3}	0.11	0.33	Treated
	200	0.46	0.43		1.1×10^{-2}	0.26	0.29	Cu excess
Ishikawa and Miyatani [6.113, 114]	100	3×10^{-2}	5×10^{-4}					
	110	3×10^{-2}	8×10^{-2}					
		4.5×10^{-1}	8×10^{-2}					

Table 6.7a. Electrical properties of Cu$_2$S at 300 K. Bulk material

Authors	σ [Ω^{-1}cm^{-1}]	ϱ [Ωcm]	p [cm^{-3}]	μ_p [cm^{2}V^{-1}s^{-1}]	Q_p [μV/°C]	E_F-E_v [meV]	E_a-E_v [eV]	m_p^*/m_0	Remarks
Hirahara [6.107]			4.4×10^{18}	9			0.102		0.15% S deficit
			2.4×10^{19}	14			0.062		Stoichiometric
Eisenmann [6.116]		200					0.6		
Abdullaev et al. [6.109]	170		7.4×10^{19}	25	90		0.064	0.58	Single crystal
Sorokin et al. [6.116]	40		4.07×10^{17}	612	180				Single crystal Cu$_{1.998}$S
Astakhov et al. [6.118]	$(1.7–2.0) \times 10^{-2}$		$(3–5) \times 10^{16}$	4.5	750		0.03–0.09	1.4–1.5	Cast polycrystals
			3.35×10^{18}						

Authors	σ [Ω^{-1}cm^{-1}]	ρ [Ωcm]	p [cm^{-3}]	μ_p [cm^{2}V^{-1}s^{-1}]	Q_p [μV/°C]	E_F-E_v [meV]	E_a-E_v [eV]	m_p^*/m_0	Remarks
Ishikawa et al. [6.114]	$(1-5)\times 10^3$								Polycrystals
Okamoto et al. [6.115]	48				125				Stoichiometric
	15				210				
	4.9				300				Treated
	0.014				811				
Gassanova [6.108]	5		10^{16}	30	348				Cu$_2$S + 1.45%Cu
Potter et al [6.119]			$10^{20}-10^{21}$	1–10					CdS crystal dipped
Dumon [6.120]	30				215				Sintered crystal Cu$_{1.995}$S
Guastavino [6.110]		0.40	3.3×10^{18}	4.75	327	0.08			Undoped
		0.122	9.2×10^{18}	5.60		0.07		1.65	Cd doped
		0.111	2.2×10^{19}	2.57	267			1.82	Undoped

Table 6.7b. Electrical properties of Cu$_2$S at 300 K. Thin layers

Authors	σ [Ω^{-1}cm^{-1}]	ρ [Ωcm]	p [cm^{-3}]	μ_p [cm^{2}V^{-1}s^{-1}]	Q_p [μV/°C]	E_F-E_v [meV]	E_a-E_v [eV]	m_p^*/m_0	Remarks
Eisenman [6.116]	100								Evaporated
Selle et al. [6.121]		2	4×10^{15}	7	200		0.43		Evaporated
Nakayama [6.122]	15		2.6×10^{19}	3.6	60				Evaporated
Martinuzzi et al. [6.142]	$10^{-1}-10^{-2}$		10^{20}	1.5					Evaporated
Miloslavskii et al. [6.123]	2×10^{-3}		10^{16}						Evaporated
Loferski et al. [6.124]		0.005–20							Cu sulfurized
Nimura et al. [6.125]	3×10^3								Evaporated
	9×10^{-1}								

I) At high temperatures (30–600 °C), the thermal variations of the conductivity, the Hall factor, and the thermoelectric coefficient show two breakdowns due to phase changes as indicated on the phase diagram. These discontinuities are well known today and have been pointed out by many authors [6.107–111].

β Cu$_2$S (103.5 °C < T < 435 °C) shows a large ionic conductivity (Table 6.6) which can be, in some cases, of the same order of magnitude as electronic type conductivity. This ionic conductivity plays an important role when the copper sulfides are in an electric field which displace the copper ions. This is an important factor accounting for the degradation of the performances of CdS–Cu$_2$S cells. *Ishikawa* et al. and *Okamoto* et al. who studied this ionic conductivity σ_i have shown that it is independent of the composition [6.114, 115]. For copper vacancy densities in the range of 3×10^{18} to 8.4×10^{19} cm^{-3} and temperatures between 150 and 400 °C, the law

$$\sigma_i = (8.9 \times 10^{14}/T)\exp(-0.24\,e/kT)$$

(where e is the electronic charge, k the Boltzmann constant, and T the absolute temperature) was found by *Okamoto* et al. [6.115]. They also show that the mobility of the cationic vacancies is inversely proportional to their density and independent of the temperature.

With regard to the Hall factor variation in the range of temperatures in which γ Cu$_2$S exists, we note that we cannot reach the intrinsic regime because of the phase change at 103.5 °C, so we are unable to determine the corresponding value of the energy gap by this method.

II) At room temperature the electrical characteristics of γ Cu$_2$S (Table 6.7) indicate a large number of carriers and imperfections showing that really we do not reach the stoichiometric composition Cu$_{2.00}$S. We note that the mobility values are generally low (3–30 cm^2 V^{-1} s^{-1}). However, *Sorokin* et al. obtain higher values. Also we note some disparities in the values of p. Some authors give values significantly lower than the average ones.

In order to decrease the high value of carrier density in Cu$_2$S some doping with Cd, Zn, and In was tried. *Okamoto* shows that In plays the same role as Cu in the stoichiometry variation. *Guastavino* confirms this result with Cd and Zn and shows that Cd displaces Cu when the amount of the latter is such that stoichiometry is reached. This is the inverse mechanism of dipping [6.126].

Abdullaev et al. give the value 0.58 m_0 for the effective mass of the holes obtained from thermoelectric power measurements [6.109]. *Guastavino* found 1.7 m_0 [6.110]. *Astakhov* et al. give a similar value [(1.4–1.5) m_0] [6.118]. *Mulder*, from optical considerations, gives 3 m_0 [6.127].

Among the parameters relevant to explain the mechanism of CdS–Cu$_2$S solar cells, the electronic affinity has been studied very little. From indirect measurements based on Si–CdS and Si–Cu$_2$S open-circuit voltage comparison, *Schewchun* et al. suggest a value of 4.05 eV [6.128]. *Pfisterer* et al. indicate 4.2 eV on their schemes. The direct measurements based on photoemission edge study are difficult because Cu_2S cannot be cleaved. *Duchemin* et al. give an electronic affinity value of 4.4 eV [6.130].

According to [6.109, 110], the Fermi level, which is an extrinsic parameter, is situated in a region at less than $3kT$ from the maximum level of the valence band because of the great number of free carriers.

III) At low temperatures, the thermal variations of the conductivity and of the Hall constant for γCu_2S are typical of a semiconductor. *Eisenmann* shows the existence of an activation energy of 0.6 eV [6.116]. *Hirahara* places this energy between 0.062 and 0.102 eV from electrical measurements performed between 250 and 400 K [6.107]. *Abdullaev* et al. [6.109] suggest a level at 0.064 eV from the valence band. *Astakhov* et al. [6.118], for samples with carrier density equal to 3.35×10^{18} and 5.9×10^{19} cm^{-3} propose the existence of two levels at 0.03 ± 0.003 and 0.09 ± 0.005 eV from the valence band. *Guastavino* also finds two activation energies, 0.08 and 0.007 eV, and shows the existence of an impurity conduction for temperatures below 50 K. The origin of these two acceptor levels is due to copper atoms. *Rau* [6.131] gives a model based on doubly ionizable copper vacancies. However, *Weiss* [6.132] suggests a model with copper vacancies and interstitial copper.

With regard to the thermal variation of μ_p, *Guastavino* shows that between 300 and 77 K the diffusion mechanism via acoustic phonons is not alone but becomes predominant at 300 K [6.110]. *Sorokin* et al. [6.117] find a $T^{-3/2}$ law between room temperature and the phase transition temperature of 100 °C, and a $T^{-\alpha}$ law, with $\alpha > 3/2$, between 100 and 200 °C. *Abdullaev* et al. [6.109] also found a $T^{-3/2}$ law between room temperature and 140 °C. *Kerimov* et al. [6.133] show a diffusion mechanism via ionized defects between 20 and 100 K and another via lattice vibrations between 100 K and room temperature. *Astakhov* et al. [6.118] obtain a $T^{-0.7}$ law and admit that there is competition between acoustic phonons and ionized impurity diffusion mechanisms between 30 and -200 °C.

6.3.3 Electrical Properties of Thin Copper Sulfide Layers Near the Cu_2S Composition

In Table 6.7b, results concerning the electrical properties of thin copper sulfide layers prepared by evaporation or the dipping process are indicated. We note some disparities between the values of carrier densities found by the different authors. However, there is a fairly good concordance between mobility values in the case of thin layers and in the case of bulk material. The main problem for thin layers is the control of the composition during the fabrication. The difference between sulfur and copper partial pressures results in a departure from stoichiometry during evaporation. The pH and the solubility of salt in solution play also a role in the completion of the dipping reaction, so it is necessary to determine the true composition obtained. *Mathieu* [6.99] studied the Cu–S phase diagram by electrochemical experiments, and *Vedel* et al. extended the results of this study to the determination of the thin Cu_xS layer

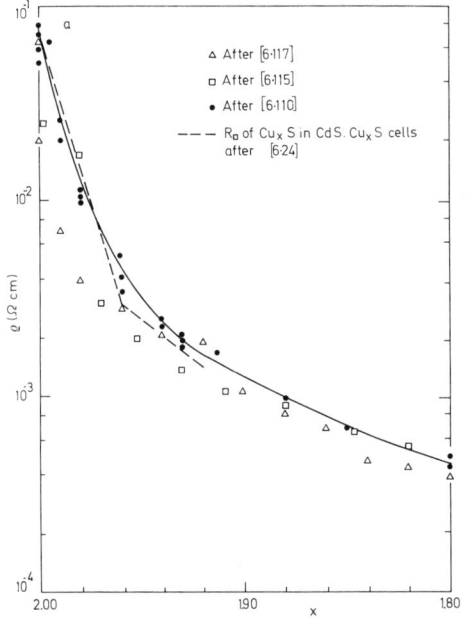
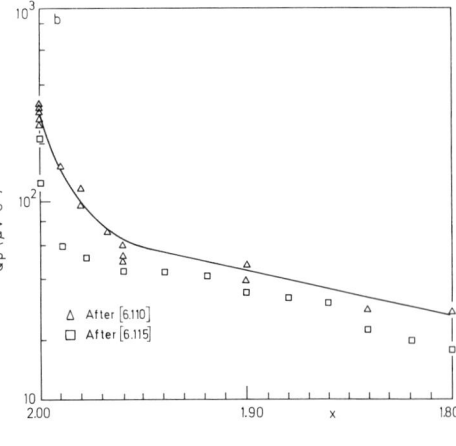

Fig. 6.19. (a) Resistivity of Cu_xS at 300 K versus composition. **(b)** Thermoelectric power variations of Cu_xS at 300 K versus composition

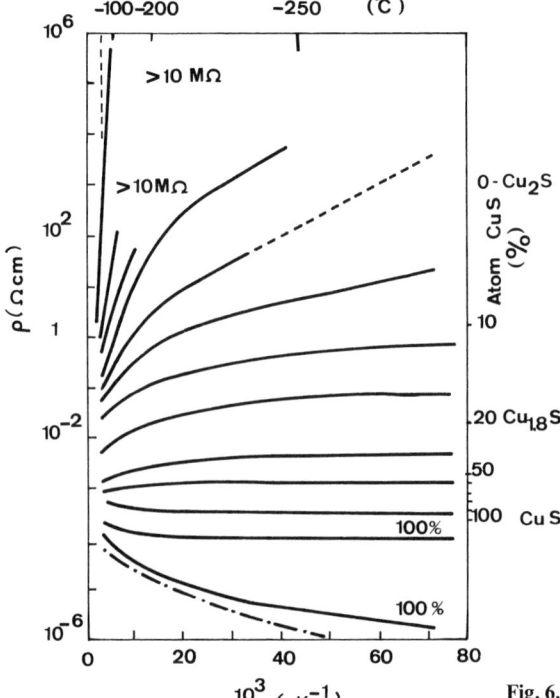

Fig. 6.20. Resistivity variations of thin Cu_xS layers after [6.116]

composition [6.134]. X-ray measurements and cathodoluminescence emission were also used by *Loferski* et al. [6.124] for characterizing the composition of his Cu_2S and $Cu_{1.96}S$ layers. *Mulder* [6.135] used microchemical analysis and atomic absorption spectroscopy methods. Differential thermal analysis (DTA) measurements also seem to be applicable in the case of cuprous sulfide layers near the stoichiometry Cu_2S [6.110].

Studies concerning the action of Cu, S or O on the electrical properties of layers of Cu_2S have also been carried out by various authors [6.136, 137].

The diffusion length L_n of minority carriers in Cu_2S involved in practical CdS–Cu_2S cells has been measured by different authors. By a light microprobe investigation, *Gill* and *Bube* measured the short-circuit photocurrent as a function of light spot distance from the junction. They obtain for L_n a value ranging from 1×10^{-5} to 4×10^{-5} cm [6.80]. From the photovoltage developed in a small CdS crystal pressed against one side of Cu_2S or $Cu_{1.96}S$ crystals as a function of the penetration depth of the light incident on the other side, *Mulder* found a value of 300–350 Å for chalcocite in equilibrium with djurleite, and $L_n < 50$ Å for djurleite [6.138]. From the decay of the short-circuit current generated by a 20 keV electron traversing the junction of a CdS/Cu_xS polycrystalline thin film solar cell, *Oakes* et al. [6.139] obtained diffusion lengths ranging from 0.11 to 0.57 μm for electrons in cuprous sulfide. Indeed, taking account of uncertainties introduced by surface recombination effects, internal fields, and experimental accuracy, they estimated the actual bulk diffusion lengths to vary between 0.09 to 1.71 μm.

6.3.4 Variation of the Electrical Properties of Cuprous Sulfides with Composition

The departure from the stoichiometry Cu_2S toward the sulfur-rich composition induces changes in the apparent electrical parameters of the copper sulfides which in fact become a phase mixture as indicated on the phase diagram. The changes are increase of the free carrier density and of the conductivity, and a decrease of the thermoelectric power coefficient Q_p. The explanation of these variations is the increase of the main defects in the copper sulfides: the copper vacancies.

In Fig. 6.19 the variation of resistivity and thermoelectric power of Cu_xS with composition for $2.00 > x > 1.79$ can be seen. The variation of the sheet resistance of copper sulfides in CdS–Cu_xS is also indicated. Notice the agreement in these variations [6.24].

In the literature we can find plots of μ_p, m^*/m_0, and Q_p versus the apparent free carrier density of the mixtures [6.110, 118].

In Fig. 6.20 we show the variation of the conductivity of thin Cu_xS layers at low temperature for different compositions according to [6.116]. In Fig. 6.21 variations of the resistivity of bulk Cu_xS at high temperature are given for the composition range: $2.00 > x > 1.85$ [6.111].

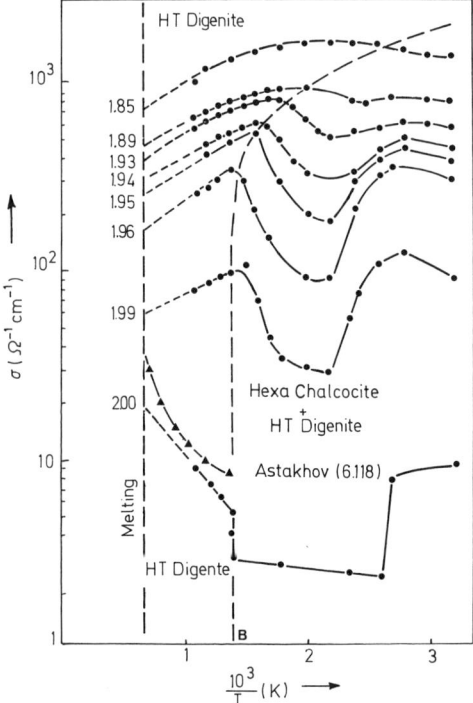

Fig. 6.21. Variations of the conductivity of bulk Cu_xS at high temperature after [6.111]

In the literature we find sparsely results on nonstoichiometric thin layers. For example, *Nakayama* [6.122] gives for $Cu_{1.96}S$ und $Cu_{1.8}S$ respectively

$\sigma = 35 \, \Omega^{-1} \, cm^{-1}$ $\mu_p = 10^{-2} \, cm^2 \, V^{-1} \, s^{-1}$ $p = 10^{22} \, cm^{-3}$
$\sigma = 230 \, \Omega^{-1} \, cm^{-1}$ $\mu_p = 0.51 \, cm^2 \, V^{-1} \, s^{-1}$ $p = 2.8 \times 10^{21} \, cm^{-3}$.

Nimura et al. [6.125] find for a mixture of chalcocite and djurleite:

$\sigma = 10^3 \, \Omega^{-1} \, cm^{-1}$ $\mu_p = 1.1 \, cm^2 \, V^{-1} \, s^{-1}$ $p = 6 \times 10^{20} \, cm^{-3}$
$\sigma = 550 \, \Omega^{-1} \, cm^{-1}$ $\mu_p = 0.6 \, cm^2 \, V^{-1} \, s^{-1}$ $p = 9 \times 10^{21} \, cm^{-3}$.

Other results are given in [6.136, 137].

6.3.5 Optical Properties of Copper Sulfides

In Table 6.8 we present different results found in the literature and concerning the value and the nature of the energy gaps in copper sulfides. Obviously there is a large disparity between these different results. Also, there is a great difference between the values of the absorption coefficients found and of the sharpness of the absorption edge (Fig. 6.22).

Table 6.8. Energy gap values found for γ Cu$_2$S

Author	Sample	Experimental method	Gap value	Gap nature
Shiozawa et al. [6.150]	Thin layers	$T+R$ $T+R$	1.21 eV (300 K) 1.83 eV (300 K)	Ind. Direct
Gassanova [6.108]	Bulk	Absorption	2 eV (300 K)	
Marshall et al. [6.140]	CdS converted	$T+R$	1.21 eV (300 K) 1.26 eV (80 K)	Ind.
Sorokin et al. [6.141]	Bulk and Thin layers	$T+R$ Photoconductivity	1.93 eV (300 K) 1.84 eV (300 K)	
Abdullaev et al. [6.109]	Bulk	Absorption	1.9 eV (300 K)	
Selle et al. [6.121]	Thin layers	Absorption Photoconductivity	1.22 eV (300 K) 1.25 eV (95 K)	
Ramoin et al. [6.142]	Thin layers	$T+R$ $T+R$	1.7 eV (300 K) 1.05 eV (300 K)	Direct Ind.
Nakayama [6.122]	Thin layers	$T+R$ $T+R$	1.05 eV (300 K) 1.08 eV (100 K)	Ind. Ind.
Mulder [6.135]	CdS converted	$T+R$ $T+R$ $T+R$ $T+R$	1.13 eV (a axis) 1.09 eV (b axis) 1.40 eV (c axis) 2.50 eV	Ind. Ind. Ind. Direct
Loferski et al. [6.124, 128]	Thin layers and bulk	CL	1.28 eV (77 K)	Direct
Miloslavskiy et al. [6.123]	Thin layers	$T+R$	1.46 eV (120 K)	Direct
Guastavino [6.110]	Bulk	$T+R$	1.21 eV (300 K)	...

T: transmission; R: reflection; CL: cathodoluminescence

In spite of these differences, what seems sure today is the existence of one edge at 1.20 eV at 300 K. This agrees with the cut-off of the spectral response of CdS–Cu$_2$S cells at this energy. However, the nature of this gap is still in discussion because the magnitude of the absorption coefficient is not sufficiently high to indicate a direct gap and not sufficiently low to indicate an indirect one. Taking into account the halfwidth of the luminescence band of Cu$_2$S (200 Å), *Schewchun* et al. [6.128] think that it is rather narrow and more indicative of a direct transition phenomenon which occurs at 1.2 eV.

Also in Table 6.8 we can see the existence of another edge in the range 1.7–2.5 eV. This edge can be shown only on thin samples as in the case on evaporated layers. This indicates a second electronic transition occuring in copper sulfides. *Mulder* suggests a band model with two types of transitions: indirect at around 1.2 eV and direct in the range of 2–2.5 eV [6.127].

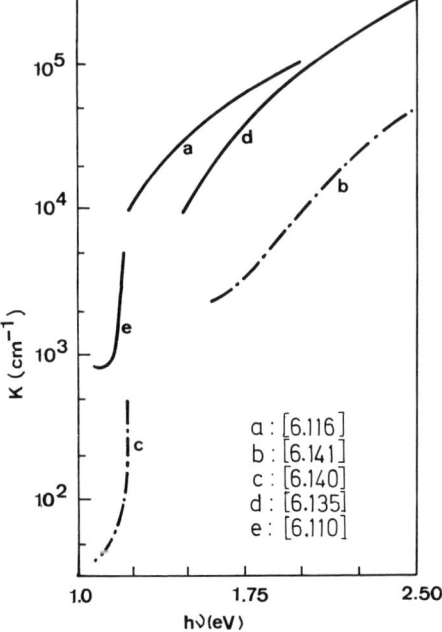

Fig. 6.22. Optical absorption curves of Cu_2S after different authors

With regard to the ir reflectivity which is directly connected with the free carrier absorption, *Guastavino* shows that the coefficient increases from 27 to 80% when the composition changes from $Cu_{2.00}S$ to $Cu_{1.79}S$. From the plasma resonance minimum, he deduces also the effective mass of the hole in digenite [6.110].

6.4 Photovoltaic Properties of Cu_2S–CdS Cells

Most efficient photovoltaic conversion has been obtained on thin film cells made by dipping the evaporated or sprayed CdS layer in CuCl. The properties given below concern these cells except when indicated.

6.4.1 Junction Structure

I) CdS: The CdS deposited by vacuum evaporation grows along the c axis of the hexagonal lattice, perpendicular to the substrate surface, and forms columns of 0.1–1 µm in diameter. For a long time, thicknesses of more than 30 µm were necessary to obtain stable cells, and also to avoid the Cu_2S going across the CdS layer during dipping and reaching the ohmic contact. In recent years, cells have been obtained on evaporated films less than 20 µm thick or on sprayed films of less than 5 µm thick [6.3, 21].

II) Cu_2S: The thickness of Cu_2S in cells illuminated from the Cu_2S side is about 0.3 μm [6.143, 144, 150, 201]; however we note the existence of copper 1 or 2 μm into the CdS [6.150, 162].

The stoichiometry of the Cu_2S in the cells was measured by an electrochemical method [6.132]. The best results were obtained on $Cu_{1.995}S$ [6.143, 50].

Massicot [6.149] conjectured the existence of a thin layer of djurleite ($Cu_{1.93}S$) between the Cu_2S and CdS by a thermal or electrochemical process, in order to explain the observed high open circuit voltages and the shifting of the low energy threshold of the spectral response from 1.2 eV (Cu_2S) to 1.8 eV ($Cu_{1.93}S$).

III) *Junction*: Today, we consider that the electric junction is situated at the Cu_2S and CdS interface. Because of the structure of the evaporated CdS surface, the area of the junction may be greater than that of the cell. The ratio of these areas is not determined precisely. A value of ten has been proposed [6.146, 147].

The surface increase gives place to a slight increase of the short-circuit current and a decrease in open-circuit voltage [6.154].

6.4.2 Current-Voltage (IV) Characteristics

a) In Darkness

Forward Bias. The variation of the current against voltage shows two modes of conduction; in general the current law is as follows:

$$I = I_{s1}[\exp a_1(V-rI) - 1] + I_{s2} \exp a_2(V-rI) + \frac{(V-rI)}{R},$$

where r is the series resistance ($\sim 1\,\Omega\,cm^{-2}$) and R is the shunt resistance (many kilohms) which for good cells can be neglected in comparison with $I_{s1}[\exp a_1(V-rI) - 1]$. The coefficient a_1 is approximately 10–15 V^{-1}, and a_2, 25–35 V^{-1} at room temperature [6.143, 163, 165, 80, 151]. The variation of a_1 and a_2 with temperature is slight and they generally do not vary as $1/kT$ [6.163, 165]. Barrier heights of 0.45 and 1.2 eV have been determined from the variations of saturation current I_{s1} and I_{s2} with temperature [6.143, 167]. A variation law $I_s = I_{s0} \exp(cT)$ has been obtained for cells from different origins [6.162, 165].

Reverse Bias. In this case, the current does not reach the saturation value. The law $I \propto V \exp(V_d - V)^{-1/2}$ has been proposed [6.162, 163] to summarize its characteristics under breakdown voltage V_{br} (between 1.6 and 4 V according to the origins of the cells). Above this value of V_{br} the current increases very rapidly with the voltage. It is not clear whether the cause of this effect is Zener or avalanche breakdown.

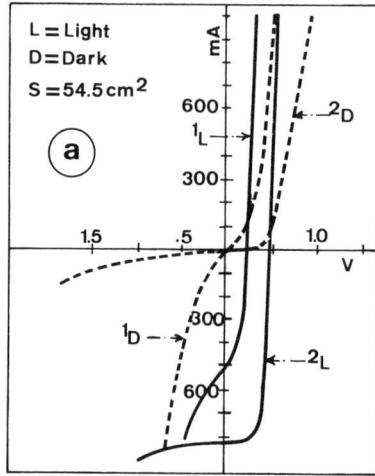
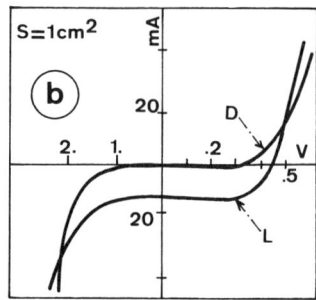

Fig. 6.23. (a) Current-voltage characteristic of one cell of I.E.C. of Delaware before ($1_L, 1_D$) and after ($2_L, 2_D$) heat treatment [6.32]. (b) Current-voltage characteristic of one cell of S.A.T. [6.165]

b) Under Illumination

Generalities. Conduction laws for the junction are altered under illumination. For certain cells, this modification is shown by the intersection of the forward bias characteristics in the dark and under illumination (Fig. 6.23a). The behavior of cells under light depends on the fabrication method and heat treatment to which cells are submitted after dipping (Fig. 6.23a, b). *Hewig* et al. [6.152] show that at forward bias greater than 350 mV, the cell current is increased by the wavelengths less than 850 nm and decreased by wavelengths between 850 and 1900 nm.

Short-Circuit Current I_{sc}. Cells of 7–8 % efficiency have a current of about 20 mA cm^{-2} at AM 1 (100 mW cm^{-2}) [6.5]. The current increases linearly with the light intensity [6.42, 175, 186]. At temperatures higher than 100 K, it varies with a positive temperature coefficient of about 10^{-3} [K^{-1}] [6.165]. It seems that the main part of this current is generated in Cu$_2$S. With the low energy threshold of the spectral response corresponding to the gap of Cu$_2$S, the variation of the short-circuit current with stoichiometry leads to this hypothesis (Fig. 6.24). The variation of I_{sc} (and V_{oc}) with the heat treatment time is explained by *Te Velde* [6.151] as a modification of the junction. For the compound cell CdS, ZnS–Cu$_2$S [6.148] the short-circuit current decreases with the Zn percentage in CdS, as if the ion exchange in CuCl did not also take place in the presence of zinc.

Open-Circuit Voltage V_{oc}. At room temperature, the open-circuit voltage is between 450 and 500 mV depending on fabrication methods. Above 100–120 K, V_{oc} decreases when the temperature increases with a negative coefficient of 2×10^{-3} [K^{-1}] close to that determined for the variation of the diffusion barrier. At low illuminations, V_{oc} is very low because of the high value of the

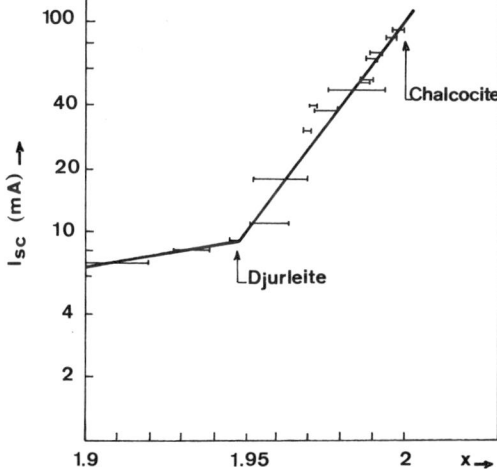

Fig. 6.24. Short-circuit current of Cu_xS–CdS cells as a function of x [6.50]

first saturation current. Due to a better lattice agreement of Cu_2S–CdZnS structure, the open-circuit voltage is greater in these cells.

Efficiency η. The highest efficiencies obtained on thin film cells are 7–8%; the illumination is from the Cu_2S side (frontwall). *Nakayama* [6.31] has obtained 9% efficiency with a ceramic CdS cell. On sprayed cells lower efficiencies are obtained [6.21]. The high value of dark currents reduces the V_{oc} for a given photoelectric current and hence limits the efficiency of this structure.

6.4.3 Capacity-Voltage (*CV*) Characteristics

Donor Concentrations in CdS – Barrier Height. Donor concentrations in CdS determined for thin film cells from the classical interpretation of characteristics are not precise because of the difference between the surface of the cell and that of the junction.

Acceptor concentrations have been determined for bulk Cu_2S samples. The values 10^{19}–10^{20} cm^{-3} obtained in [6.110] suggest placing the space-charge region inside the CdS.

Lindmayer et al. [6.155] determine two donor concentrations in CdS: 1.8×10^{17} cm^{-3} far from the junction, and 3.5×10^{15} cm^{-3} near it. These values vary from 10^{17} cm^{-3} (Delaware [6.143]) to 10^{19}–10^{20} cm^{-3} (SAT [6.163, 165]) far from the junction, and from 10^{15} to 10^{17} cm^{-3} near it, depending on the origin of the cell. Copper diffusion in CdS during dipping or heat treatment explains the lower value of donor concentration in the CdS near the junction.

Nakayama [6.31] explains the existence of a capacitance independent of bias by the formation of an insulating CdS layer.

The barrier heights determined from capacitance variation for forward bias are between 0.6 and 1 V (for single crystals the values are between 0.8 and 1 V).

Table 6.9. Spectral response[a]

Reference	Structure Frontwall F Backwall B	Peaks [μm]	Cutoff Short [μm]	Cutoff Long [μm]
Cu_2S/single crystal CdS				
Reynolds et al. [6.173]	B	0.550, 0.620	0.525	
Hammond and *Shirland* [6.174]	B	0.700	0.525	1.0
	F	0.700	0.265	1.0
Woods and *Champion* [6.175]	B	0.510		1.1
	F	0.700		
Bockemuehl et al. [6.176]	B	0.520		
Williams and *Bube* [6.38]	B	0.520		
	F	0.480		
Shitaya and *Sato* [6.177]		0.600, 0.800		1.1
Gill et al. [6.178]	F before heat treatment			1.0
	after			0.75
Miya [6.179]	Cd plane			1.1
	S plane			0.85
Cu_2S/evaporated CdS				
Moss [6.180]	B	0.570		
Shirland [6.181]	F	0.520		1.0
Drozdov et al. [6.182]	B Cu_2O	0.520, 0.420		
Spakowski et al. [6.183]	F	0.480		1.0
Pastel [6.184]	B thin Cu	0.730		
	B thick Cu	0.620		
Shirland [6.42, 185]	F	0.520, 0.650–0.700	0.250	1.0
Balkanski et al. [6.186]	F	0.650	0.450	0.9
Bernard [6.48]	F	0.500–0.600		1.0
Potter et al. [6.187]	F			
Mytton [6.188]	F	0.500		
Anshon et al. [6.189]	FB	0.500–0.580–0.900		
Cu_2S/sprayed CdS				
Chamberlin et al. [6.22, 190]	B	0.700		
	F	0.495, 0.580, 0.700		
Jordan [6.21]	B	0.500–0.700		
Besson [6.4], *Martinuzzi* [6.28]	B	0.600, 0.700		
Sputtered films				
Pavelets et al. [6.191]	F	0.520		
		0.520, 0.600		
Yefremenkova et al. [6.18]	F	0.420, 0.450–0.460, 0.480		
Egorova [6.192]		0.490–0.500		
	B	0.580–0.650		
	F	0.420, 0.450, 0.490–0.500		0.8–0.85
	B	0.480–0.600	0.510	0.8–0.85

[a] See [6.5].

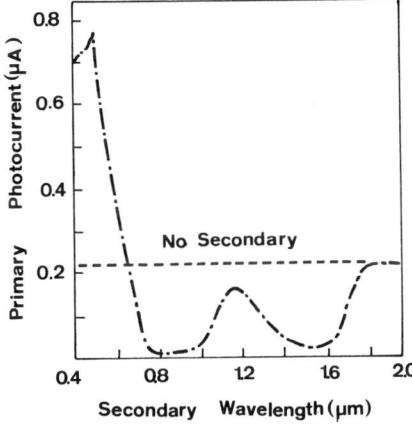

Fig. 6.25. Enhancement and quenching of photocurrent by secondary light. Both the 0.655 μm primary and the secondary intensity were 300 μW cm^{-2} [6.80]

For reverse bias, the intersection of the curve of C^{-2} against V with the V axis give greater values in agreement with the existence of an interface layer between Cu$_2$S and CdS [6.31, 165, 167].

Photocapacitance. The capacitance is increased with illumination near of about 0.5 μm wavelength [6.155]. *Lindquist* et al. [6.156–159] explain the same phenomena observed on single crystals by trapping holes by deep traps. The decrease in the capacitance with illumination of a greater wavelength than 0.85 μm allows them to position the traps in the middle of the band gap of CdS. The observations and interpretation have been carried out on thin film cells [6.160, 161].

6.4.4 Spectral Response

In Cu$_2$S–CdS cells, carriers are generated in the Cu$_2$S by photons of energy greater than 1.2 eV. In CdS, they are created by intrinsic excitation at energies greater than 2.45 eV and by extrinsic excitation at lower energies. When the cell is illuminated from the CdS side (backwall), the spectral response appears between the wavelengths of 0.5 and 1 μm. When illuminated from the Cu$_2$S side (frontwall), we increase the absorption in the Cu$_2$S, and the maximum will shift to the less energy side and the response will spread out to the short wavelengths (0.25 μm). Spectral response data are summarised in Table 6.9.

At the present time, the choice of the illuminated side is determined by the method of fabrication. *Rothwarf* [6.146], using a theoretical calculation, links this choice to the reflexion conditions on the rear contact. The existence of a high density of traps in the CdS band gap having a long relaxation time modify the spectral response following the direction of wavelength scan [6.143, 150]. The existence of these traps leads *Lindquist, Gill, Fahrenbruch*, and *Bube* [6.80, 156–159] to explain the increase in the spectral response by a secondary illumination of wavelengths near 0.5 μm and the decrease by wavelengths between 0.8 and 1.1 μm (Fig. 6.25).

6.4.5 Stability

The decrease (reversible or not) in the performance of the Cu_2S–CdS cells which appears during use are due to the structure and the components of these cells.

Changes Due to the Structure. Due to the different expansion coefficients of each component during thermal cycling, cells split and delaminate resulting at the beginning in a decrease of the I_{sc} current without affecting V_{oc} or the fill factor. Improvement has been made by coating with Kapton and using new processes for fixing the grid. These tests have been carried out for use in space; for use on earth, the outside temperature is much less than in space, which raises the possibility of using glass as substrate.

Changes Due to the Components. The cells must be protected against moisture either by the use of transparent epoxy, or, in the case of use on earth, between two sheets of glass. Moisture diffuses to the junction and forms recombining centers which lead to a deterioration of the characteristics of the cells and an increase in the series resistance.

Copper Sulfide Diffusion. It is generally agreed upon that optimal efficiency is obtained with $Cu_{1.995}S$ [6.141, 143]. When the copper concentration decreases, the number of free electron-hole pairs generated in the Cu_2S decreases as well. At temperatures higher than 100 K, the high diffusion velocity of Cu [6.73–75, 144] leads to a decrease of the amount of Cu in Cu_2S. The diffusion of Cu into CdS lowers its conductivity. This diffusion may be reduced by doping with CdS and reducing the neutral Cu density in Cu_2S.

Ionic Conduction. Palz et al. [6.53] have shown that I_{sc} decreases with temperature near 100 °C, simultaneously with the decrease of the intensity of X-ray diffraction of chalcocite γ, i.e., when chalcocite γ becomes chalcocite β. Chalcocite β shows high ionic conduction. Also, when the cell is kept under illumination and bias at these temperatures, a change in its characteristics is observed.

Electrochemical Decomposition. Electrochemical decomposition [6.34, 171, 172] of Cu_2S at voltages of 0.35–0.4 V explains some damage to the cell while functioning in open circuit. In fact, a failure at the junction lowers the CdS potential to within that of the Cu_2S; at this point, with voltages around V_{oc}, the Cu_2S decomposes, and copper filaments are formed. These filaments first decrease the shunt resistance, then the short-circuit resistance of the cell. This can be avoided by adjusting the stoichiometry of the copper sulfide to 1.995, and by decreasing the number of failures at the junction level. Nowadays, cells do not show this kind of damage.

6.5 Conduction Mechanisms in Cu_2S–CdS Cells

Since the discovery of the photovoltaic effect in Cu_2S–CdS cells, many models have been proposed to explain the conduction mechanisms in light and in

Table 6.10. Models and mechanisms[a]

1. Impurity band conduction
   Reynolds et al. [6.173]
   Woods and *Champion* [6.175]
2. Photoemission across Cu–CdS barrier
   Williams and *Bube* [6.38]
   Paritskii et al. [6.193]
   Fabricius [6.194]
   Mead and *Spitzer* [6.195]
3. CdS p–n junction
   Grimmeiss and *Memming* [6.196]
   Backemuehl et al. [6.176] Two junctions produced by photo-cond. layer in CdS

   Duc Cuong and *Blair* [6.197] Unspecified p–n junction
4. Cu$_2$S–CdS heterojunction
   Cusano [6.198]
   Keating [6.199]
   Hill et al. [6.200] Mobile ions
   Balkanski and *Choné* [6.186] Absorption at interfaces states
   Pavelets and *Fedorus* [6.191]
   Spakowski et al. [6.183, 187] Barrier height 0.85 eV
   Gill et al. [6.178, 180] Tunneling through a spike
   Lindmayer and *Revesz* [6.155] Interfaces states and electron traps – Barrier height 0.95 eV
5. Cu$_2$S–CdS heterojunction with intrinsic layer of CdS
   Shiozawa et al. [6.201] Barrier height 0.85 eV in light
                                                           1.2 eV in dark

   Shitaya and *Sato* [6.177]
   Mytton [6.188] Barrier height 0.8 eV
   Nakayama [6.31]
   Böer et al. [6.202]

[a] See [6.5].

darkness. These models have been summarized by *Van Aerschodt* et al. [6.204] and, more recently, by *Stanley* [6.5] (Table 6.10).

Except in the first models, there is agreement that the absorption and thus the electron photocurrent is situated in Cu$_2$S. *Rothwarf* [6.146, 147] and *Boer* [6.145] have calculated the maximum theoretical value (of 36 mA cm^{-1} at AM 1) for the short-circuit current I_{sc} by the use of values found in the literature (cf. Sect. 6.4) for absorption, the diffusion length of electrons, and the refractive index, and by taking for the mobility of electrons, the same mobility as for holes. They also supposed a uniform concentration of the Cu$_2$S layer, though the Cu$_2$S layer obtained by dipping may, in fact, be transversally inhomogeneous.

During ion exchange, we can expect to find a higher Cu concentration near the surface rather than beyond it. Also, the Cu$_2$S obtained must contain a high percentage of cadmium, which acts as a donor impurity and compensates or

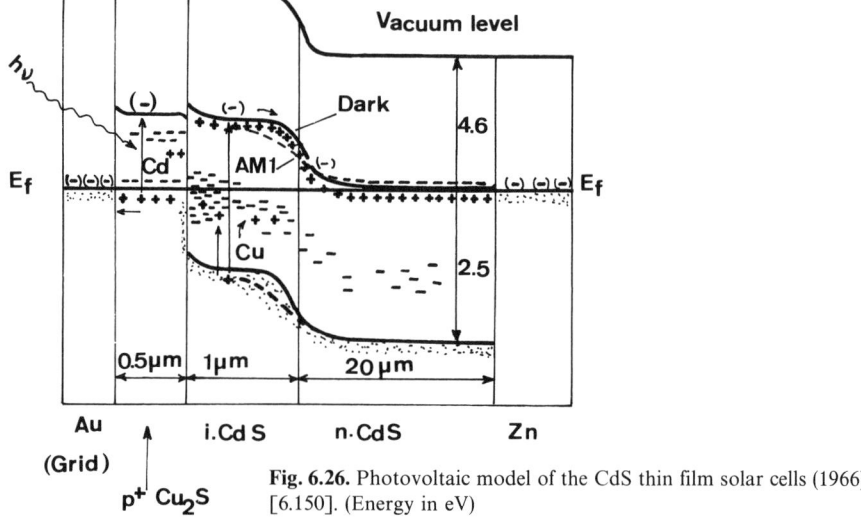

Fig. 6.26. Photovoltaic model of the CdS thin film solar cells (1966) [6.150]. (Energy in eV)

decreases the Cu vacancies by establishing a bond with the S in Cu_2S. In both cases there will be a weaker hole concentration in the neighborhood of the interface than that usually found in Cu_2S crystals. After the different annealing the cell undergoes and because of the high mobility of copper, it is very difficult to foresee the variation of stoichiometry perpendicular to the surface.

Massicot [6.149] has noted that by a thermal or photochemical process, a djurleite layer ($E_g = 1.8$ eV) may appear at the interface. This may explain certain high values of V_{oc} (~1 V) found on some cells [6.23].

A high enough concentration gradient in Cu_2S may modify the apparent diffusion length of electrons by the appearance of a non-negligible electric field.

In the model proposed by *Shiozawa* et al. [6.150] (Fig. 6.26), the diffusion of copper gives place to a 1 µm layer of very resistive and photoconductive CdS. This layer becomes very conductive under light and thus explains the intersection of the *IV* characteristics in the dark and under illumination. The electric junction is situated in the CdS. This model was slightly modified, as shown in Fig. 6.27. For cells obtained from ceramic CdS, *Nakayama* [6.31] explains the existence of a constant capacity with bias by the appearance of a Mott barrier due to copper diffusion.

According to *Bube* et al. [6.38, 80, 156, 157], and *Hewig* et al. [6.152], the conduction band in CdS at the interface is above that of Cu_2S (spike) (Fig. 6.28). Illumination of the heterojunction, and according to the wavelength, empties the interface traps situated in the middle of the CdS gap, decreasing the spike width.

Te Velde [6.151] explains the evolution of *IV* characteristics of monocrystals under illumination by the appearance of the spike during air annealing

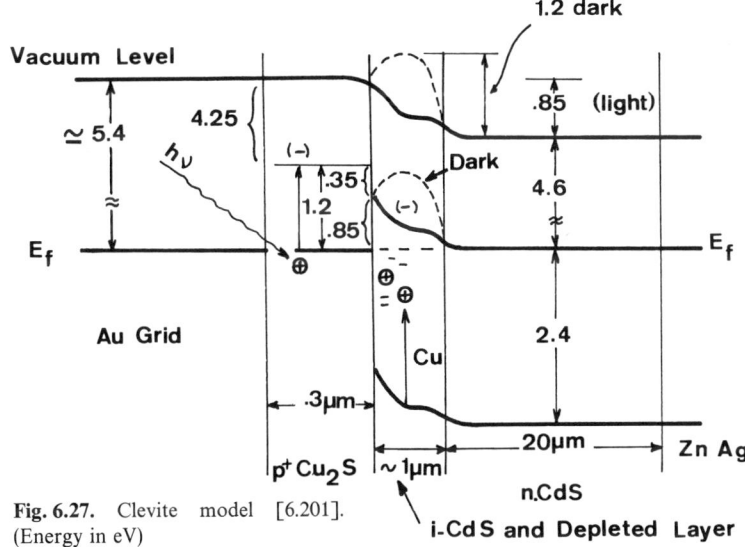

Fig. 6.27. Clevite model [6.201]. (Energy in eV)

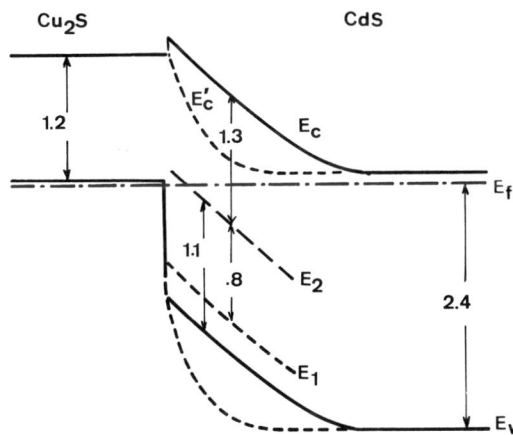

Fig. 6.28. Stanford model (solid lines). Illumination decreases the spike width (dotted lines) [6.178]. (Energy in eV)

(Fig. 6.29), the oxygen acting on CdS band, bending it at the interface. He concludes that the highest efficiency is obtained by a direct coupling of the conduction band.

The Delaware group [6.143] and *Martinuzzi* et al. [6.164] propose an energy band diagram without spike (Fig. 6.30) where ΔE_c varies from 0.14 to 0.35 eV according to the author. The ΔE_c at the junction increases with the duration of heat treatment, i.e., with the quantity of copper diffused in the CdS.

Deb et al. [6.167] have proposed the existence of an interface layer with interface states (caused by O_2, according to the assumption of *Te Velde* [6.151]),

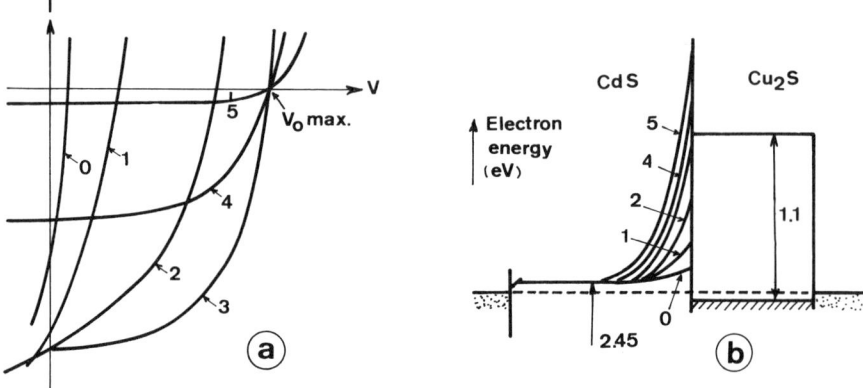

Fig. 6.29a and b. Modification of the current-voltage characteristic (**a**) and of the band picture (**b**) as a function of the duration of the heat treatment in air [6.151]. (Energy in eV)

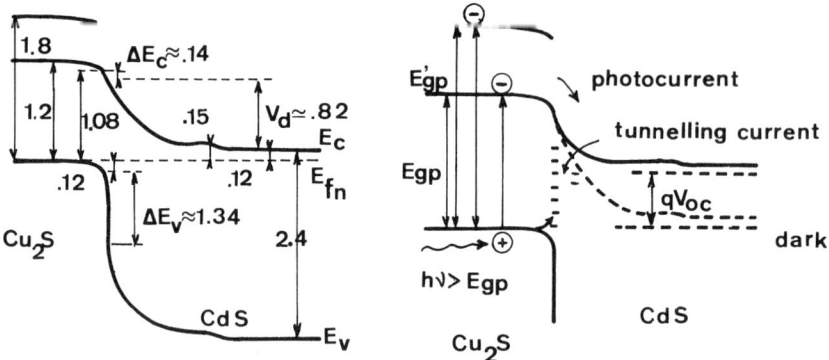

Fig. 6.30. Energy band profile and model for the photovoltaic mechanism proposed by *Martinuzzi* et al. [6.164]. (Energy in eV)

to explain the improvement of the performance with the different heat treatments. The model is similar to that already proposed to explain certain results obtained on Schottky diodes [6.166, 168, 169].

In thin film cells the dark current is too high to be a simple diffusion or/and emission current. *Bube* et al. [6.80, 157] have proposed conduction across the spike by tunneling, this current being increased under illumination by the reduction of the spike width. *Hewig* et al. [6.152] calculate the transmission coefficient in the spike. With an acceptor concentration of $\sim 10^{20}$ cm^{-3} in the Cu_2S, and donor concentration of $\sim 10^{19}$ cm^{-3} in CdS, which is near enough to that obtained experimentally, they find a width of the space-charge region compatible with tunnel conduction.

Martinuzzi et al. [6.164, 165] have used the model of the multistep tunnel conduction via interfaces states proposed by *Riben* and *Feucht* [6.170] to explain experimental results obtained on these cells. However, this model does not use a strict calculation of the probability of jumping from one state to the other. The current thus obtained may be greater than that obtained by rigorous calculation.

The Delaware group [6.143] keeps the hypothesis of a tunnel current at low forward bias; above a certain value, the current is controlled by the recombination velocity at the interface and by the electric field in the same region. This recombination velocity is coupled to the number of existing defects at the interface, i.e., with the mismatch between the Cu_2S and CdS lattices.

Boer [6.144, 145], in analogy with his observations on photoconductive CdS single crystals, suggests the existence of a "high field domain" controlling the current where the copper has diffused. Thus he explains the appearance of a saturation current after heat treatment, while, before annealing, the cells rectify only to a small extent.

6.6 Conclusion

The present state of development in the field of Cu_2S/CdS solar cells is characterized by the following two approaches.

In the first approach, which presently obtains the best results, the CdS layer is formed by evaporation and the Cu_2S layer by dipping, with the final cell being formed thermally. Recent attempts of the group at the Institute of Energy Conversion at Delaware, Newark, to optimize the performance of this type of cell have resulted in efficiency values of 9.15%, a very interesting result indeed for terrestrial applications [6.205–208].

In the second approach, the CdS layer is sprayed on, and the Cu_2S layer and the cell formation are obtained as in the first method. This approach has recently been emphasized on account of its economic attractiveness.

The spray methods used by SAT [6.28], by Photon Power [6.209] and the group GEMCES – CEES of the USTL, Montpellier [6.25, 210] have recently raised the efficiency of previous cells of this type considerably, obtaining values between 4 and 5%. It is presently not clear whether these studies will ultimately lead to cell efficiencies of the order of 8% which are required to compete with silicon cells in the manufacturing stage. The price of packaging these cells may play a role.

Parameters presently not known in both types of cells include the donor concentration in CdS, the acceptor concentration in Cu_2S, as well as the surface and junction geometry. Knowledge of these parameters is not only required for the improvement of existing cells, but also for the interpretation of the conduction mechanism in the heterojunction.

Conduction mechanism as well as the photovoltaic properties of Cu_2S/CdS cells in general have been shown to depend on the structure of the interface and the presence of traps. The influence of the interfacial states depends on the level of illumination and the direction and size of the bias. Some authors assume that for small forward bias, the conduction mechanism corresponds to a tunneling transition between states on either side of the interfacial zone. Under high illumination and reverse bias, however, the conduction proceeds by tunneling from interfacial into band states, or by tunneling between band states.

Attempts to develop a model for the Cu_2S/CdS heterojunction are limited by a lack of definitions of the parameters available in the literature. Some are only insufficiently known, such as the electron affinity, the mobility, the absorption coefficient, and the lifetime and diffusion length of the minority carriers. In addition, all these parameters may also depend on the method of fabrication.

A novel trend was observed during the recent Thirteenth IEEE Photovoltaics Specialists Conference, Washington, DC (June 1978), by placing greater emphasis on the properties of Cu_2S to the extent that the resulting cells are called Cu_2S cells only. Recent theoretical work [6.211] predicts that for the ideal case of vanishing surface and interface recombination and without absorption in the CdS, backwall structures should obtain 18% efficiency. This result is in agreement with work by *Rothwarf* et al. [6.203]. Thin films of Cu_2S have been fabricated by vacuum evaporation [6.212, 213], by sputtering [6.214] or even by spraying [6.39], with oxygen strictly excluded in order to prevent the formation of copper oxides. At present, different methods are being developed for the characterization of thin Cu_2S films by relating the electrical and optical properties to changes in the stoichiometry and the fabrication parameters.

If successful, these studies will remove the CdS as the optical window from the Cu_2S solar cells, and the dipping step will no longer be necessary to obtain Cu_2S. Transparent and conductive materials such as SnO_2, ITO or others will take the place of the optical window, the ultimate choice being dictated by an optimum barrier height which does not affect the photoelectric current.

References

6.1 A.E. Carlson: "Research on Semiconductor Films" WADC Tech. Rep. 56–62 Clevite Corp. (1956)
6.2 A.E. Carlson, L.R. Shiozawa, J.D. Finegan: US Patent **2**, 820–841 (1958)
6.3 J.F. Jordan: *Proc. Int. Conf. on Photovoltaic Power Generation*, ed. by H.R. Losch (Deutsche Gesellschaft für Luft- und Raumfahrt e.V., Köln 1974) pp. 221–228
6.4 J. Besson, T. Nguyen Duy, A. Gauthier, C. Martin: *Photovoltaic Spec. Conf. Rec. 11th* (IEEE, New York 1975) pp. 468–476
6.5 A.G. Stanley: "Cadmium Sulfide Solar Cells", in *Applied Solid State Science*, Vol. 5, ed. by R. Wolfe (Academic Press, New York 1975) pp. 251–366
6.6 R.E. Aitchison: Nature (London) **167**, 812 (1951)
6.7 M.G. Miksic, E.S. Schlig, R.R. Haering: Solid-State Electron. **7**, 39 (1964)
6.8 N.F. Foster: Proc. IEEE **53**, 1400 (1965)
6.9 J.C. Schaefer, E.R. Hill, T.A. Griffin: Final Rep. Contract NAS 3-7631, Harshaw Chem. Co (1966)
6.10 J.C. Schaefer, J. Evans, T.A. Griffin: Final Rep. Contract NAS 3-8515, Harshaw Chem. Co (1967)
6.11 R.J. Miller, C.H. Bachman: J. Appl. Phys. **29**, 1277 (1958)

6.12 C. Pastel: J. Phys. (Paris) **26**, 127 (1965)
6.13 G.O. Müller, H. Peibst: Phys. Status Solidi **8**, K51 (1965)
6.14 F.A. Shirland, F. Augustine, W.K. Bower: 2nd Quart. Rep. Contract NAS 3-6461, NASA-CR-54413, Clevite Corp. (1965)
6.15 J.P. David, S. Martinuzzi, F. Cabane-Brouty, J.P. Sorbier, J.M. Mathieu, J.M. Roman: *Proc. Int. Colloq. Solar Cells 1970* (CNES, Toulouse 1971) p. 81
6.16 J. Dresner, F.V. Shallcross: J. Appl. Phys. **34**, 2390 (1963)
6.17 M. Balkanski, M. Renata Chaves: J. Phys. (Paris) **34**, 173 (1966)
6.18 V.M. Yefremenkova, I.V. Egorova, V.E. Yurasova: Izv. Akad. Nauk SSSR, Ser. Fiz **32**, 1242 (1968)
6.19 D.B. Fraser, H. Melchior: J. Appl. Phys. **43**, 3120 (1972)
6.20 K.R. Albrand, E.W. Justi, W. Möhle, G.H.A. Schneider, D. Ullrich: "Coopération méditerranéenne Energie Solaire", Bull. 13–5 (1967)
6.21 J.F. Jordan: *Proc. Int. Conf. Solar Electricity* (CNES, Toulouse 1976) pp. 57–81
6.22 R.R. Chamberlin, J.S. Skarman: J. Electrochem. Soc. **113**, 86 (1966)
6.23 R.R. Chamberlin, J.S. Skarman: Solid-State Electron. **9**, 819 (1966)
6.24 W. Palz, J. Besson, T. Nguyen, J. Vedel: *Photovoltaic Spec. Conf. Rec. 9th 1972* (IEEE, New York 1972) p. 91
6.25 J. Bougnot, M. Perotin, J. Marucchi, M. Sirkis, M. Savelli: *Photovoltaic Spec. Conf. Rec. 12th 1976* (IEEE, New York 1976) p. 519
6.26 J. Bougnot, M. Perotin, J. Marucchi, M. Sirkis, M. Savelli: *Proc. Int. Conf. Solar Energy* (Barcelona 1977) [in Spanish]
6.27 M. Savelli: Rapport Final Contract C.E.E. Bruxelles, Belgium, No. 159-76-7-ES F, 31 déc. (1977)
6.28 S. Martinuzzi, F. Cabane-Brouty, J. Oualid, J. Gervais, A. Mostavan, J.L. Granier: *Proc. of Int. Conf. on Solar Energy*, ed. by A. Strub (Reidel, Dordrecht, Holland 1977) p. 581
6.29 F.B. Micheletti, P. Mark: Appl. Phys. Lett. **10**, 136 (1967)
6.30 Chen Ho Wu, R.H. Bube: J. Appl. Phys. **45**, 648 (1974)
6.31 N. Nakayama: Jpn. J. Appl. Phys. **8**, 450 (1969)
6.32 J. Philips: *Proc. Int. Workshop on CdS Solar Cells and Other Abrupt Junctions*, ed. by K.W. Böer, J.D. Meakin (University of Delaware, Newark, Delaware 1975) p. 475
6.33 H.W. Schock, G. Bilger, G.H. Hewig, F. Pfisterer, W.H. Bloss: *Proc. Int. Conf. Solar Electricity* (CNES, Toulouse 1976) p. 285
6.34 L. Clark, R. Gale, K. Moore, R.S. Mytton, R.S. Pinder: *Proc. Int. Colloq. Solar Cells 1970* (CNES, Toulouse 1971) p. 241
6.35 R.A. Mickelsen, D.D. Abbott: Final Rep. Contract NAS 3-13232, NASA-CR-120812 (Boeing Co, Seattle, Washington 1971)
6.36 J. Dieleman: *Proc. Int. Workshop on CdS Solar Cells and Other Abrupt Heterojunctions*, ed. by K.W. Böer, J.D. Meakin (University of Delaware, Newark, Delaware 1975) p. 92
6.37 P.A. Crossley, G.T. Noel, M. Wolf: Final Rep. Contract NASW 1427. RCA Astro Electron Div. Hightstour New Jersey (1968)
6.38 R. Williams, R.H. Bube: J. Appl. Phys. **31**, 968 (1960)
6.39 J. Vedel, M. Soubeyrand, P. Cowache, G. Leduc: *Proc. of the Int. Photovoltaic Solar Energy Conf.*, ed. by A. Strub (Reidel, Dordrecht, Holland 1977) p. 601
6.40 R.H. Bube, W. Gill, P. Lindquist: Prog. Rep. No. 1, Grant NGR-05-020-214, Stanford University (1967)
6.41 K. Bogus, S. Matts: *Photovoltaic Spec. Conf. Rec. 9th 1972* (IEEE, New York 1972) p. 106
6.42 F.A. Shirland: Adv. Energy Convers. **6**, 201 (1966)
6.43 F.A. Shirland, J.R. Hietanem: *Proc. 19th Ann. Power Sources Conf. 1965* (IEEE, New York 1965) p. 177
6.44 F.A. Shirland, J.R. Hietanem: *Photovoltaic Spec. Conf. Rec. 5th 1965* (IEEE, New York 1966) Sect. II C3
6.45 F.A. Shirland, F. Augustine: *Photovoltaic Spec. Conf. Rec. 5th 1965* (IEEE, New York 1966) Sect. II C4
6.46 J.R. Hietanem, F.A. Shirland: *Photovoltaic Spec. Conf. Rec. 6th 1967* (IEEE, New York 1966) Vol. 1, p. 179

6.47 J.C. Schaefer, R.L. Slater: *Photovoltaic Spec. Conf. Rec. 4th 1964* (IEEE, New York 1966) Vol. 2, p. A6-1
6.48 J. Bernard: Techn. Note NT 02.2, Centre National d'Etudes Spatiales, Toulouse (1967)
6.49 W. Palz, G. Cohen-Solal, J. Vedel, J. Fremy, T.N. Duy, J. Valerio: *Photovoltaic Spec. Conf. Rec. 7th 1968* (IEEE, New York 1968) p. 54
6.50 W. Palz, J. Besson, T.N. Duy, J. Vedel: *Photovoltaic Spec. Conf. Rec. 10th 1973* (IEEE, New York 1974) p. 69
6.51 W. Palz, J. Besson, J. Fremy, T.N. Duy, J. Vedel: *Photovoltaic Spec. Conf. Rec. 8th 1970* (IEEE, New York 1970) p. 16
6.52 G. Coste, J. Fremy, T.N. Duy: *Proc. Int. Colloq. Solar Cells 1970* (CNES, Toulouse 1971) p. 187
6.53 W. Palz, J. Besson, T.N. Duy, J. Vedel: *Photovoltaic Spec. Conf. Rec. 9th 1972* (IEEE, New York 1973) p. 91
6.54 M. Daspet, J. Besson, M. Lacroix: Coopération Méditerranéenne Energie Solaire, Bull. **14**, 117 (1968)
6.55 K. Bogus, H. Fisher, S. Mattes, N. Peters: *Proc. Int. Coll. Solar Cells 1970* (CNES 1971) p. 121
6.56 K.W. Boer, C.E. Birchenall, I. Greenfield, H.C. Hadley, T.L. Lu, L. Partain, J.E. Phillips, J. Schultz, W.F. Tseng: *Photovoltaic Spec. Conf. Rec. 10th 1973* (IEEE, New York 1974) p. 77
6.57 A.E. Middleton, D.A. Gorski, F.A. Shirland: Prog. Astronaut. Rocketry **3**, 275 (1961)
6.58 L.R. Shiozawa, F. Augustine, W.R. Cook: 1st Quart. Prog. Rep. Contract F33615, 68-6-1732, Clevite Corp. (1969)
6.59 E. Konstantinova, S. Kanev: J. Appl. Phys. **42**, 5861 (1971)
6.60 B. Ray: *II–VI Compounds*, Monographs in the Science of the Solid State, Vol. 2 (Pergamon, Oxford 1969)
6.61 M. Aven, J.S. Prener (eds.): *Physics and Chemistry of II–VI Compounds* (North-Holland, Amsterdam 1967)
6.62 V. Rossler, M. Lietz: Phys. Status Solidi **17**, 597 (1966)
6.63 M. Cardona, G. Harbeke: Phys. Rev. **137**, A1467 (1965)
6.64 M. Balkanski: J. Phys. (Paris) **28** C3, 36 (1967)
6.65 M. Balkanski, Y. Petroff: *Proc. 7th Int. Conf. Physics Semiconductors* (Dunod, Paris 1964) p. 244
6.66 M. Balkanski, J. des Cloiseaux: J. Phys. Radium **21**, 825 (1960)
6.67 T.K. Bergstresser-Cohen: Phys. Rev. **164**, 1069 (1967)
6.68 S.S. Delvin: "Transport properties", in *Physics and Chemistry of II–VI Compounds*, ed. by M. Aven, J.S. Prener (North-Holland, Amsterdam 1967) Chap. II, pp. 551–609
6.69 M. Fujite et al.: J. Phys. Soc. Jpn. **20**, 109 (1965)
6.70 Ko Bayashi: *Int. Conf. II–VI Semiconducting Compounds* (Benjamin, New York 1967) pp. 755–785
6.71 M. Onuki, K. Shiga: J. Phys. Soc. Jpn. **21**, Suppl. 427/30 (1966)
6.72 R.H. Bube: *Photoconductivity of Solids* (Wiley and Sons, New York 1960)
6.73 J. Zmija, M. Demianiuk: Acta Phys. Pol. A**39**, 539 (1971)
6.74 W. Szeto, G.A. Somorjai: J. Chem. Phys. **44**, 3490 (1966)
6.75 G.A. Sullivan: Phys. Rev. **184**, 736 (1969)
6.76 R.L. Clarke: J. Appl. Phys. **30**, 957 (1959)
6.77 H.H. Woodbury: J. Appl. Phys. **36**, 2287 (1965)
6.78 R.K. Purohit, B.L. Sharma, A.K. Sleedhar: J. Appl. Phys. **40**, 4677 (1969)
6.79 J. Mott, W.E. Spear: Phys. Rev. Lett. **8**, 314 (1962)
6.80 W.D. Gill, R.H. Bube: Appl. Phys. **41**, 1694 (1970)
W.D. Gill, R.H. Bube: J. Appl. Phys. **41**, 3731 (1970)
6.81 F.V. Shalcross: Trans. AIME **236**, 309 (1966)
6.82 J.I.B. Wilson, J. Woods: J. Phys. Chem. Sol. **34**, 171 (1973)
6.83 Y. Terasaki, T. Murakami, H. Toyoda: Rev. Elect. Comm. Lab. **14**, 425 (1966)
6.84 N.F. Foster: J. Appl. Phys. **38**, 149 (1967)
6.85 I. Lagnado, M. Lichtensteiger: J. Vac. Sci. Technol. **7**, 318 (1970)
6.86 R.L. Petritz: Phys. Rev. **104**, 1508 (1956)

6.87   L.L. Kazmerski, W.B. Berry, C.W. Allen: J. Appl. Phys. **43**, 3515 (1972)
6.88   L. Partain, J.J. Oakes, I.G. Greenfield: *Proc. Int. Workshop on CdS Solar Cells and Other Abrupt Heterojunctions*, ed. by K.W. Böer, J.D. Meakin (University of Delaware, Newark, Delaware 1975) p. 346
       J.J. Oakes, I.G. Greenfield, L.D. Partain: *Photovoltaic Spec. Conf. Res. 11th 1975* (IEEE, New York 1975) p. 454
6.89   H. Rau: J. Phys. Chem. Sol. **28**, 903 (1967)
6.90   E. Jensen: Avh. Nor. Vidensk-Akad. Oslo Mat.-Naturvidensk. Kl. **6**, 3 (1947)
6.91   M. Hansen, K. Anderko: *Constitution of Binary Alloys* (McGraw-Hill, New York 1958) p. 620
6.92   G. Kullerud: Ann. Report of Director of Geophys. Lab., Carnegie Inst. Washington Yearb. **56**, 195 (1957); **57**, 215 (1958); **59**, 110 (1960)
6.93   G.H. Moh: Ann. Report of Director of Geophys. Lab., Carnegie Inst. Washington Yearb. **63**, 208 (1964); **62**, 214 (1963)
6.94   E.H. Roseboom: J. Econ. Geol. **61**, 641 (1966)
6.95   W.R. Cook, Jr.: Thesis, Case Western University USA (1971)
6.96   N. Morimoto, K. Koto: Am. Mineral. **55**, 106 (1970)
6.97   V. Wehefritz: Z. Phys. Chem. (Frankfurt am Main) **26**, 339 (1960)
6.98   R.W. Potter: J. Econ. Geol. **72**, 1524 (1977)
6.99   H.S. Mathieu: Ph.D. Dissertation, Dortmund, Germany (1971)
6.100 E.N. Eliseev, L.E. Rudenko, L.A. Sinev, B.K. Koshurnilov, N.I. Solovov: Mineralog. Sb. (L'vov.) Gos. Univ. **18**, 383 (1964)
6.101 N. Morimoto, G. Kullerud: Am. Mineral. **48**, 110 (1963)
6.102 S. Djurle: Acta Chem. Scand. **12**, 1415 (1958)
6.103 H. Takeda, J.D.H. Donnay, E.H. Roseboom, D.E. Appelman: Z. Krist. **125**, 404 (1967)
6.104 A. Janosi: Acta Crystallogr. **17**, 311 (1964)
       A. Janosi: Ph.D. Dissertation, Louvain, Belgium (1959)
6.105 N. Morimoto, K. Koto, Y. Shimazaki: Am. Mineral. **54**, 1256 (1969)
6.106 A. Etienne: J. Electrochem. Soc. **117**, 870 (1970)
6.107 E. Hirahara: J. Phys. Soc. Jpn. **6**, 422, 428 (1951)
6.108 N.A. Gassanova: Izv. Akad. Nauk Az. SSR, Ser. Fiz. Tekh. Mat. Nauk **3**, 91 (1963)
6.109 G.B. Abdullaev, Z.A. Aliyarova, E.A. Zamanova, G.A. Asadov: Phys. Status Solidi **26**, 65 (1963)
6.110 F. Guastavino: Ph.D. Dissertation, Montpellier, France (1974)
6.111 F. Guastavino, H. Luquet, J. Bougnot, M. Savelli: J. Phys. Chem. Sol. **36**, 621 (1975)
6.112 I. Yokota: J. Phys. Soc. Jpn. **8**, 595 (1953)
6.113 S. Miyatani: J. Phys. Soc. Jpn. **11**, 1059 (1956)
6.114 T. Ishikawa, S. Miyatani: J. Phys. Soc. Jpn. **42**, 159 (1977)
6.115 K. Okamoto, S. Kawai: Jpn. J. Appl. Phys. **12**, 1132 (1973)
6.116 L. Eisenmann: Ann. Phys. (Leipzig) **10**, 129 (1952)
6.117 G.P. Sorokin, I.Ya. Andronik, E.V. Kovtun: Izv. Akad. Nauk SSSR, Neorg. Mater. **11**, 2129 (1975)
6.118 O.P. Astakhov: Izv. Akad. Nauk SSSR, Neorg. Mater. **11**, 1506 (1975)
       O.P. Astakhov, V.V. Lobankov, I.V. Sgibnev, B.M. Surkov: Teplofiz. Vys. Temp. **10**, 654 (1972)
       O.P. Astakhov, V.V. Lobankov: Teplofiz. Vys. Temp. **10**, 905 (1972)
6.119 A.E. Potter, R.L. Schalla: *Photovoltaic Spec. Conf. Rec. 6th* (IEEE, New York 1967) p. 24
6.120 A. Dumon: C.R. Acad. Sci., Sér. B**269**, 835 (1969)
6.121 B. Selle, J. Maege: Phys. Status Solidi **30**, K153 (1968)
6.122 N. Nakayama: J. Phys. Soc. Jpn. **25**, 290 (1968)
6.123 V.K. Miloslawskii, B.I. Perekrestov: Ukr. Fiz. Zh. T**14**, 7, 1160 (1969)
6.124 J.J. Loferski, J. Schewchun, E.A. de Meo, R. Arnote, E.E. Crisman, R. Beaulieu, H.L. Hwang, C.C. Wu: *Photovoltaic Spec. Conf. Rec. 12th* (IEEE, New York 1976) p. 496
6.125 H. Nimura, A. Atoda, T. Nakau: Jpn. J. Appl. Phys. **16**, 403 (1977)
6.126 F. Guastavino, S. Duchemin, J. Bougnot, M. Savelli: *Photovoltaic Spec. Conf. Rec. 12th* (IEEE, New York 1976) p. 508

6.127 B.J. Mulder: Phys. Status Solidi a **18**, 633 (1973)
6.128 J. Schewchun, J.J. Loferski, A. Wold, R. Arnote, E.A. Demeo, R. Beaulieu, C.C. Wu, H.L. Hwang: *Photovoltaic Spec. Conf. Rec. 11th* (IEEE, New York 1975) p. 482
6.129 F. Pfisterer, G.H. Hewig, W.H. Bloss: *Photovoltaic Spec. Conf. Rec. 11th* (IEEE, New York 1975) p. 461
6.130 S. Duchemin, F. Guastavino, C. Raisin: Solid State Commun. **26**, 187 (1978)
6.131 H. Rau: J. Phys. Chem. Sol. **28**, 903 (1967)
6.132 K. Weiss: Ber. Bunsenges. Phys. Chem. **73**, 338 (1969)
6.133 I.G. Kerimov, A.M. Musaev, F.Yu Aliev, A.G. Rustamov, E.I. Manafli: Izv. Akad. Nauk SSSR, Ser. Fiz. Tech. Mat. 6 (1970)
6.134 J. Vedel, E. Castel: *Int. Conf. Photovoltaic Power and Its Applications in Space and on Earth* (CNES, Paris 1973) pp. 199–207
6.135 B.J. Mulder: Phys. Status Solidi a **13**, 79 (1972)
6.136 F. Guastavino, H. Luquet, J. Bougnot: C.R. Acad. Sci., Sér. B**269**, 831 (1969)
6.137 V.P. Kryzhanvoskii: Opt. Spectrosc. **24**, 135 (1968)
6.138 B.J. Mulder: Phys. Status Solidi a **13**, 569 (1972)
6.139 J.J. Oakes, I.G. Greenfied, L.D. Partain: J. Appl. Phys. **48**, 2548 (1977)
6.140 R. Marshall, S.S. Mitra: J. Appl. Phys. **36**, 3882 (1965)
6.141 G.P. Sorokin, Yu.M. Papshiev, P.T. Oush: Sov. Phys. Solid State **7**, 1810 (1966)
6.142 M. Ramoin, J.P. Sorbier, J.F. Bretzner, S. Martinuzzi: C.R. Acad. Sci., Sér. B **268**, 1097 (1969)
S. Martinuzzi: Phys. Status Solidi **2**, K9 (1970)
6.143 K.W. Böer, J.D. Meakin: Final Report Contract No. NSF/RANN/AER 72-03478 AO3/FR/75 (Institute of Energy Conversion, University of Delaware, Newark, Delaware 1971)
6.144 K.W. Böer: Phys. Rev. B**13**, 5373 (1976)
6.145 K.W. Böer: Phys. Status Solidi a **40**, 355 (1977)
6.146 A. Rothwarf: *Proc. Int. Conf. on Solar Electricity* (CNES, Toulouse 1976) p. 273
6.147 A. Rothwarf: *Photovoltaic Spec. Conf. Rec. 12th* (IEEE, New York 1976) p. 544
6.148 L.C. Burton, B. Baron, W. Devaney, T.L. Hench, S. Lorenz, J.D. Meakin: *Photovoltaic Spec. Conf. Rec. 12th* (IEEE, New York 1976) p. 526
6.149 P. Massicot: Phys. Status Solidi a **11**, 531 (1972)
6.150 L.R. Shiozawa, G.A. Sullivan, F. Augustine: Contract No. AF33 (615) 5224, Clevite Corporation, Cleveland, Ohio (1967)
6.151 T.S. Te Velde: Solid State Electron. **16**, 1305 (1973)
6.152 G.H. Hewig, F. Pfisterer, W.H. Bloss: *Proc. Int. Conf. on Photovoltaic Power Generation*, ed. by H.R. Lösch (Deutsche Gesellschaft für Luft- und Raumfahrt e.V., Köln 1974) p. 255
6.153 G.H. Hewig, W.H. Bloss: *Photovoltaic Spec. Conf. Rec. 12th* (IEEE, New York 1976) p. 483
6.154 M.K. Mukherjee, F. Pfisterer, G.H. Hewig, H.W. Schock, W.H. Bloss: J. Appl. Phys. **48**, 1538 (1977)
6.155 J. Lindmayer, A.G. Revesz: Solid State Electron. **14**, 647 (1971)
6.156 P.F. Lindquist, R.H. Bube: J. Appl. Phys. **43**, 2839 (1972)
6.157 P.F. Lindquist, R.H. Bube: J. Electrochem. Soc. **119**, 936 (1972)
6.158 A.L. Fahrenbruch, R.H. Bube: J. Appl. Phys. **45**, 1264 (1974)
6.159 A.L. Fahrenbruch, R.H. Bube: *Photovoltaic Spec. Conf. Rec. 10th 1973* (IEEE, New York 1974) p. 85
6.160 J. Bernard, T. Amand: *Int. Conf. on Solar Electricity* (CNES, Toulouse 1976) p. 309
6.161 J. Bernard, J.P. Vormus: *Proc. of the Int. Photovoltaic Solar Energy Conf.*, ed. by A. Strub (Reidel, Dordrecht, Holland 1977) p. 570
6.162 S. Martinuzzi, F. Cabane-Brouty, J.F. Bretzner: *Photovoltaic Spec. Conf. Rec. 9th* (IEEE, New York 1972) p. 111
6.163 S. Martinuzzi, O. Mallem: Phys. Status Solidi a **16**, 339 (1973)
6.164 S. Martinuzzi, O. Mallem, T. Cabot: Phys. Status Solidi a **36**, 227 (1976)
6.165 H. Luquet, L. Szepessy, J. Bougnot, M. Savelli, F. Guastavino: *Photovoltaic Spec. Conf. Rec. 11th* (IEEE, New York 1975) p. 445
H. Luquet: Thesis, Montpellier, France (1977)

6.166 B. Lepley, S. Ravelet, J.M. Hess, P. Renard: *Int. Conf. on Solar Electricity* (CNES, Toulouse 1976) p. 543
6.167 S. Deb, H. Saha: *Photovoltaic Solar Energy Conf.* (Reidel, Dordrecht, Holland 1977) p. 570
6.168 S.J. Fonash: J. Appl. Phys. **47**, 3597 (1976)
S.J. Fonash: *Photovoltaic Spec. Conf. Rec. 11th* (IEEE, New York 1975) p. 376
6.169 R.J. Stirn, Yea-Chuan M. Yeh: *Photovoltaic Spec. Conf. Rec. 10th* (IEEE, New York 1973) p. 15
6.170 A.R. Riben, D.L. Feucht: Solid State Electron. **9**, 1055 (1966)
A.R. Riben, D.L. Feucht: Int. J. Electron. **20**, 583 (1966)
6.171 H.J. Mathieu, H. Rickert: Z. Phys. Chem. **79**, 315 (1972)
6.172 H.J. Mathieu, K.K. Reinhartz, H. Rickert: *Photovoltaic Spec. Conf. Rec. 10th* (IEEE, New York 1973) p. 93
6.173 D.C. Reynolds, G. Leies, L.L. Antes, R.E. Marburger: Phys. Rev. **96**, 533 (1954)
D.C. Reynolds: *Encyclopedia of Chemical Technology*, Suppl. (Wiley-Interscience, New York 1957) p. 667
R.E. Marburger, D.C. Reynolds, L.L. Antes, R.S. Hogan: J. Chem. Phys. **23**, 2448 (1955)
D.C. Reynolds, S.J. Czyzak: Phys. Rev. **96**, 1705 (1954)
6.174 D.A. Hammond, F.A. Shirland: *Proc. Electron. Components Conf.* (1959) p. 98
F.A. Shirland: ARL Tech. Rep. 60.293, Harshaw Chem. Co. (1960)
6.175 J. Woods, I.A. Champion: J. Electron. Control. **7**, 243 (1959)
6.176 R.R. Bockemuehl, J.E. Kauppila, D.S. Eddy: J. Appl. Phys. **32**, 1324 (1961)
6.177 T. Shitaya, H. Sato: Jpn. J. Appl. Phys. **7**, 1348 (1968)
6.178 W.D. Gill, P.F. Lindquist, R.H. Bube: *Photovoltaic Spec. Conf. Rec. 7th* (IEEE, New York 1968) p. 47
W.D. Gill, R.H. Bube: *Proc. Int. Conf. Photocond. 3rd 1969* (1971) p. 395
P.F. Lindquist, R.H. Bube: *Photovoltaic Spec. Conf. Rec. 8th* (IEEE, New York 1970) p. 1
6.179 N. Miya: Jpn. J. Appl. Phys. **9**, 768 (1970)
6.180 H.I. Moss: RCA Rev. **22**, 29 (1961)
6.181 F.A. Shirland, G.A. Wolff, J.C. Schaefer, G.H. Dierssen: ASD-TDR 62 69, Vol. II, Harshaw Chem. Co. (1962)
6.182 V.A. Drozdov, Sh.D. Kurmashev, A.L. Rvachev: Radiotekh. Elektron. **10**, 1358 (1965) [Radio Eng. Electron. Phys. (USSR) **10**, 1172 (1965)]
6.183 A.E. Spakowski, A.E. Potter, R.L. Schalla: *Photovoltaic Spec. Conf. Rec. 5th 1965*, PIC SOL 209/6.1, Sect. C-7 (IEEE, New York 1966)
A.E. Spakowski, A.E. Potter, R.L. Schalla: NASA Tech. Memo NASA TM X-52 144 (1965)
6.184 C. Pastel: Ph.D. Thesis, University of Paris (1962)
6.185 F.A. Shirland: Clevite Corp. Eng. Memo 66-17, 1966, presented at 2nd OAR Res. Appl. Conf. (1967)
6.186 M. Balkanski, B. Choné: Rev. Phys. Appl. **1**, 179 (1966)
6.187 A.E. Potter, R.L. Schalla: *Photovoltaic Spec. Conf. Rec. 6th 1967* (IEEE, New York 1967) p. 24
A.E. Potter, R.L. Schalla: NASA Tech. Note NASA TN D 3849 (1967)
A.E. Potter: NASA Tech. Note NASA TN D 4333 (1967)
A.E. Potter, R.L. Schalla, H.W. Brandhorst, L. Rosenblum: *Photovoltaic Spec. Conf. Rec. 7th 1968* (IEEE, New York 1968) p. 62
6.188 R.J. Mytton: Br. J. Appl. Phys. **1**, 721 (1968)
6.189 A.V. Anshon, I.A. Karpovich: Fiz. Tekh. Poluprovodn. **3**, 503 (1969); Sov. Phys. Semicond. **3**, 503 (1969)
6.190 J.S. Skarman, A.B. Budinger, R.R. Chamberlin: Summary Rep. Contract AF 33 (615) 1578 National Cash Register Co, Dayton Ohio (1965)
R.R. Chamberlin, J.S. Skarman: ASD TDR 63 223, Part II, National Cash Register Co, Dayton Ohio (1963)
R.R. Chamberlin, J.S. Skarman, D.E. Koopman, L.E. Blakely: ASD-TDR-63-323, Part I, National Cash Register Co, Dayton Ohio (1963)
6.191 S. Yu Pavelets, G.A. Fedorus: Ukr. Fiz. Zh. **11**, 686 (1966)
6.192 I.V. Egorova: Fiz. Tekh. Poluprovodn. **2**, 319 (1968); Sov. Phys. Semicond. **2**, 266 (1968)

6.193 L.G. Paritskii, A.G. Rogachev, S.M. Ryvkin: Fiz. Tverd. Tela (Leningrad) **3**, 1613 (1961); Sov. Phys.-Solid State **3**, 1170 (1961)
6.194 E.D. Fabricius: J. Appl. Phys. **33**, 1597 (1962)
6.195 C.A. Mead, W.G. Spitzer: Appl. Phys. Lett. **2**, 74 (1963)
W.G. Spitzer, C.A. Mead: J. Appl. Phys. **34**, 3061 (1963)
6.196 H.G. Grimmeiss, R. Memming: J. Appl. Phys. **33**, 2217 (1962)
6.197 N. Duc Cuong, J. Blair: J. Appl. Phys. **37**, 1660 (1966)
6.198 D.A. Cusano: Solid State Electron. **6**, 217 (1963)
6.199 P.N. Keating: J. Appl. Phys. **36**, 564 (1965)
6.200 J.C. Schaefer, E.R. Hill, T.A. Griffin: Final Rep. Contract NAS 3, 7631, Harshaw Chem. Co (1966)
E.R. Hill, B.G. Keramidas: *Photovoltaic Spec. Conf. Rec. 5th 1965*, PIC-SOL 20916.1, Sect. C6 NASA CR 70169 (1966)
E.R. Hill, B.G. Keramidas: Rev. Phys. Appl. **1**, 189 (1966); IEEE Trans. Electron Devices **14**, 22 (1967)
E.R. Hill, B.G. Keramidas, D.J. Krus: *Photovoltaic Spec. Conf. Rec. 6th 1967*, Vol. 1 (IEEE, New York 1967) p. 35
6.201 L.R. Shiozawa, G.A. Sullivan, F. Augustine, J.M. Jost: Interim Tech. Rep. Contract AF 33 (615) 5224, Clevite Corp. (1967)
L.R. Shiozawa, F. Augustine, G.A. Sullivan, J.M. Smith, W.R. Cook: Final Rep. Contract AF 33 (615) 5224, Clevite Corp. (1969)
L.R. Shiozawa, F. Augustine, G.A. Sullivan: 8th Quart Prog. Rep. Contract AF 33 (615) 5224, Clevite Corp. (1968)
L.R. Shiozawa, G.A. Sullivan, F. Augustine: *Photovoltaic Spec. Conf. Rec. 7th 1968* (IEEE, New York 1968) p. 39
L.R. Shiozawa, F. Augustine, W.R. Cook, Jr.: Final Rep. Contract F 33 (615) 69C, 1732, Clevite Corp. (1970)
6.202 K.W. Böer: NASA CR 129675, GRANT NGR-08-001-028, University of Delaware, Newark (1971)
K.W. Böer, J. Phillips: *Photovoltaic Spec. Conf. Rec. 9th 1972* (IEEE, New York 1973) p. 125
6.203 A. Rothwarf, A.M. Barnett: IEEE Trans. Electron Devices **24**, 381 (1977)
A.M. Barnett, J.D. Meakin, A. Rothwarf: *Photovoltaic Solar Energy Conf.*, ed. by A. Strub (Reidel, Dordrecht, Holland 1977) p. 535
6.204 A.E. Van Aerschodt, J.J. Capart, K.H. David, F. Fabbricotti, K.H. Heffels, J.J. Loferski, K.K. Reinhartz: *Photovoltaic Spec. Conf. Rec. 7th 1968* (IEEE, New York 1968) p. 22; IEEE Trans. Electron Devices **18**, 471 (1971)
6.205 S.P. Shea, L.D. Partain: *Photovoltaic Spec. Conf. Rec. 13th 1978* (IEEE, New York) p. 393
6.206 A. Rothwarf, J. Phillips, N. Convers Wyeth: *Photovoltaic Spec. Conf. Rec. 13th 1978* (IEEE, New York) p. 399
6.207 B. Baron, A.W. Catalano, E.A. Fagen: *Photovoltaic Spec. Conf. Rec. 13th 1978* (IEEE, New York) p. 406
6.208 J.A. Bragagnolo: *Photovoltaic Spec. Conf. Rec. 13th 1978* (IEEE, New York) p. 412
6.209 V.P. Singh: *Photovoltaic Spec. Conf. Rec. 13th 1978* (IEEE, New York) p. 507
6.210 J. Marucchi, M. Perotin, Oudeacoumar, J. Bougnot, M. Savelli: *Photovoltaic Spec. Conf. Rec. 13th 1978* (IEEE, New York) p. 298
6.211 M.O. Henry: Thesis, Montpellier, France (1978)
6.212 J. Shewchun, R.A. Clarke, J.J. Loferski, K. Rajkanan, D. Burk, T. Vanderwel, J.P. Marton, A. Kazandjian: *Photovoltaic Spec. Conf. Rec. 13th 1978* (IEEE, New York) p. 378
6.213 F. Guastavino, S. Duchemin, B. Rezig, B. Girault, M. Savelli: *Photovoltaic Spec. Conf. Rec. 13th 1978* (IEEE, New York) p. 303
6.214 G. Armantrout, J. Yee, E. Fischer-Colbrie, D. Miller: *Photovoltaic Spec. Conf. Rec. 13th 1978* (IEEE, New York) p. 383
6.215 W. Arndt, G. Bilger, W.H. Bloss, G.H. Hewig, F. Pfisterer, H.W. Schock: *Photovoltaic Solar Energy Conference*, ed. by A. Strub (Reidel, Dordrecht, Holland 1977) p. 547

7. Heterojunction Phenomena and Interfacial Defects in Photovoltaic Converters

A. L. Fahrenbruch and J. Aranovich

With 36 Figures

The overriding criterion for widespread terrestrial use of photovoltaics is the cost of solar cells in large quantities. If low cost is to be achieved by the use of thin films of the active materials, then we are led directly to the use of high absorption constant, direct band-gap materials. This follows because the small crystallite sizes ordinarily seen in thin films on amorphous substrates provide an upper bound to the minority carrier diffusion length and hence on the absorption length necessary to obtain reasonable quantum efficiencies. Photons must be absorbed within a diffusion length for useful collection of the photogenerated carriers. Since loss by surface recombination generally becomes dominant for carriers generated close to the surface, we are further led to utilizing the heterojunction window effect to admit light to the active region without suffering nonuseful attenuation.

Unlike the situation for homojunctions, the carrier transport properties of heterojunctions (HJs) are generally dominated by phenomena at or near the metallurgical interface. The current transport has been variously attributed to recombination in the depletion layers, tunneling, and a tunneling/recombination process involving interface states. In some treatments transport is lumped into an effective interface recombination velocity. In very few heterojunctions is the transport dominated by the properties of the quasi-neutral regions outside the junction field region. In addition the picture is complicated by the presence of conduction and valence band discontinuities arising from the difference in electron affinities of the two materials or from interface dipole layers. The distribution of interface states may be electrically charged which further distorts the junction fields.

Heterojunctions have previously been studied principally for luminescent devices, solid-state lasers, and somewhat less for transistor applications. Almost all of these are single crystal applications. The parameters useful for solar converters are usually quite different; high conductivities are required and the ultimate aim is extension to all thin film, polycrystalline devices of very large area. In this chapter we attempt to gain a physical perspective in the domain of solar cell parameters. Section 7.1 circumscribes that domain, focusing on how the HJ parameters affect the ultimate solar efficiency of the devices. A brief review and critique of the pertinent theories of HJ transport is given in Sect. 7.2. The interface plays a crucial role in the properties of HJs, and in Sect. 7.3, we focus on interface effects and the diagnostics currently in use to study these phenomena.

7.1 The Relation of Solar Conversion Efficiency to Heterojunction Parameters

In this section we establish the connections between the observable electrical parameters of a heterojunction (HJ) and the conversion efficiency of an HJ solar cell. These connections follow directly from the application of the carrier transport equation to a system in which the law of superposition holds. We do this not so much to derive these relationships but to examine the assumptions on which they are based, especially with regard to superposition. To avoid complexity that might distract from the physical principles involved, we concentrate on simple solutions. In the latter part of this section we discuss the collection function approach as a first-order departure from the superposition approach. Finally, we turn to questions about the diode parameters themselves and the effect of light on them. For the most part, the considerations in this section are equally applicable to homojunctions and heterojunctions since we focus on the quasi-neutral regions and the adjoining depletion layer on one side of the junction only.

7.1.1 The Ideal Solar Cell and Calculation of Light-Generated Current

The operation of the usual photovoltaic cell involves: I) the generation of electron-hole pairs by band-to-band photon absorption in the bulk, II) diffusion transport of the photogenerated minority carriers to the edge of the depletion region, III) separation by the junction field, and IV) collection of the current via ohmic contacts and grid structures. This problem has been discussed by many authors [7.1–4]. It is generally treated by linear superposition assuming thermal equilibrium. These are good approximations except for high intensity illumination and for certain HJs as we shall see. Several steps are involved in the calculation of the solar efficiency: I) calculation of the total light-generated current J_L as a function of the incident spectral intensity and the bulk semiconductor parameters [the absorption constant $\alpha(\lambda)$ and the minority carrier diffusion length L_d], and II) determination of the maximum power P_m using the current-voltage relationship of the diode and the J_L value. The latter step is usually broken up into: a) calculation of the open-circuit voltage V_{oc} from the diode parameters; b) determination of the fill factor $\text{ff} \equiv V_{oc} J_{sc}/P_m$, again from the diode parameters by maximization of the JV product with the device resistance included; and c) calculation of the solar efficiency $\eta_S = V_{oc} J_{sc} \text{ff}/S_0$, using the total solar input power S_0. As discussed later, these steps are not always independent.

a) Calculation of the Light-Generated Current

Without any real loss of generality, we restrict ourselves to the HJ shown in Fig. 7.1 where we assume that all the useful generation takes place in the quasi-neutral region ($x_p < x < x_{p2}$) of the lower band-gap, p-type absorber layer and

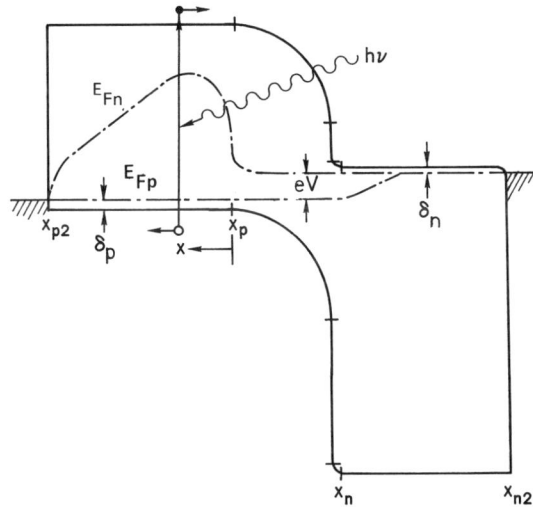

Fig. 7.1. Schematic band diagram of elementary heterojunction solar cell at forward bias V showing quasi-Fermi levels with $n_p = n_{p0} \exp(eV/kT)$ at $x = x_p$ assumed

that the large band-gap material may be thought of as a passive window layer with regard to photogeneration.

The transport equation which expresses the diffusion transport is obtained from the diffusion equations for electrons

$$J_n = e\mu_n \mathscr{E} + eD_n \nabla n$$

and for holes

$$J_p = e\mu_p \mathscr{E} - eD_p \nabla p$$

with $e = |e|$ the magnitude of the charge on the electron, and the charge-conservation relations

$$\partial \varrho / \partial t + \nabla \cdot \boldsymbol{J} = 0$$
$$\partial n / \partial t - \nabla \cdot \boldsymbol{J}_n / e = G_n - U_n$$
$$\partial p / \partial t + \nabla \cdot \boldsymbol{J}_p / e = G_p - U_p$$

and

$$\boldsymbol{J} = \boldsymbol{J}_n + \boldsymbol{J}_p.$$

G_n and G_p and U_n and U_p are the bulk generation and recombination rates for electrons and for holes. It can be shown that since $G_n = G_p = G$ then $U_n = U_p = U$ in steady state. These equations result in two steady-state transport equations such as, for electrons in one dimension,

$$D_n d^2 n/dx^2 + \mu_n n d\mathscr{E}/dx + \mu_n \mathscr{E} dn/dx - U(x) + G(x) = 0 \tag{7.1}$$

with a similar equation for holes. The implicit assumptions are:

I) Thermodynamic equilibrium holds within the bands and the carrier relaxation time $\tau_r \approx \mu m^*/e \approx 10^{-13}$ sec is much smaller than the effective minority carrier lifetime τ_e.

II) The recombination rate can be represented by a linear function such as $U(x)=(n_p-n_{p0})/\tau_e$ where n_p and n_{p0} are the illuminated and dark equilibrium carrier densities in the p-type layer.

III) The carrier mobility μ_n and lifetime τ_e are not functions of n, $\mathscr{E}$, or position.

In most cases the condition that $n_p \ll p_{p0}$ enables us to eliminate the hole equation to obtain

$$D_n d^2 n_p/dx^2 + n_p \mathscr{E} \mu_n dn_p/dx - (n_p - n_{p0})/\tau_e + G(x) = 0. \quad (7.2)$$

Thus the transport is controlled by the minority carrier gradient. We further assume that:

IV) The electric field $\mathscr{E}=0$ in the quasi-neutral region.

V) The thickness of the absorber $x_{p2} \gg 1/\alpha$ so that $n_p \to n_{p0}$ as $x \to \infty$ and that the generation rate is given by $G(x)=\alpha \Gamma_0 \exp[-\alpha(x-x_p)]$ where Γ_0 is the incident photon flux.

VI) The carrier density at the depletion layer edge is given by

$$n_p(x=x_p) = n_{p0} \exp(eV/kT). \quad (7.3)$$

The last assumption (to be examined later) is that the electron quasi-Fermi level is constant throughout the depletion layer. At this stage we ignore generation in the depletion layer.

The solutions can easily be obtained for this semi-infinite absorber case (with x measured from x_p, see Fig. 7.1) (e.g., see *Sze* [7.5]):

$$n_p(x) = \alpha \Gamma_0 [\exp(-x/L_n) - \exp(-\alpha x)]/D_n(\alpha^2 - 1/L_n^2)$$
$$+ n_{p0}[\exp(eV/kT) - 1] \exp(-x/L_n) + n_{p0}$$
$$J_n(x) = eD_n dn_p/dx$$
$$J_n(x=x_p) = e\alpha \Gamma_0(\lambda)/[1 + 1/\alpha(\lambda) L_n] - (eD_n n_i^2/L_n N_A)[\exp(eV/kT) - 1] \quad (7.4)$$

with $L_n = (D_n \tau_e)^{1/2}$ and $n_{p0} = n_i^2/N_A$.

Because of the large barrier for holes, the hole current through the junction is assumed negligible so that $J_p = J - J_n = 0$ at $x = x_p$ and the total monochromatic light current J_L is given by (7.4). The first term of this expression is just the total light generated current in the absorber layer which makes its way to the depletion layer edge. The second term is the familiar forward-bias injected electron current for one side of a homojunction. This was implicit in our boundary condition (assumption VI). Note that the only influence on $|J_{sc}| = J_L$ from the diode parameters arises from this boundary condition and that the light-generated current is independent of bias voltage.

We are assuming superposition only in the quasi-neutral region. At this point we require only that the quasi-Fermi level for electrons is the same at the depletion layer edges at both sides of the junction and we are unconcerned with its behavior inside the depletion layer. Note also that the dark diode current is added linearly to the light current; this is an example of an "ideal" solar cell: $J = J_0[\exp(eV/kT) - 1] - J_L$.

Bulk recombination in the quasi-neutral region determines the quantum efficiency η_Q of this device as well as its dark forward-bias current; for maximum efficiency, we want the largest αL_n product possible. But now we wish to focus on the depletion layer. Unlike the indirect band gap homojunction solar cell where transport is usually dominated by the quasi-neutral regions, the direct band-gap HJ collects an appreciable fraction of the available photons in the depletion layer. For instance, for a GaAs HJ with a depletion layer width $W_d = 0.1\,\mu\text{m}$, about 25% of AM 2 (air mass two) photons are absorbed in the depletion layer assuming no absorption in the wide band-gap window component. Moreover, the dark forward-bias current is usually dominated by recombination in the depletion layer. The common assumption with respect to the light current is that the photogenerated minority carriers suffer no recombination in the depletion region because they are swept out rapidly by the large electric field there. For a diffusion voltage V_d, $\mathscr{E}_{depl} \approx 2V_d/W_d \approx 10^4$ Vcm^{-1} compared with the diffusion field $\mathscr{E}_{diff} = kT/eL_n \approx 250$ Vcm^{-1}. With this assumption we can immediately write an approximation for J_L:

$$J_L = e\Gamma_0(\lambda)\{1 - \exp[-\alpha W_{dp}(V)]\} + e\Gamma_0(\lambda)\exp[-\alpha W_{dp}(V)]/(1 + 1/\alpha L_d), \quad (7.5)$$

with $W_{dp} = x_p - x_0$, i.e., the portion of the depletion layer on the p-type side. The total light current J_L is just the sum of that generated in the quasi-neutral region and that from the p-region depletion layer. Note that J_L is now a slowly varying function of bias voltage – our first departure from ideality. We return to these considerations in Sect. 7.2 but now we complete our discussion of the solar efficiency for an ideal cell. The total light-generated current J_L in sunlight is obtained by integration of the monochromatic quantum efficiency $\eta_Q(\lambda) = J_L(\lambda)/e\Gamma_0(\lambda)$ over the solar spectrum:

$$J_L = e \int_0^\infty \eta_Q(\lambda)(d\Gamma/d\lambda)d\lambda.$$

The resulting JV curves for an ideal cell are shown in Fig. 7.2 (for series resistance $R_s = 0$). The dark JV curve is translated downward by J_L without

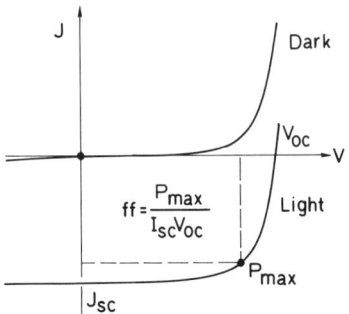

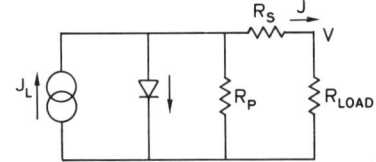

Fig. 7.3. Simplified equivalent circuit for a solar cell

Fig. 7.2. Light and dark JV curves for an ideal solar cell

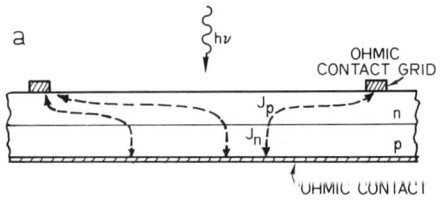

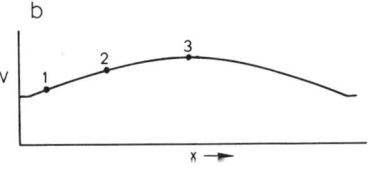

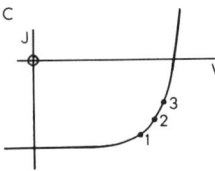

Fig. 7.4a–c. Current collection in a front-gridded solar cell: (a) current paths, (b) voltage distribution between grid lines for operation near P_{max}, and (c) corresponding voltage values on JV curve

change in shape. This ideal solar cell characteristic is the result of superposition and in this case $J_{sc} = -|J_L|$. The implications of this model are:

I) $\log J_{sc}$ vs V_{oc} is identical with $\log J$ vs V dark.

II) J_0 and the slope of the $\log J$ vs V curve are not functions of illumination intensity or wavelength.

III) J_L is not a function of bias.

In general R_s and R_p are present giving an equivalent circuit as shown in Fig. 7.3. The current collection paths for a typical cell are illustrated in Fig. 7.4. Generally we are concerned only with the effective lumped R_s. We note in passing, however, that the distributed nature of R_s in the thin window layer between the collecting grid lines can give rise to complex, nonlinear effects and introduce serious error in measurements (e.g., such as lowering V_{oc}) of devices with larger values of film resistivity ($\gtrsim 10\,\Omega$ cm for a 2 μm film). The calculation of η_s was treated by *Sze* [7.5] for lumped R_s and by *Wolf* and *Rauschenbach* [7.6] and *Fang* and *Hauser* [7.7] for the distributed model.

We could easily extend our treatment to absorber layers of finite thickness, the effects of surface recombination, and collection in the window layer but this would add little to our understanding of the HJ problem.

Once J_L is determined, V_{oc}, the maximum power P_m, and the solar efficiency η_s can be readily determined from the JV relationship of the diode (including R_s and R_p) by maximizing the JV product and multiplying by the appropriate factors: $\eta_s = P_m/S_0 = J_{sc}V_{oc}\text{ff}/S_0$ where $J_{sc} \simeq J_L$.

7.1.2 Critique of Assumptions

In this section we examine several of the assumptions underlying our derivation of the photogenerated current J_L. We briefly mention those pertaining to the quasi-neutral layer, but we wish to concentrate our attention on the assumptions which concern transport in the depletion layer, thinking of the quasi-neutral absorber more as a source that is modified only by conditions at the boundary of the depletion layer.

a) $\mathscr{E} = 0$ in the Quasi-Neutral Region

The electric field in the quasi-neutral region is given by the IR drop there (or $\Delta V = J\varrho$ where ϱ is the bulk resistivity). For an efficient solar cell the series resistance in the p and n layers should be less than $\approx 0.1\,\Omega$ for each cm² of cell area. For a 4 μm thick cell at AM 2, $\mathscr{E} \approx J_L RA/(x_{p2}-x_{n2}) \approx 5$ V cm^{-1} which is small compared with the diffusion field (≈ 250 V cm^{-1}) or the junction field ($\approx 10^4$ V cm^{-1}).

b) $U = (n_p - n_{p0})/\tau_e$

The use of a minority carrier lifetime which is independent of n_p is based on the Shockley-Read model for a one-level recombination center when $p_p \gg n_p$. The approximation is valid in the quasi-neutral regions of most systems of interest at AM 0 illumination levels providing that the τ_e used is an effective lifetime which is measured under similar conditions of carrier density, illumination, and temperature. (See [7.8, 9] for a discussion of high-intensity effects.)

In the depletion layer, $p_p \gg n_p$ no longer holds, and the full Shockley-Read expression must be used. The recombination rate in the depletion layer then depends on the position of the recombination levels and the quasi-Fermi levels for electrons and holes in a complex way, generally reaching a maximum value for $E_{Fn} > E_r > E_{Fp}$ where E_{Fn} and E_{Fp} are the electron and hole quasi-Fermi levels and E_r is the energy level of the recombination centers. This is particularly important for the direct band-gap cells where a large fraction of the light is absorbed in the depletion region. Under forward bias, the total recombination current is given by integration of the recombination rate across the junction:

$$J = \int_{x_n}^{x_p} eU(x)dx.$$

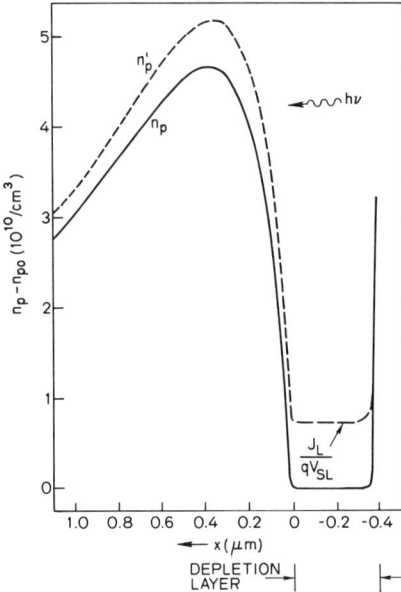

Fig. 7.5. Calculated carrier density in a CdTe pn junction at zero bias with (---) and without (——) velocity limited by scattering [the solid $n_p(x)$ curve is "forced" to fit the boundary condition $n_p = n_{p0}$ at $x = x_p$]

However, the fact that the rate reaches a strong maximum within a limited region of the depletion layer allows some simplifying assumptions.

c) The Boundary Condition at the Depletion Layer Edge

For a zero-bias boundary condition we previously assumed that *all* the photogenerated carriers are swept away by the junction field, driving the minority carrier concentration at the depletion layer edge to its dark, thermal equilibrium value. Although this is a great convenience in calculation, we know intuitively that this cannot be true. If we assume that the photogenerated carriers traversing the junction are traveling at a scattering-limited velocity v_{sl}, the limiting value under high fields, then the minimum carrier density in the depletion layer is J_L/ev_{sl} which is considerably greater than the dark, equilibrium value. For example, for GaAs, μ_n begins to saturate at $\mathscr{E} \approx 2 \times 10^3$ V cm^{-1} and $v_{sl} \approx 10^7$ cm s^{-1}. The carrier density in these types of junctions must be approximately as shown in Fig. 7.5. For a silicon homojunction the minimum carrier density as limited by v_{sl} is much smaller than the light-generated carrier density, so the effect is not important, and $n_p = n_{p0}$ at $x = x_p$ is a good assumption. However, for direct band-gap materials which commonly have much shorter carrier lifetimes τ_e, the maximum photogenerated carrier density in the quasi-neutral region is much smaller, and the effect of the scattering-limited velocity carrier density is relatively larger. If we assume a v_{sl} throughout the depletion layer to obtain the electron density there, then the reduction factor for J_L is approximately $(1 - D_n/L_n v_{sl})$ with D_n and L_n taken in the quasi-neutral region. This is a *lower* bound on the reduction of J_L and in the example of Fig. 7.5 for CdTe, J_L is reduced by a factor of approximately 2% assuming $L_n = 1\,\mu\text{m}$ and $v_{sl} = 1.3 \times 10^7 \text{ cm s}^{-1}$.

Consideration of the carrier density in the junction region is further complicated by the fact that the carriers are "hot" with respect to the dropping conduction band edge. In fact the thermalization of hot carriers in the depletion layer is a major loss mechanism for solar cells, reducing the output power by a factor of $\{1 - [V_{max}/(E_{g1} - \delta_n - \delta_p)]\} \approx 40\%$, where V_{max} is the voltage at the maximum power point, E_{g1} is the absorber band gap, and δ_n and δ_p are the energy differences between the Fermi-levels and the conduction and valence band edges in the n- and p-type quasi-neutral regions. The diffusion of hot electrons in the junction field is discussed by *Seeger* [7.10], *Stratton* [7.11], and *Persky* [7.12]. The usual treatment is to use the "diffusive approximation" to the distribution function $[f = f_0 - \tau_m \mathscr{E}_z(e/m^*) \partial f_0 / \partial v_z$, for example] where τ_m is the momentum relaxation time and v_z and $\mathscr{E}_z$ are the components of the velocity and electric field in the z direction. The Boltzmann approximation is used, but with an electron temperature T_e which is higher than that of the lattice, $f_0 \approx \exp(-mv_z^2/2kT_e)$. Equivalent T_e values of $\approx 1000\,\text{K}$ are predicted [7.13] and observed [7.14] for junction fields of $\approx 10^4 \text{ V cm}^{-1}$. The general effects of these high fields are: I) the Einstein relation in its simple form $(D = kT\mu/e)$ no longer holds, II) the carrier velocity saturates toward the thermal velocity at high fields, and III) the current-voltage relation for the diode is changed somewhat, and must include T_e which is itself a function of the electron current [7.10, 15].

This consideration introduces the realization that thermal equilibrium may not hold in the dark depletion layer under forward bias and that it certainly does not hold in the light. A question that remains to be answered by the solar HJ theorist is how to describe the density of hot electrons thermalizing in the depletion layer and further how to describe the recombination of the photo-excited carriers traversing the junction in the nonequilibrium situation.

d) Position of the Quasi-Fermi Levels Within the Depletion Layer

We have considered briefly the effect of the carrier density in the depletion layer on the J_L boundary condition. Now we go on to consider the relation of the carrier density at one edge of the depletion layer to that at the other. In our simplified treatment of the calculation of J_L we assumed that $n_p = n_{p0} \exp(eV/kT)$ at $x = x_p$ to obtain the first-order boundary condition. In the homojunction case we can show that the quasi-Fermi level is almost constant within the depletion layer (i.e., $\partial E_{Fn}/\partial x \approx 0$) in forward bias in the dark. The demonstration follows from that outlined by *Sah* et al. [7.16]. If thermodynamic equilibrium is assumed (as is commonly done) then the electron current is given in terms of the electron quasi-Fermi level by

$$J_n(x) = \mu_n n(x) (dE_{Fn}/dx)$$

with $n = n_i \exp\{[E_{Fn}(x) - E_i(x)]/kT\}$ where $E_i(x)$ is the intrinsic level. These relations lead directly to a differential equation with $\exp[(E_{Fn} - E_i)/kT]$ as the variable and with the solution

$$\exp[E_{Fn}(x)/kT] = C + \int [J_n(x)/n_i\mu_n kT] \exp[E_i(x)/kT] dx.$$

The constant C can be eliminated by choosing the limits of integration and, with the zero of energy at $x = x_n$ [i.e., $E_{Fn}(x_n) = 0$], we obtain

$$E_{Fn}(x_p) = kT\ln\left\{1 + \int_{x_n}^{x_p} [J_n(x)/n_i\mu_n kT] \exp[E_i(x)/kT] dx\right\}. \tag{7.6}$$

Note that $J_n(x)$ is negative for forward bias.

We can evaluate this integral approximately by using the value of $J_n(V)$ for a real diode characteristic, letting J_n be constant throughout the junction, and using a linear approximation for $E_i(x)$ across the junction. For example, consider a rather poor Si pn junction with $J_0 = 10^{-9}$ A cm^{-2}, $A = 1$, a depletion layer width W_d of 1 μm, and $\mu_n \sim 10^3$ cm^2 V s^{-1}. For this example we find that the electron quasi-Fermi level differs by less than $0.4 kT$ from one side of the junction to the other. In general, for the forward biased Shockley diode, the depletion layer behaves like an infinite source of carriers and E_{Fn} is almost constant as long as $L_n \geq kTW_d/e(V_d - V)$ where V_d is the diffusion voltage and L_n is the minority carrier diffusion length in the p-type quasi-neutral region. For a recombination/generation diode, the variation in E_{Fn} is even smaller due to the variation of J_n with x. These results should also hold in the HJ case. However, the result is not true for appreciable reverse bias [7.16], in the light, or for very high forward bias when resistance drops in the depletion layer must be accounted for.

Under illumination the current is opposite in sign and (7.6) can again be used to calculate the position of the quasi-Fermi level. This is done for the case of Fig. 7.5 (where the carrier density was scattering-velocity limited) and the result is shown in Fig. 7.6.

This result has important consequences, allowing us to calculate, at least to first order, the recombination rate within the depletion layer, because E_{Fn} and E_{Fp} are known to within a small error.

e) On the Constancy of the Photogenerated Current Across the Junction

Implicit in our derivation for J_L was that the photogenerated electron current was constant across the depletion layer and that once the carriers passed the depletion layer edge they were fully collected. The current of photogenerated carriers raises the quasi-Fermi level within the depletion layer above its equilibrium value (as in Figs. 7.5, 6), implying that more recombination will take place and that some of the carriers will be lost in the depletion layer, even considering the increased effective diffusion length due to the high junction

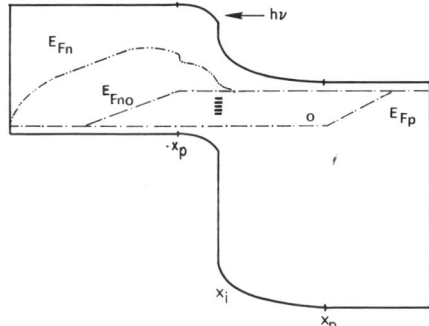

Fig. 7.6. Schematic band diagram of an HJ solar cell under forward bias with absorption in the depletion layer [dark (—·—) and light (—··—) quasi-Fermi levels]

field. A simplified treatment, assuming that we can superimpose J_L on the dark, forward-bias current is given here for the case where the important recombination is assumed to occur within a thin layer at the metallurgical interface of the HJ. The recombination loss is lumped into an effective interface recombination velocity $S_I = \delta U(V)$ where δ is the thickness of the layer and $U(V)$ is the applied-voltage-dependent recombination rate. The continuity of current across the interface is $J_{in} = J_{out} + enS_I$, and the carrier density at the interface can be approximated by $n(x_I) = J_{out}/ev(\mathscr{E})$ where $\mathscr{E} \approx 2(V_d - V)/W_d(V)$. For low carrier mobility near the interface $v(\mathscr{E})$ can be replaced by $\mu\mathscr{E}$ and for high μ or $\mathscr{E}$, $v(\mathscr{E}) \to v_{sl}$. Combining, we have

$$J_{out}/J_{in} = [1 + S_I(V)/v(\mathscr{E})]^{-1} \tag{7.7}$$

where, for example, $U(V) \approx U_0[\exp(eV/2kT) - 1]$ for a single level of recombination centers midway between E_{Fn} and E_{Fp} in a symmetrical junction. This illustrates a way in which it is possible to treat the recombination loss of J_L in the junction region rather independently of the dark junction current.

7.1.3 Bias Voltage Dependence of Collection Efficiency

The use of collection functions is a means of gaining perspective on the processes of photogenerated carrier transport without resorting to the solution of a nonequilibrium transport equation in the depletion layer. Their use assumes superposition of J_{dark} and J_L in the depletion layer, however, and they are therefore good only to first order. The collection function H is a product of two factors, the first $g(\lambda, V)$ representing the bulk processes of absorption and recombination, and the second $h(V)$ representing the bias voltage dependence of interfacial recombination loss. An example of $g(\lambda, V)$ is given by $g(\lambda, V) = J_L/e\Gamma_0$ where J_L is in turn given by (7.5). We can obtain $g(\lambda, V)$ from the shape of the experimentally measured spectral response of J_L curve, particularly the long λ edge. The factor $h(V)$ is independent of λ, providing the junction shape or the interface recombination velocity S_I doesn't change significantly

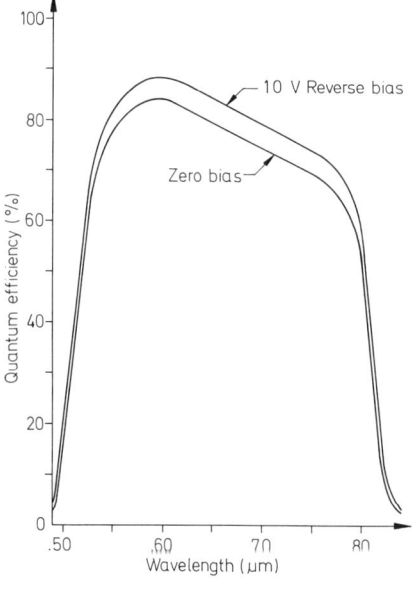

Fig. 7.7. Measured spectral response for CdTe/CdS HJ at zero and 1 V reverse bias (corrected for 13% reflection loss) [7.17]

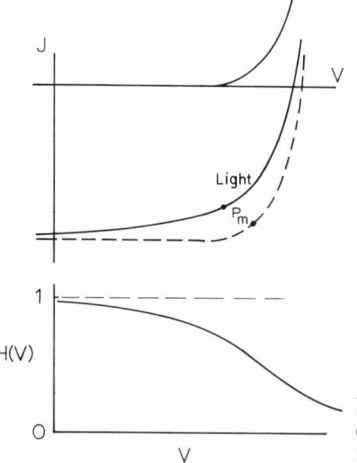

Fig. 7.8. Schematic JV curves showing the effect of a collection function (dashed curve is dark curve displaced by $-J_L$)

with illumination, since where the carriers are generated in the bulk does not matter. This fact provides a means of separation of h and g, and $h(V)$ multiplies the entire spectral response curve by a constant factor $\gtrsim 1$ for each applied bias voltage as shown in Fig. 7.7. An example of $h(V)$ is (7.7) for a CdTe/CdS HJ solar cell measured by *Mitchell* et al. [7.17]. In this case, S_I was considered constant and a value of 2×10^6 cm s^{-1} was deduced. The product $H = gh$ is identical with the internal quantum efficiency of a device such as that of Fig. 7.7 where collection in the wide band-gap material is neglected.

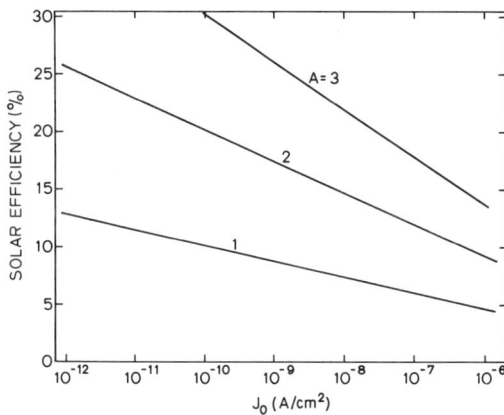

Fig. 7.9. Ideal solar efficiency vs $\log J_0$ for various quality factors (calculated for $J_L = 25$ mA cm^{-2}, solar input $S_0 = 100$ mW cm^{-2}, and $R_s = 0$)

The effect of an $H < 1$ on the JV characteristic is shown in Fig. 7.8. The value of light-generated current actually collected is reduced with respect to that generated in the absorber, particularly for forward bias. As a result V_{oc} is reduced a small amount, and there is a considerable reduction in ff. J_{sc} is usually changed only a small amount in efficient cells. The collection function H can be measured by comparison of light and dark JV curves or by ac methods, but the experimenter must first verify that A and J_0, the diode parameters, are relatively unaffected by illumination or at least account for this effect.

The collection-function approach gives us a first-order means for separation of interface and bulk recombination effects in a real device. It was used by *Fahrenbruch* et al. [7.18] and *Rothwarf* [7.19] with considerable success in the analysis of the Cu$_2$S/CdS HJ cell.

7.1.4 The Solar Efficiency of Heterojunction Converters

So far we have largely ignored the actual JV characteristics of the HJ, concentrating instead on background aspects. At this point we show explicitly the relation of solar efficiency to diode parameters. Most HJ solar cell JV characteristics can be represented by an expression such as

$$J = J_0[\exp eV/(AkT) - 1] - H(\lambda, V)J_L \tag{7.8}$$

at least over the range of J values which determine the solar efficiency. The efficiency of an ideal solar cell is plotted versus J_0 and A in Fig. 7.9. This relationship can be expressed formally as:

$$\eta_s = [J_L AkT(1-\gamma)/eS_0] \ln(\gamma J_L/J_0), \tag{7.9}$$

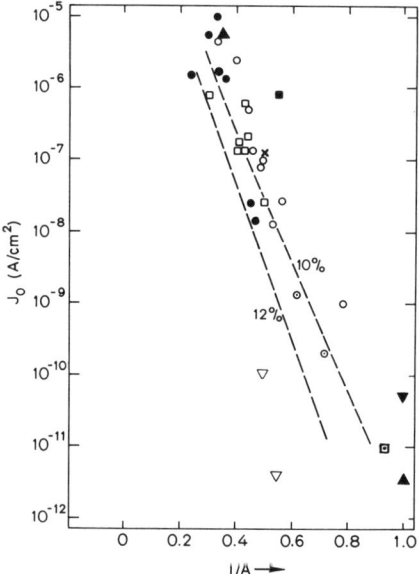

Fig. 7.10. Diode parameters J_0 and A for a variety of junctions at ≈ 300 K. Two constant solar efficiency curves are also plotted for $J_L = 17$ mA cm^{-2}, $S_0 = 75$ mW cm^{-2}, and $R_s = 0$.

Symbols: GaAs p/n △ [7.121, 122], ZnSe/GaAs ⊙ [7.23], ITO/InP □ [7.24], CdS/InP crystal ● [7.24–27], CdS/InP poly ■ [7.26], InP p/n ⊡ [7.26], CdS/CdTe ○ [7.17], ZnO/CdTe × [7.28], Si p/n ▲ [7.6], and Cu$_2$S/CdS, illuminated, ▼ [7.19].

All data shown appear to include recombination/generation except ITO/InP and Si p/n for $A = 1$. Note that solar efficiency values are for *one* specific value of J_L

where γ is a slowly varying constant given (for $R_s = 0$) by

$$\gamma = [1 + \ln(\gamma J_L/J_0)]^{-1} \approx 0.05\text{–}0.10 \tag{7.10}$$

by *Mitchell* et al. [7.20]. In this example $H(V) = 1$.

In most HJ solar cells of interest, the quantum efficiency already lies in the 80–95% range so that the largest potential for improvement in efficiency is from increases in A and decreases in J_0 which depend chiefly on the properties of the interface and depletion layer in the HJ rather than on the bulk collection properties. However, A and J_0 are somewhat loosely dependent on each other as is indicated by a plot of $\log J_0$ vs $1/A$ in Fig. 7.10 for a wide variety of HJs. By assuming the form

$$J_0 = J_{00} \exp(-\Delta E/AkT) \tag{7.11}$$

one can obtain an activation energy from this plot of about $E_g/2$ for many of the junction types.

So as we examine the various models for HJ transport we have the questions: just what is the physical significance of A, and what parameters determine J_0?

7.2 Present Theories of Heterojunction Transport

To begin this section we explore the meaning of the diode parameters J_0 and A in a general way, drawing primarily on homojunction theory. The section goes on to examine and contrast several of the present HJ theories. Since these transport models are presented in numerous papers and in excellent review articles by *Tansley* [7.30] and *Van Ruyven* [7.31] and the book by *Milnes* and *Feucht* [7.32], we set our examination at a rather general level, attempting to seek out the essence of the models with particular attention focused on the solar cell domain. A map for the exploration follows:

– Extension of Shockley diode theory to HJs with no interface states [7.33];
– Interpretation of the HJ in terms of back-to-back Schottky diodes with a high recombination, "metallic" interface [7.34];
– Introduction of charged interface states [7.35];
– Introduction of interfacial dipoles [7.31]; and
– Stepwise recombination/tunneling through states adjacent to or at the interface [7.21, 36].

Finally, metal/insulator/semiconductor (MIS) and semiconductor/insulator/semiconductor (SIS) structures and the connections with Schottky barriers are considered briefly.

7.2.1 The Diode Parameters J_0 and A

As we have seen in the previous section, the quantities A and J_0 are perhaps the most direct connections between the solar efficiency and the electrical properties of the junction. The quality factor A for *homo*junctions arises from the type of transport involved: $A=1$ for injection and diffusion into the quasi-neutral regions where n_i^2 enters the J_0 expression [e.g., $J_0=(D_n/\tau_n)^{1/2}n_i^2/N_A$ for the p region], and $A \approx 2$ for the mechanism in which recombination/generation in the depletion layer dominates the current transport and n_i enters the J_0 expression (e.g., $J_0 \approx W_d n_i/2\tau_{\text{eff}}$).[1] In the diffusion case the electron (hole) quasi-Fermi level $E_{\text{F}n}(E_{\text{F}p})$ moves directly with the applied bias with respect to the recombination energy level E_r while $E_{\text{F}p}(E_{\text{F}n})$ remains relatively constant in the highly doped p-type (n-type) quasi-neutral region. In the recombination/generation case *both* $E_{\text{F}n}$ and $E_{\text{F}p}$ move with respect to the recombination energy level in the depletion layer. The applied voltage is split between $(E_{\text{F}n} - E_i)$ and $(E_i - E_{\text{F}p})$ and, viewed on a simplistic level, the voltage splitting introduces the factor of 2. The total current is the sum of the diffusion and recombination/generation components and the latter generally dominates HJ transport, especially at low to moderate bias, in the range of solar cell operation.

In a basic paper on the recombination/generation mechanism, *Sah*, *Noyce*, and *Shockley* (SNS) [7.16] have shown that A depends on the energy level of the effective recombination centers with respect to the intrinsic level E_i as well as on the magnitude of the applied bias, ranging from 1 for shallow recom-

---
1 τ_{eff} is the effective carrier lifetime in the depletion layer.

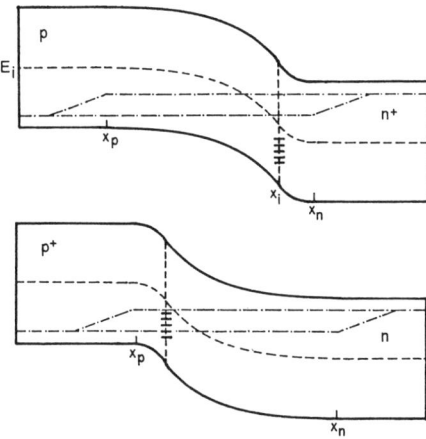

Fig. 7.11. Asymmetrical homojunctions with interface states

bination centers close to the band edges, to almost 2 for centers with levels within a few kT of E_i. The determining factor for A in these symmetrical junctions appears to be how far the effective recombination centers are from the intrinsic level (for moderate forward bias).

If we now consider an asymmetrically doped *homo*junction with a high recombination region at the interface between p- and n-type regions (perhaps formed when an n layer was grown epitaxially on a "dirty" p substrate) we see that the positions of E_{Fn} and E_{Fp} at the interface change considerably with respect to E_r depending on the magnitude and direction of the asymmetry as shown in Fig. 7.11. The relative positions of E_{Fn}, E_{Fp}, E_r, and E_i affect the recombination rate $U(x)$ and hence the total junction current considerably.[2] This introduces an asymmetry dependence of the diode factor which is readily extended, in principle, to the HJ. Since recombination centers of a given kind (donorlike, for example) usually have very different values for σ_n and σ_p, a further asymmetry is introduced. The recombination/generation current for the asymmetric homojunction is derived by *Choo* [7.37] and the effect of the SNS and Choo theories on solar efficiency is discussed by *Hovel* [7.38].

Using the general expression (7.11), we can associate J_{00} for the junctions of Fig. 7.11 with the number and cross section of the effective recombination centers and A with the location of their energy levels with respect to E_i. The complexity of these relationships for HJs, coupled with the lack of specific knowledge about the centers involved in the depletion layer, particularly at the interface, has made A and J_0 largely into curve fitting parameters.

---

[2] Recall that the calculation of the forward bias recombination/generation current involves: I) assuming that E_{Fn} and E_{Fp} are both almost constant in the depletion layer and that $eV = (E_{Fn} - E_{Fp})$, II) calculation of $U(x) = U[E_{Fn}(x), E_{Fp}(x), E_i(x)]$ using the $E_r(x)$ and the σ_n and σ_p for the recombination region, and III) integration across the depletion layer (assuming the junction shape is unchanged) $J_{rg} = e \int U(x)dx$. For a treatment for three recombination center levels see [7.39].

7.2.2 Anderson's Model for the Heterojunction

a) The Basic Model

The model of *Anderson* [7.33] forms the basis and starting point for most other HJ theories. The essence of the model is to incorporate the discontinuities in the material properties ε_r, χ, and E_g across an abrupt metallurgical interface into the Shockley diode theory. The discontinuity in the relative dielectric constant ε_r is easily dealt with by requiring that the electric displacement be continuous across the interface, i.e., $\varepsilon_1 \mathscr{E}_1 = \varepsilon_2 \mathscr{E}_2$. The differences in E_g and the electron affinities χ require discontinuities in the conduction and valence band edges, ΔE_c and ΔE_v, which are easy to visualize but rather discomforting theoretically. The result of bringing the two semiconductors together and lining up the Fermi levels is shown in Fig. 7.12. Anderson's model assumes that no interface states are present and that current transport is via injection into the quasi-neutral regions or by recombination/generation in the depletion layer. The relation between the various quantities is given here:

$$\Delta E_c = \chi_1 - \chi_2 \qquad V_d = V_{d1} + V_{d2}$$
$$\Delta E_v = \chi_2 - \chi_1 + E_{g2} - E_{g1} \qquad V = V_1 + V_2, \tag{7.12}$$

$$V_{d2}/V_{d1} = V_2/V_1 = (V_{d2} - V_2)/(V_{d1} - V_1) = \varepsilon_1 W_{d2}/\varepsilon_2 W_{d1} = \varepsilon_1 N_{A1}/\varepsilon_2 N_{D2}, \tag{7.13}$$

$$\left.\begin{array}{l} W_{d1}^2 = (x_i - x_p)^2 = 2N_{D2}\varepsilon_1\varepsilon_2(V_d - V)/eN_{A1}(\varepsilon_1 N_{A1} + \varepsilon_2 N_{D2}) \\ W_{d2}^2 = (x_n - x_i)^2 = 2N_{A1}\varepsilon_1\varepsilon_2(V_d - V)/eN_{D2}(\varepsilon_1 N_{A1} + \varepsilon_2 N_{D2}) \end{array}\right\}. \tag{7.14}$$

V_{d1} and V_{d2} are the diffusion voltages on either side of the junction and V_1 and V_2 are the portions of the applied voltage dropped on either side of the junction. Of course other values of δ_n, δ_p, χ_1, χ_2 permit a wide variety of other profiles to be obtained, some of which are schematized in Fig. 7.13. We examine a few of these profiles in greater detail in later sections. Turning to the HJ of Fig. 7.13a which represents the usual configuration for a solar cell, we want $E_{g2} > E_{g1}$ to maximize the window effect so that in most cases the transport of one of the carrier types *across* the metallurgical interface will dominate. In the case shown, hole transport across the interface to recombine in W_{d2} is negligible with respect to the electron current. In addition, the photogeneration in a wide band-gap material is usually neglected because its thickness is much greater than $1/\alpha_2$ so that hole transport to the left may also be neglected in this example. With these assumptions, the J vs V relationship for this diode (considering diffusion dominated transport for example) is very similar to that for a homojunction, and is given by

$$J \approx (D_{n1}/\tau_{e1})^{1/2}(N_{c1}/N_{D2}/N_{c1})\exp(-eV_b/kT)[\exp(eV/kT) - 1] \tag{7.15}$$

$$= (D_{n1}/\tau_{e1})^{1/2}(n_{i1}^2/N_{A1})[\exp(eV/kT) - 1],$$

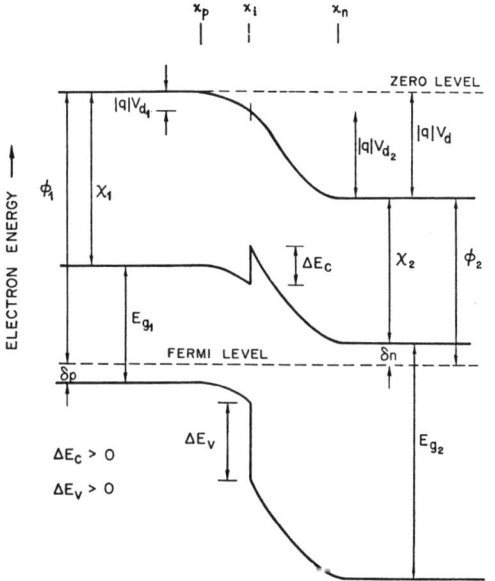

Fig. 7.12. Schematic Anderson heterojunction

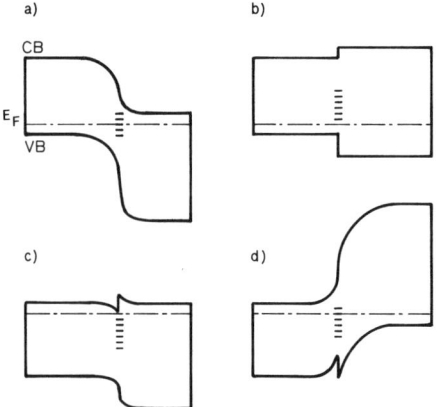

Fig. 7.13. Various heterojunction configurations

where V is the applied bias, $V = V_1 + V_2$, and $V_b = V_{d1} + V_{d2} - \Delta E_{c/e}$. The model is somewhat successful in predicting the configuration of the bands at the junction but in many cases the measured JV relations are much different, both in quality and magnitude, from those predicted. Other transport mechanisms are needed. In general, many HJs admit the empirical formulation:

$$J \approx J_{00} \exp(-\Delta E/kT)[\exp(eV/AkT) - 1]. \tag{7.16}$$

b) Examination of Assumptions

The constancy of the electron and hole quasi-Fermi levels across the depletion layer follows the same arguments as for the homojunction case in Sect. 7.1.2d.

The Anderson model predicts band-edge discontinuities depending only on the electron affinities and the band gaps of the two materials. Their presence is clear from experimental measurements [7.21, 33] although their magnitude, as well as their direct dependence on the electron affinities, is uncertain because of our inability to separate out interface effects. The Schottky barrier is of course the simplest example of such a discontinuity. When $\Delta E_c < 0$ as shown in Fig. 7.13a, the effect is merely to reduce V_d and ultimately to reduce the V_{oc} obtainable from an HJ solar cell. On the other hand, $\Delta E_c > 0$ implies a conduction band spike or notch (Fig. 7.12) which tends to impede photo-generated carrier transport and, in some cases, may change the dark JV characteristics considerably.

Formally, the electron affinity χ is the energy required to raise an electron from the conduction band edge to the vacuum level. Intuitively we feel that it is a bulk property and in many cases it has been treated as such in the calculation of ΔE_c and ΔE_v. The value of χ is usually inferred from measurements of the electron work function ϕ. All such measurements involve a surface and unfortunately the value of χ is very surface sensitive. The principal means for measuring the work functions are listed below.

I) In the Kelvin method of contact potential measurement, a reference surface is brought into close proximity with the semiconductor surface and their spacing is varied sinusoidally. The contact difference of potential across the vibrating capacitor so formed produces an ac voltage which may be nulled to measure $\Delta \phi$.

II) In photoemission, electrons within the valence band are photoexcited with enough energy to be ejected through the surface of the material into the vacuum. The kinetic energy spectrum of the ejected electrons is measured and the data reduced to obtain the work function and hence the value of χ. A similar sort of experiment may be done utilizing thermionic emission. When a junction is made with another semiconductor or a metal, the internal photoemission can be used to determine barrier heights (on both sides of the junction for an HJ) so that the effective $\Delta \chi$ can be inferred.

III) The capacitance-voltage characteristics of a metal/semiconductor junction or of an HJ can be used to obtain V_d from which ΔE_c can be inferred with certain assumptions about interface states (Sect. 7.3.3a).

With all of these methods one is measuring with respect to a particular interface (vacuum in 1 or 2 and a metal or semiconductor in the last case). There is ample evidence that each pair of materials forms a unique case with considerable influence on the effective electron affinity difference obtained due to surface states, pinning of the Fermi level at the surface, and/or a variation of E_g at the surface due to reconstruction, strain, or adsorbed foreign atoms (see

Sect. 7.3). Each of the above types of measurement gives slightly different results for the same pair, and all are highly surface sensitive.

Kroemer [7.40] discussed the use of electron affinities to predict ΔE_c and ΔE_v and suggested that they be decomposed into a polarization part (corresponding to work done against the image force and equal to the electron affinity of a hypothetical perfect crystal surface) and an electrostatic part (corresponding to work done against any surface dipole layers). In this way the bulk and surface properties are separated. *Kroemer* went on to suggest that the polarization component be measured relative to some reference point on the periodic potential of the lattice, thus making it accessible to theoretical and experimental verification (e.g., by XPS) as a bulk constant.

In whatever manner the χ values are measured for an HJ pair, it is likely that the band profile will be modified in forming the junction according to the particular interface that is formed (depending on, for example, the chemical reactivity of the components). Published values of χ can serve only as a rough guide to the direction and magnitude of the ΔE_c and ΔE_v.

An example of the possible effects of a conduction band spike on a solar cell is the Cu_2S/CdS junction. *Te Velde* [7.41] and *Lindquist* et al. [7.42] independently proposed that a conduction band spike controls photogenerated carrier transport from the Cu_2S to the CdS. *Lindquist* et al. proposed that the thickness of the spike was the medium of control as modulated by the depletion layer width in the CdS while *Te Velde* proposed that heat treatment varied ΔE_c per se, thus controlling V_{oc} and J_{sc}. *Fahrenbruch* and *Bube* [7.18], and *Rothwarf* et al. [7.43] later showed that a spike was not required to explain the observed variations, suggesting instead that the local electric field controlled the recombination at the interface. *Böer* has recently resurrected the spike concept to explain the dependence of the J_{sc} and V_{oc} on y in the $Cu_2S/Zn_yCd_{1-y}S$ junction [7.44].

c) The Question of Abruptness

While Anderson's model is based on the assumption of a perfectly abrupt HJ, many junctions are fabricated at high enough temperatures so that at least some interdiffusion occurs, forming solid solutions of the junction components in some cases. *Oldham* and *Milnes* [7.45], and *Cheung* et al. [7.46] have shown that junction grading over distances of approximately 100 Å can smear out and lower the conduction band spike to the point where carrier flow is unimpeded. Photoexcitation and transport properties in such HJs with thin, graded layers are discussed in [7.22, 47].

d) Difficulties with the Anderson Model

There is considerable disagreement between the Anderson model and experiment. The measured values of J_0 obtained by extrapolation of the log J vs V curves to $V=0$ are generally much higher than expected. For example, in the

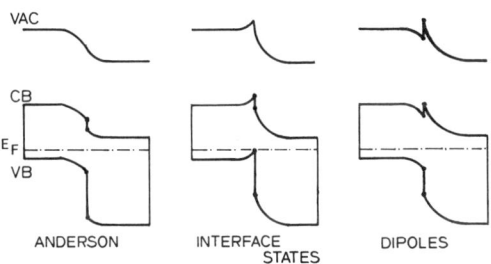

Fig. 7.14. Energy band profile of a *pn* heterojunction according to three models: Anderson (without interface states or dipoles), including charged interface states, and including dipoles – all with identical doping levels

Ge/GaAs HJ studied by *Riben* and *Feucht* [7.21], J_0 was 6 decades higher than predicted theoretically at 300 K. In addition, the observed slope at low forward bias was constant with temperature, whereas the model of (7.15) predicts a slope change by a factor of 4 over the temperature range used. The values of ΔE_c and ΔE_v, as measured by capacitance and other techniques, are not as predicted by published χ values. In fact, *Newman* [7.48] found that many junction characteristics appear to fit the relation $J = J_{00} \exp(\beta T) \exp(\alpha V)$ where α and β are constants, in many cases showing two branches of different slope. For these and other reasons the Anderson model has been modified as described in the following sections.

7.2.3 Sophistication for the Simple Heterojunction

The modification of the Anderson model to include the effects of interface states and dipoles is illustrated in Fig. 7.14 where the three band profiles are compared. These changes, plus the introduction of various tunneling mechanisms, bring the theory of HJs more into agreement with experimental observations.

a) Introduction of Interface States

The presence of a large density of electrically active interface states provides for two mechanisms in the HJ: I) the charge stored in the states distorts the band profile, raising or lowering the conduction bands at the interface with respect to the equilibrium Fermi-level, and I) the states provide a large density of recombination centers needed to explain the high J_0 values observed [7.34]. In this section, we discuss Fermi-level pinning and the density and energy distributions of interface states. In the following section, the recombination behavior of the interface states is discussed.

With the advent of highly sophisticated measurement techniques for surface properties, it is now quite natural to think in terms of surfaces and interfaces at which the Fermi-level is pinned by the charge on states localized there. Such pinning has long been observed at dislocations and grain boundaries, sometimes to the extent of inversion as in the *p–n–p* behavior of grain boundaries

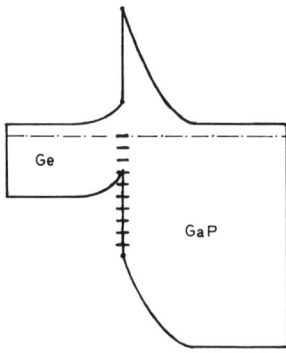

Fig. 7.15. Isotype Ge/GaP heterojunction showing effects of charged interface states (lattice mismatch 3.8%). (Redrawn from [7.52])

in Si observed by *Weinreich* et al. [7.49]. A considerable portion of the excellent and very thorough book by *Mataré* [7.50] is devoted to such effects in bicrystals.

One of the most convincing arguments for Fermi level pinning in HJs is in the isotype heterojunctions Ge/Si [7.51], Ge/GaP [7.52], and Ge/Si [7.53]. By observation of several reversals in sign of the photogenerated current as wavelength was varied, together with capacitance and current-voltage measurements, these authors deduced a band profile such as that shown in Fig. 7.15, where the pinning of E_F at the interface adds substantially to the barrier seen by conduction band electrons by raising the bands at the interface.

Although pinning of E_F at surfaces and isotype HJs has been widely studied, much less is known about E_F pinning in anisotype HJs since the effects are much more subtle in the latter. The properties would be expected to be considerably different since the states responsible for pinning the Fermi level at a free surface are different depending on the majority carrier type (acceptorlike for n-type and donorlike for p-type materials in general) and some neutralization might occur. It should be noted, however, that *Van Ruyven* et al. [7.52] do not share this view, holding instead that the surface pinning of the Fermi level retains its identity for each of the two components when they are joined, creating an interfacial dipole to account for the difference.

The state distribution in density and energy at "free" surfaces has been extensively studied. Similarly, these properties of interface states at grain boundaries, dislocations, and to a lesser extent, isotype HJs, have been researched a fair amount. However, these quantities are more difficult to obtain in anisotype junctions, and results tend to be very model dependent. The energy and spatial distribution of states in the depletion layer adjacent to the interface have been measured by the use of capacitance transient spectroscopy (e.g., Deep Level Transient Spectroscopy, DLTS) in connection with photo- and thermal excitation which is reviewed in [7.54]. *Sah* [7.55] also compares the thermal and photostimulated capacitance methods. Unfortunately, the sensitivity of these techniques falls off as one probes closer to the interface so that distributions at or near the interface cannot be obtained unless an insulating layer is present. Analysis of the capacitance-voltage data gives considerably more information about the energy levels, densities, and optical and thermal cross sections of states at the interface. The intercept of $1/C^2$ vs V curves as $1/C^2 \to 0$ gives the diffusion voltage V_d modified by the charge stored at the interface Q_{ss} and the dipole voltage ϕ_d.

Consider a heterojunction consisting of a p-type material of acceptor density N_{A1}, an n-type material of donor density N_{D2}, and an intermediate layer of width W_i and donor density N_{i2}. The new, modified voltage intercept V_d'' is given by

$$V_d' = V_d + \frac{e(N_{i2}+N_{A1})(N_{i2}+N_{D2})W_i^2}{2\varepsilon(N_{A1}+N_{D2})} + \phi_d \qquad (7.17)$$

with $\varepsilon_1 = \varepsilon_2 = \varepsilon$. For a region of zero thickness, the charge stored at the interface Q_{ss} can be expressed as the limit of $W_i N_{i2}$ as $W_i \to 0$. Equation (7.17) then becomes

$$V_d' = V_d + \frac{Q_{ss}^2}{2e(\varepsilon_1 N_{A1}+\varepsilon_2 N_{D2})} + \phi_d . \qquad (7.18)$$

In this case all the states near the interface behave as though they were *at* the interface, the spatial dependence is lost, and it is difficult (if not impossible) to discriminate between charged interface states and dipoles without additional information. Measurements as a function of frequency can be used to separate time-dependent interface state population and ΔE_c (time-independent) contributions to the V_d offset, as well as the bias voltage dependence of the occupation of interface states. These aspects are discussed in [7.56].

Interface state densities were determined for the Au/n-GaAs Schottky diode in [7.57] using a combination of J vs V vs T and C vs V data. Transient capacitance measurements of the DLTS type were used to measure electron and hole capture cross sections for n- and p-type Si MOS structures in [7.58]. Very few, if any, measurements on anisotype HJ appear to have been done.

b) Direct Recombination Through Interface States

Direct recombination through states at the interface appears to be a dominant mechanism of transport in many HJs. *Dolega* (pn junctions) [7.34] and *Oldham and Milnes* (isotype junctions) [7.35] described the HJ as two Schottky diodes back-to-back on either side of an almost metallic layer of interface states where recombination was very strong. This model predicts essentially (7.16) with $1 < A < 2$ depending on $\varepsilon_1 N_{A1}/\varepsilon_2 N_{D2}$. However, since those models imply thermal excitation over barriers they cannot explain the temperature-independent slopes of the log$J$ vs V curves seen for many HJs.

With regard to the recombinative aspects of interface states it is again difficult to measure their spatial distribution. One would wish to do so with a resolution of 20–50 Å in order to establish whether tunneling is involved in the recombination pathway. In addition, one would wish to discriminate between recombination in the region *close* to the interface (within approximately 100 Å, due to lattice mismatch dislocations, interdiffusion, and other "bulk" effects) and recombination in the thin (2–5 atomic layers) region *at* the interface (due to electrically active dangling bonds, oxide layers, compound formation, and the like). In most cases these are both lumped into an effective surface recombination velocity S_I. In some treatments (such as the excellent book on surface properties by *Many* et al. [7.59], p. 194) the Shockley-Read recombination model is applied to obtain the details of possible recombination models.

Several of the experimental methods of determining interface recombination are briefly summarized below:

I) *Photogenerated Current Loss.* The relationship between J_{sc} and the junction field as described in Sect. 7.1.3 and (7.7) can be used to obtain an effective S_I although the results can be quite model dependent.

II) *Recombination in Double Heterostructures.* By studying recombination kinetics in the thin center layer of a *p*-AB/*p*-CD/*n*-AB or *n*/*p*/*n* type HJ structure, one can obtain approximate values for S_I since the recombination current is given approximately by $e(2S_I + Uw)$ where the variable w is the intermediate layer thickness. The effective lifetime in the slab can be measured by a variety of techniques including electroluminescence time decay [7.60, 61], current decay from pulsed laser excitation (for $E_{gAB} > E_{gCD}$) [7.62], and pulsed reverse-bias recovery [7.63].

III) *$1/C^2$ Versus V Analysis.* Using the C vs V analysis outlined in Sect. 7.2.3a with variation of T and/or illumination intensity and wavelength, the energies, densities, and thermal and optical cross sections of the states involved can be obtained which, in principle, can be used to calculate the recombination kinetics (given a thorough knowledge of the junction structure). Such measurements have been done using these and DLTS techniques on *p*-Si MOS structures [7.64] and on GaAs and InP MOS devices [7.65].

IV) *Surface Photovoltage Spectroscopy.* The decay of photoexcited surface voltage of a semiconductor, measured by capacitance probe for a free or insulator/semiconductor surface or by V_{oc} decay in an HJ, can be used to obtain information about recombination losses close to the surface of interest (in the narrow band-gap material for an HJ). The recombination current is, schematically, $\int U dx + S_I$ so that S_I can be ferreted out by careful analysis particularly if one wishes to compare S_I for otherwise identical devices.

Separation of the charging and recombinative effects can be somewhat artificial, depending on how well these interface states communicate with the conduction and valence bands of the particular side of the junction of interest. A complete description would require simultaneous solution with respect to charging and recombination.

c) Combined Tunneling and Recombination

A mechanism which includes *interband* tunneling through a thin portion of the barrier, in addition to the previously mentioned thermal currents, was used by *Riben* [7.66] and extended by *Riben* and *Feucht* [7.36] to include stepwise tunneling into, and subsequent recombination through, a staircase of closely spaced states in the depletion layer. These theories follow from similar "excess current" mechanisms for tunnel diodes, for example as discussed by *Chynoweth* et al. [7.67]. *Riben* and *Feucht*'s model gives

$$J_{tr} = BN_t \exp[-4(2m^*e)^{1/2}(V_D - k_2 V)/3\hbar H] \tag{7.19}$$

with $H = (2eN_A/\varepsilon_2)^{1/2}$ and $k_2 = (1 + \varepsilon_1 N_A/\varepsilon_2 N_D)^{-1}$ where N_t is the density of tunneling/recombination centers.

Equation (7.19) fits data for an n-Ge/p-GaAs diode with N_{A1}, ε_1; N_{D2}, ε_2 like that shown in Fig. 7.13d. The J vs V relation of (7.19) predicts that the slope of the log J vs V curves be temperature independent with the small remaining temperature dependence arising from the variation in band gaps with temperature. This results in a simplified version of (7.19):

$$J_{tr} = B' \exp(\beta T) \exp(\alpha V). \tag{7.20}$$

A number of tunneling steps are required for thicker barriers and, in a succeeding work by *Riben* and *Feucht* [7.21], an additional parameter depending on the number of tunneling steps and on the temperature is introduced into (7.19). Although the form of this parameter is postulated by forcing the model to fit the experimental data, its form and the number of steps predicted appear to be physically reasonable.

A qualitatively similar process, *intraband* tunneling (such as through spikes in the conduction or valence bands, Fig. 7.12), was used to explain results for GaAs/Ge and GaAs/GaSb HJs [7.68]. The intraband tunneling mechanism is discussed in [7.30] and used to explain transport in the Cu_2S/CdS solar cell in [7.42].

Many HJs exhibit a combination of two mechanisms: tunneling/recombination as described above, and thermal activation to higher energies where the barrier is thin enough for direct tunneling or where direct access to the interface recombination centers obtains. A large portion of the recombination in both cases occurs at the interface. The combined situation is reviewed rather elegantly by *Owen* and *Tansley* [7.69]. An example of this sort of mechanism is the n-CdS/p-CdTe HJ studied by *Mitchell* et al. [7.17] who found tunneling like transport of dark diode currents at temperatures < 290 K following (7.20) with $\alpha = 22 \text{ V}^{-1}$ and $\beta = 0.015 \text{ K}^{-1}$. For $T > 290$ K, a recombinationlike mechanism operated following (7.16) with $A = 1.89$ and an activation energy for J_0 of $\Delta E = 0.59$ V. The zero-bias depletion-layer width of this junction is 0.19 µm, predominantly in the CdTe. Another example is that of the n-Ge/p-Si junction [7.70] which shows thermally activated currents ($A = 1.86$ at $T \approx 300\,\text{K}$) at low bias, but tunneling-dominated transport ($\alpha = 12.7\,\text{V}^{-1}$) at high bias (>0.1 to 0.3 V, depending on temperature).

d) Interfacial Dipoles

Interfacial dipoles were introduced by *Van Ruyven* to explain the observed conduction band discontinuities in isotype HJs which were not as predicted by the electron affinities [7.52]. The presence of such dipoles follows from the pinning of the Fermi-level at the surface of each semiconductor by a large number of surface states forming an almost metallic layer at the surface. The joining of these metallic layers on each of the semiconductors on the creation of an HJ is accompanied by the formation of a dipole of atomic dimensions, much as it occurs in the joining of two bulk metals with different work functions.

Van Ruyven's model [7.31] and *Anderson*'s model [7.33] of the HJ represent two extremes. In the former case a dipole is established and the depletion layers are unaltered by connection of the two component semiconductors, regardless of the doping levels in the bulk of the semiconductors. In the Anderson model, the depletion layers are formed by the connection of the two semiconductors. Most HJs probably lie somewhere between these two extremes.

The physical nature of such dipoles in semiconductors is discussed in Sect. 7.3. Suffice it to say here that dipoles could arise from the mechanism discussed above, from charge on opposite sides of a thin insulating layer at the interface, or from a spatial distribution of oppositely charged defects on either side of the interface in the bulk of the semiconductors. In the latter case the charge might be spread over several hundreds of angstroms; e.g., at dislocations propagating away from a lattice-misfit interface.

e) Junction Grading

Grading of composition of the components of a HJ (thus grading the band-gap) may be used for a number of reasons:

I) to enhance the efficiency of conversion of photons of a distribution of wavelengths;

II) to introduce an effective electric field in order to increase the diffusion length of minority carriers;

III) to spread the distribution of lattice mismatch states over a larger volume in order to decrease tunneling between them and thus reduce tunneling/recombination diode current; and

IV) to eliminate the effect of conduction or valence band discontinuities as discussed in Sect. 7.2.2c.

Van Ruyven [7.31] discusses the effects of grading due to perturbation of the electron wave functions by the changing periodic potential for rather large band-gap gradients.

Smaller band-gap gradients (over distances of 0.1 to 2 µm) are of more interest for solar converters. These were first considered by *Tauc* [7.71], and later *Wolf* [7.72]. *Emtage* [7.73] estimated that maximum solar efficiencies of roughly 35% could be obtained by judicious combination of grading and band gaps. Theoretical analysis and experimental confirmation in the HgTe/CdTe graded structure were done by *Marfaing* and *Chevallier* [7.74], and the work was continued by *Bouazzi* et al. [7.75] who showed that such graded gap structures could considerably exceed their *pn* HJ counterparts in solar efficiency.

The graded gap design seems simpler, in principle, than various tandem-junction approaches now being widely considered for concentration applications. It remains to be seen whether adequate electronic properties can be obtained in the graded region in order to complete with the high solar efficiencies already obtained in the tandem cells {e.g., 28.6% in a (GaAl, AsP) system [7.76]}.

f) The Effect of Illumination on Junction Parameters

In many cases the diode parameters A and J_0 are changed by illumination of the junction, complicating analysis of HJ solar converters. This is particularly true when deep trapping centers at or near the junction are not in good thermal communication with the bands ("slow states") and can be optically pumped. By plotting $\log J_{sc}$ vs V_{oc} as the illumination level is varied and comparing with dark $\log J$ vs V data, an estimate of the extent of these effects may be found. Plotting $\log J_{sc}$ vs V_{oc} also removes the effect of simple, nondistributed series resistance from the plotted characteristic. Such analysis cannot discriminate between changes in A and J_0 however.

Perhaps the extreme example of such light-induced changes is the Cu_2S/CdS HJ where the diode quality factor A changes from $A \gtrsim 2$ in the dark to $A \approx 1$ under illumination accompanied by large changes in J_0 as discussed in [7.19, 77, 78]. This HJ is particularly sensitive to the ratio of red to blue illumination especially at low illumination levels. These variations are due to charging of states in the CdS in the vicinity of the interface which in turn modifies the junction profile (as determined by capacitance-voltage data) [7.18].

7.2.4 Metal-Insulator-Semiconductor Junctions

It has been shown in the previous section that significant departures from "ideal" behavior are observed in most practical HJs. This is especially true for the HJs considered for solar cells. This departure from ideality is usually associated with defects in the interface region and space-charge region. In most cases, the nonideal junction behavior has deleterious effects on the solar cell's performance, specifically reducing the open-circuit voltage and the fill factor.

The design of useful devices demands the control of the bulk properties of the two materials as well as the properties of the interface region. Interface related phenomena need not be regarded as a factor in the degradation of performance, however, and the interface phenomena might be controlled by tailoring the properties of the interface region. Furthermore, interface-related effects can lead to a major improvement in device characteristics as compared with the "ideal", bulk-determined system. Perhaps the most vivid example of this fact is the Metal-Insulator-Semiconductor (MIS) solar cell. We briefly review some of the mechanisms which have been proposed to account for such an improvement.

a) Basic Schottky Barriers

Schottky barrier diodes have been under practical consideration since the early thirties. The transport mechanisms which determine the JV characteristics have been the object of much theoretical research [7.79–97], and they are covered in a recent review paper [7.98]. The most important forward transport mechanisms are (considering only n-type semiconductors): I) thermionic emission of electrons from the semiconductor over the top of the barrier into the metal; II) recombination in the depletion region; III) quantum-mechanical tunneling through the barrier; and IV) hole (minority carrier) injection.

The most important of these mechanisms is thermionic emission (TE), and metal/semiconductor (MS) diodes where this mode dominates are usually referred to as "ideal". Actually, the TE mechanism is in series with diffusion and drift in the depletion region. If the transport of carriers were limited by the diffusion mechanism, the JV characteristics in forward bias would be of the form

$$J = eN_c \mu_e \mathscr{E}_{max} \exp(-e\phi_b/T)[\exp(eV/kT)-1], \tag{7.21}$$

where $\mathscr{E}_{max}$ is the maximum electric field in the barrier, ϕ_b is the barrier height, and the rest of the parameters have been previously defined. This expression was the result of the diffusion theory of Schottky barriers [7.79, 80, 99]. If the process is limited by thermionic emission, the JV relationship will be

$$J = A^* T^2 \exp(-e\phi_b/kT)[\exp(eV/kT)-1], \tag{7.22}$$

where $A^* = 4m^*\pi e k^2/h^3 = eN_c v_R/T^2$ is the Richardson constant corresponding to the effective mass of the semiconductor, v_R being an effective recombination velocity at the potential energy maximum. These mechanisms can be combined as in [7.83] to arrive at the expression

$$J = \frac{eN_c v_R}{1+v_R/v_d} \exp\left(-\frac{e\phi_b}{kT}\right)\left[\exp\left(\frac{eV}{kT}\right)-1\right], \tag{7.23}$$

where v_d is an effective diffusion velocity defined in [7.83]. This expression has as limiting cases (7.21) or (7.22), depending on how v_d compares with v_R. Considering optical phonon scattering and quantum-mechanical reflection effects, *Crowell* and *Sze* [7.83] redefined an effective Richardson constant A^{**} such that

$$J = A^{**}T^2 \exp(-e\phi_b/kT)[\exp(eV/kT)-1] \qquad (7.24)$$

with $A^{**} = A^* f_p f_Q/(1 + f_p f_Q v_R/v_d)$; f_p is a scattering related probability and f_Q is the average transmission coefficient of the barrier.

The experimental forward characteristics are usually abbreviated as

$$J = J_0[\exp(eV/nkT)-1]. \qquad (7.25)$$

The Schottky diode quality factor[3] n shows deviations from unity even in the ideal case due to a small lowering of the actual barrier by the image force effect, which is field dependent.

Nonideal transport mechanisms such as recombination in the depletion region and tunneling have been observed (see, for example, [7.98]) but we shall not consider them here. These mechanisms have similar effects in the saturation current, J_0, and the diode quality factor n as observed in HJ devices.

Minority carrier injection is usually disregarded in the general case since it accounts only for a negligible contribution to the forward current. It is essentially proportional to $\exp(-E_{Fp}/kT)$, where E_{Fp} is the hole quasi-Fermi level measured from the valence band at the edge of the depletion region. The majority carrier current is proportional to $\exp(-\phi_b/kT)$. The injection ratio J_h/J_e is, therefore, generally much less than unity in most practical junctions [7.98].

b) Consideration of the Barrier Height

A most important result of (7.21–24) is the dependence of the saturation current on ϕ_b. As a consequence the magnitude of the barrier height determines the achievable values for the open circuit voltage. If the interface did not play a role in the determination of ϕ_b, its value would be given by

$$\phi_b = \phi_m - \chi_{sc}, \qquad (7.26)$$

where ϕ_m is the metal work function and χ_{sc} is the semiconductor electron affinity. The actual barrier heights observed experimentally are lower than predicted and do not follow an expression like (7.26). Instead, the barrier heights are almost independent of the nature of the metal for low polarity semiconductors like Si, Ge, and most III–V compounds (cf. Sect. 7.3). With increasing polarity, there is an increasing dependence of ϕ_b on ϕ_m, but not as

---

[3] It is conventional to use the quality factors A for HJ and n for MS and MIS junctions.

strong, nor even linear as suggested by (7.26). This independence of the experimentally determined barrier heights from the nature of the metal expresses the lack of reaction of the Fermi level to the deposition of metal layers under practial conditions, as if it were pinned by interface states. Barrier formation and the determination of the final barrier height are governed by truly microscopic interface processes which are discussed in Sect. 7.3.

These processes can be represented macroscopically as resulting in net interface charges and dipoles which affects the "true" metal and semiconductor work functions and the resulting barrier height. This has been the traditional way of attacking the problem by device physicists. A simple model assumes a constant distribution of interface states at the semiconductor side of the interface, separated from the metal by a thin insulating layer, leading *Crowell* and *Sze* [7.82], for example, to derive the expression

$$\phi_0 = \{C_2(\phi_m - \chi_{sc}) + (1 - C_2)[(E_g/e) - \phi_0^*]\} + C_0 , \qquad (7.27)$$

where $C_2 = \varepsilon_i/(\varepsilon_i + e^2 \delta D_{ss})$, with ε_i and δ as the permittivity and thickness of the interfacial layer respectively, and D_{ss} as the density of interface states. The position of the Fermi level if the metal were absent is ϕ_0^*, measured from the valence band, and C_0 is a small term that for practical considerations can be disregarded.

If $D_{ss} \to \infty$, then $C_2 \to 0$ and

$$\phi_b \cong [(E_g/e) - \phi_0^*] . \qquad (7.28)$$

In this case the Fermi level is pinned by interface states and the barrier height is independent of the metal and entirely determined by the interface properties.

If $D_{ss} \to 0$, then $C_2 \to 1$ and

$$\phi_b \cong (\phi_m - \chi_{sc}) \qquad (7.29)$$

as in (7.26).

It has been customary to fit experimental data with expressions such as

$$\phi_b = a\phi_m + b \qquad (7.30)$$

for a given semiconductor. From the experimental values of a and b the magnitudes of D_{ss} and ϕ_0^* could be determined. For a large variety of low polarity semiconductors [7.82], this analysis indicates an effective pinning of the Fermi level at $(E_c - e\phi_0^*) \sim \frac{2}{3} E_g$, the well-known "two-thirds rule".

The inclusion of fixed charge trapped within the insulating layer Q_{fix} can in turn be represented as modifying the metal work function ϕ_m to $\phi_m^{eff} = \phi_m - Q_{fix} d_{eff}/\varepsilon_i$ where d_{eff} is an effective distance from the trapped charge sites to the metal [7.105]. The final experimental barrier height in this case would be modified by a similar amount.

The two types of charge distributions mentioned above are expected to be dependent on the actual material systems and experimental conditions. The control of the characteristics of the interface and insulating layer finally determine the achievable barrier heights.

If we neglect any localized charge effects, the presence of an interfacial insulating layer has the following effects [7.98]: I) the passive layer acts as a dielectric separation between the metal and the semiconductor, lowering ϕ_b; II) the flow of carriers is limited by tunneling through the layer, reducing the current for a given bias; and III) the applied bias is sustained partially by the insulating layer and the barrier height will not be constant, resulting in values of n greater than unity.

c) MS and MIS Solar Cells

The utilization of MS devices for solar cell applications is very attractive because of the inherent simplicity of the system, its applicability to thin-film polycrystalline materials, and its inexpensive technology. The dark and light behavior of MS solar cells is described in [7.3], where the solar efficiencies for a number of different metals on Si and GaAs are calculated using empirical parameters. The calculated efficiencies are considerably lower than for Si and GaAs pn junctions because of the lower open-circuit voltages developed by MS devices. These low values of V_{oc} are a consequence of the large saturation currents which are, in turn, the result of the majority carrier transport mechanism and the low values observed for ϕ_b. Controlling this transport mechanism would be a way to improve the solar performance. More importantly, increasing the barrier height, a difficult task as it was shown above, would result in a major increase in the open-circuit voltage. If barrier heights close to the band gap could be obtained, the solar efficiencies for MS solar cells would approach those of the pn junction.

As a consequence of the study of MS solar cells, it has been found that the insertion of a thin insulating layer between the metal and the semiconductor can have an advantageous effect, mainly by increasing the open-circuit voltage of the devices [7.102]. The enhancement of V_{oc} can be achieved without appreciably decreasing the short-circuit current J_{sc} or the fill factor ff. These effects are usually observed up to a certain thickness of the insulating layer, of the order of 30 Å. Above this critical thickness, a strong reduction of J_{sc} and ff occurs, reducing the solar efficiency rapidly to zero. Improvements of the open-circuit voltage of over 50% have been reported for MIS structures based on n-GaAs [7.100], with V_{oc} values up to 0.7 V and solar efficiencies in the 15% range. Similar enhancement of the open-circuit voltage has been observed in MIS devices based on n-Si [7.101]. In Fig. 7.16 we show the effect of a properly controlled thin oxide layer on the light IV characteristics of an Au–GaAs Schottky barrier cell [7.108].

The role of the interfacial layer has been the subject of an increasing amount of literature recently [7.102]. The proposed mechanisms are not coincidental,

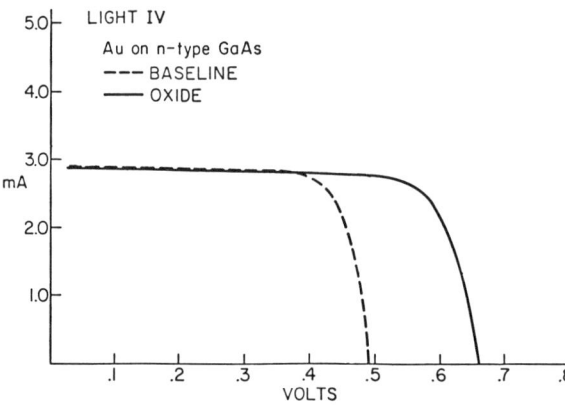

Fig. 7.16. Au/thin-oxide/n-GaAs and Au/n-GaAs (baseline) device *IV* characteristics under illumination. (Redrawn from [7.108]. Courtesy of the American Institute of Physics)

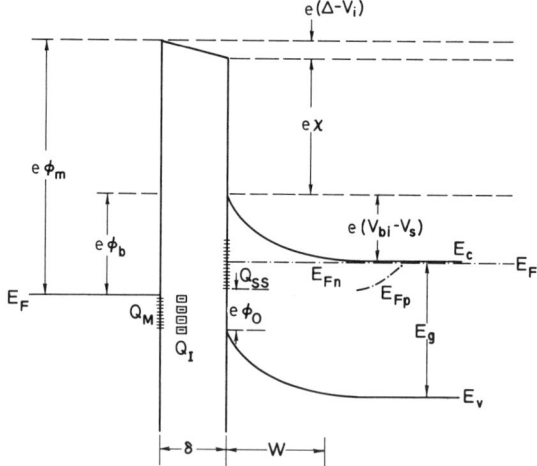

Fig. 7.17. Schematic diagram of an MIS solar cell under illumination. The applied voltage is partially supported by the interfacial layer (V_i) and partially by the semiconductor (V_s). The potential drop across the insulator is Δ for the unbiased device. The net charge at the metal surface is Q_M and the trapped charge within the insulating layer is Q_I. The potential ϕ_0 characterizes the distribution of interface states. (After [7.105], courtesy of the American Institute of Physics)

although based in similar concepts. It might well be the case that several processes must be taken into consideration to explain the variety of data, either singly or simultaneously.

Fonash et al. [103–107] has proposed a number of very basic mechanisms which can lead to V_{oc} enhancement. The schematic MIS energy band diagram shown in Fig. 7.17 illustrates the case. The junction is under illumination, developing a voltage V, a portion of which, V_s, is supported by the semiconductor, and the remainder V_i by the insulating layer. Assuming that the insulating layer interfaces an otherwise ideal MS junction, the total current density for the device under illumination can be expressed as [7.105]

$$J = \mathcal{T}_e A^{**} T^2 \exp(-e\phi_b/kT)[\exp(eV_s/kT)-1]$$
$$+ (\mathcal{T}_h p_{n0} D_h/L_n)[\exp(eV/kT)-1] - J_L. \tag{7.31}$$

The first term represents the majority carrier thermionic emission transport mechanism, modulated by a transport coefficient $\mathcal{T}_e$ which expresses tunneling processes across the insulating layer. The second term, the hole (minority carrier) ideal diffusion current, has been included since a small value for $\mathcal{T}_e$ would reduce the first term eventually to levels where minority carrier effects become important. The factor $\mathcal{T}_h$ represents hole tunneling through the insulating layer. The third term is the light-generated current which for simplicity is assumed to be independent of interface characteristics and bias. Fine details [7.105] have been omitted for simplicity. The voltage drop $V_i = V - V_s$ depends on the characteristics of the interface layer trapped charges, interface charges and dipoles, and the electrical characteristics of the semiconductor. By electrostatic analysis such as that by *Fonash* [7.105], an expression for $V_i(V)$ can be deduced. The voltage drop on the semiconductor side V_s can be written as $V_s = V - V_i = nV$, or $n = 1 - V_i(V)/V$, and (7.31) takes the form

$$J = J_{0e}[\exp(eV/nkT) - 1] + J_{0h}[\exp(eV/kT) - 1] - J_L. \tag{7.32}$$

This simple equation shows, within the limitations of the assumptions, the parameters that can be varied to increase V_{oc}: I) most important, the barrier height which determines the majority carrier saturation current J_{0e}; II) the diode quality factor n which in this formulation expresses the ratio (V_s/V), the amount of field shaping; and III) the ratio J_{0e}/J_{0h} which is controlled by $\mathcal{T}_e$ and $\mathcal{T}_h$, the transport tunneling coefficients.

I) *Barrier Height Modification.* As it was shown previously, the factors determining the ultimate barrier height ϕ_b are far from being understood. However, we can extract some operative conclusions from our previous description. Firstly, pinning of the Fermi level at the semiconductor interface is very effective in low polarity semiconductors, making ϕ_b practically independent of the metal. Secondly, while early researchers assumed that pinning was an intrinsic property of the semiconductor, it now appears to be the result of interfacial interactions between the semiconductor and its immediate overlayer (cf. Sect. 7.3). These two points leave us with an apparent paradox: pinning is induced by the overlayer, but the Fermi level position does not depend on its nature. In any event, slight variations of ϕ_b with the metal work function are expected. Furthermore, in a true MIS structure, the semiconductor is in contact with the insulating layer rather than the metal, a fact that can have an effect, at least in principle, on the ultimate barrier height. Additionally, the insulating layer is expected to be electrically active (if not, its presence would result in a reduction of ϕ_b) and trapped charge can, therefore, modify the effective barrier height. Fixed charge within the insulating layer produces a change in the effective ϕ_b proportional to $(-Q_{fix})$, where positive charge reduces it and negative charge leads to an increase. In such a case, the diode quality factor remains practically equal to unity but the saturation current changes with the amount of trapped charge. An increasing negative charge reduces J_{0e} exponentially and increases V_{oc}.

Only properly chosen insulating layers can lead to increases in ϕ_b [7.107, 109]. There is a tendency, however, for the presence of an insulating layer interfacing an *n*-type semiconductor to improve the MIS photovoltaic behavior. Experimental observation of Schottky diodes of Al, Cu, Ag, Pd, and Au on GaAs and Si [7.107] have shown higher barrier heights for the MIS configurations. It has been suggested that the presence of the interfacial oxide might satisfy dangling bonds at the semiconductor surface, reducing the density of interface states. This process would relieve the pinning of the Fermi level, resulting in higher ϕ_b values which would show a greater dependence on the metal work function. The experiments reported in [7.107, 108] show that MIS barrier heights in GaAs and Si are in fact higher than for the corresponding MS structures but have the same lack of correlation with the metal work function.

II) *Field Shaping.* The insulating layer partially sustains the total bias V across the device. The V_s/V ratio, and therefore the diode factor n, depend on the characteristics of the insulating layer and especially on the interfacial energy states at the insulator-semiconductor interface. This mechanism, called "field shaping" by *Fonash* [7.103], permits the possibility of higher V_{oc} values, even higher than the built-in potential, V_{bi}, since $V_s(V)$ is smaller than V. The V_s/V ratio depends on the actual distribution of interface states and the relative capture cross section for electrons (which should be high for the process to be effective) as compared with that for holes. This mechanism has a small effect on the saturation current but leads to n factors which are higher than unity and probably bias dependent. Several possible interface level distributions, with populations governed by the electron quasi-Fermi level, and their effects on the electrical characteristics of MIS devices have been examined in [7.103–108]. In this case, the improvement of V_{oc} follows from the increase in the n factor.

III) *Transport Control.* The mechanism of transport control can be effective for the improvement of V_{oc} if tunneling through the insulating layer is a limiting process in the overall transport of majority carriers. The effect of an increasing insulator thickness on the *IV* characteristics of a practical MS structure can be seen for example in [7.98]. Basically, the thicker the insulating layer, the smaller the forward current at a given bias. This process is lumped in the parameter $\mathcal{T}_e$ in (7.31). For solar cell applications it is, however, desirable that the insulating layer be transparent to minority carriers (i.e., $\mathcal{T}_h \cong 1$). Decreasing $\mathcal{T}_e$ would reduce the majority carrier saturation current, therefore increasing V_{oc}. Thus we wish to decrease $\mathcal{T}_e$ while holding $\mathcal{T}_h \approx 1$ if possible. The lower bound for the forward-current density is then imposed by the minority carrier diffusion current. This mechanism reduces the forward saturation current, leaving the diode factor equal to unity.

The three basic mechanisms discussed above have been contrasted with experimental evidence by *Fonash* et al. [7.107] using forward and reverse *CV*, dark and light *JV*, and infrared photoemission measurements. Certainly, additional research is needed to disclose the significant processes. At this stage it seems that the first two processes are the ones than can describe the majority of the results.

The role of tunneling across the insulating layer has been stressed in an alternative model proposed by *Shewchun* et al. [7.109–111] which compares successfully with experimental evidence of *p*-type Si MIS structures.

d) Stability of the MIS Interface

Finally, we briefly consider the electrochemical stability of the insulating layer and its interface with the semiconductor. Thus far we have considered electronic effects associated with a chemically stable MIS structure. Diffusion of oxygen through the metal into the insulating layer might produce aging effects, for example, showing a change with time of the dark JV characteristics and the photovoltaic response.

Ponpon and *Siffert* [7.112] studied aging effects on Au/n-Si Schottky diodes, finding that the electrical properties of these devices are directly correlated to oxygen diffusion through the metal. The devices were produced by vacuum evaporation of the metal at intermediate vacuum ($p \cong 10^{-6}$ Torr) where a naturally occurring native oxide is expected to be present. When tested under vacuum, the diodes showed no rectification. Exposure to air, however, resulted in increasing rectification ratios. The $\log J$ vs V characteristics showed that when the diodes are exposed to air, there is a reduction of the reverse saturation currents by several orders of magnitude which can be correlated with exposure time. The diode quality factor remains unchanged. A consistent increase in V_{oc} was observed. Both of these related effects are associated with an increase in the barrier height values with exposure time, from a small initial value for the unexposed diodes to a final value consistent with the reported ones in the literature. The increase of ϕ_b with exposure was evaluated from V_{oc}, J_0, and ir photoemission measurements and the barrier height values shown by these different measurements agreed. This fact supports the hypothesis that the main effects are the enhancement of ϕ_b towards "normal" values due to oxygen diffusion and accumulation at the insulating layer and its interface with the semiconductor.

Ponpon and *Siffert* suggested that the initial low values of ϕ_b are to be associated with a positive charge Q_{fix} trapped in the insulating layer. Oxygen diffusion through the metal neutralizes this charge restoring ϕ_b to its normal value. Metals with low oxide heat of formation, like Au, will allow O_2 to diffuse easily and aging effects will appear. Conversely, metals with high oxide heat of formation, like Al, will prevent oxygen diffusion and small aging effects will be observed. This is actually the case, as shown in [7.112], where a thorough model of the processes involved is proposed. The same authors studied Schottky barriers of a variety of metals on n-CdTe [7.113]. Unlike the case of Si, the Au barrier on CdTe is formed immediately after metal evaporation. An increase in V_{oc} is observed when the diode is exposed to air. By studying the exposure-time dependence of the forward JV characteristics, it was found that the main change arises from the factor n, while values of the saturation current remained practically the same. This result suggests that field shaping entirely due to interface states is the actual improvement mechanism.

e) The SIS Structure

Practical HJs investigated for solar cell applications often show large values of J_0 and severe limitations on the open-circuit voltage. This is the result of nonideal transport through the junction. They show the same kind of limitations found in Schottky barrier devices. One is tempted to consider, then, the possibility that the introduction of a thin insulator interfacing the two semiconductors might lead also to improved photovoltaic behavior. It can be argued that the nature of the problem is different. However, HJs are a large class of devices of which Schottky barriers represent a limit. At least close to that limit, an insulating layer may lead to enhanced characteristics. In the general case, though, there is probably no universal answer. It would depend on the relative doping of the semiconductors and the particular interface characteristics. *De Visschere* and *Pauwels* [7.114] analyzed theoretically the photovoltaic behavior of the SIS structure. Due to the complexity of the problem, restrictive assumptions had to be made. Even so, different cases could be distinguished. It was found that when collection of photogenerated carriers takes place in the more heavily doped semiconductor, the insulating layer is disadvantageous. When collection takes place in the weakly doped semiconductor, the insulating layer has an advantageous effect if the band discontinuity has the appropriate sign; in the other case, the layer has no influence. These conclusions would indicate that an insulating layer would have deleterious effects on solar cells of the Cu_2S/CdS type but might be advantageous on conductive metal-oxide/absorbing-semiconductor systems depending on the particularities of the interface.

Many practical HJs are in fact SIS structures since native insulating oxides are present at the interface. *Schewchun* et al. [1.115] have proposed an SIS model to describe the ITO/Si system, where experiments indicate the presence of a thin interfacial insulating layer. They extend the analysis of MIS structures of [7.109] to the SIS structure where one of the semiconductors is a degenerate wide-band-gap oxide. The effect of several parameters was examined and a maximum theoretical efficiency up to 20% for the ITO/thin insulator/Si system is predicted. Other than these papers, very little work appears to have been done on analysis of SIS structures.

7.2.5 Summary and Application to Real Heterojunctions

So far in this section we have examined the factors that determine characteristics of HJ diodes. In particular, the Anderson model was extended to include interface states and dipoles in order to explain experimental results. It appears that HJs can be divided into two broad classes. First, there are those with near zero lattice mismatch and extremely careful preparation, giving negligible interface state densities. The transport properties are dominated by bulk properties. Despite the expense of their preparation, such HJs show great promise for concentrating systems with already realized solar efficiencies in the 20–30% range. As we see in the following sections, the zero lattice mismatch and careful preparation may not be sufficient to insure the absence of interface states except for fortuitous choices of HJ components. In the second class are the HJs with small to large mismatch in which the transport is almost completely dominated by the presence of interface states, both by distortion of the bands and by recombination and/or tunneling.

Recently it has been found that the "$A=2$" current in many HJs, particularly those with small areas, is controlled by interfacial recombination at the periphery of the heterojunction plane. *Stringfellow* [7.116] found that the recombination/generation current in GaAsP and GaP HJ LEDs was directly related to the area-to-perimeter ratio and that it could be reduced temporarily by CF_4 plasma etching. While *Sah* et al. [7.16] found negligible perimeter current in the Si p/n junctions used to substantiate their theory, later work by *Sah* [7.117] showed that the perimeter current in Si p/n diodes could be substantial unless a surface oxide was present. *Henry* et al. [7.118], in a very thorough work, showed (using luminescence techniques) that the $A=2$ current in AlGaAs pn junctions is due to surface recombination at the junction perimeter for a wide variety of surface treatments. The "$A=2$" behavior is a consequence of a nearly constant ratio of electron to hole densities at the surface which is, in turn, required to maintain overall charge neutrality at the surface in the presence of Fermi-level pinning.

Although these measurements are for small area devices ($\gtrsim 0.04\,mm^2$), they clearly indicate that perimeter currents cannot be ignored in experimental cells with small areas. The results might also be extended to polycrystalline cells where a large number of large angle grain boundaries penetrate the junction plane.

We have briefly summarized the modes of HJ transport which have received the most attention. Most HJs exhibit a combination of two or more of these modes, depending on temperature and bias voltage. In Table 7.1 data on several of the more common HJs are presented, with values of A, J_0, and solar efficiency taken from the literature. An indication of the major mode of current transport at 300 K is also given where possible. The implications of these data are the following:

  I) Near zero lattice mismatch is beneficial but by no means required for small J_0 and large solar efficiency.

  II) The values of J_0 are in almost all cases many orders of magnitude larger than predicted by simple theory.

The foregoing discussion on HJ, MS, and MIS structures has led us to focus on the interface properties as determining factors for J_0 and A (or n for Schottky type junctions). Control of interface properties is thus a major pathway toward increased solar efficiency. Only in fortuitous cases does current

Table 7.1. Transport parameters and solar efficiency of selected HJs[a]

HJ	Ref.	$\varepsilon_{mf}[\%]$[b]	$J_0[\text{A cm}^{-2}]$	A[c]	$\eta_s[\%]$	Transport mechanism
n-CdS/p-InP[d]	[7.26]	0.3	3.0×10^{-8}	2.38	12.8	Recombination
n-CdS/p-InP[d]	[7.25]	0.3	1.5×10^{-6}	2.93	12.5	Recombination
n-CdS/p-InP[d]	[7.27]	0.3	1.3×10^{-8}	2.14	14.4	Recombination
n-CdS/p-InP[d]	[7.24]	0.3	1×10^{-5}	2.99	12	Recombination/tunneling
n-ITO/p-InP	[7.24]	large	1.2×10^{-7}	2.35	12.4	Tunneling
	[7.24]	large	2.5×10^{-8}	2.02	8.3	Tunneling
n/p InP	[7.25]	0	2×10^{-8}	2.36[e]	?	Recombination/diffusion
			$\approx 1 \times 10^{-14}$	1.08[e]		
n-CdS/p-CdTe	[7.17]	9.7	1.7×10^{-8}	1.89	8.0	Recombination
n-ZnO/p-CdTe	[7.28]	29.1	1.2×10^{-7}	1.98	≈ 3	Recombination
p/n CdTe	[7.119]	0	5×10^{-9}	≈ 2	?	Recombination
p-Cu$_2$S/n-CdS	[7.19]	4.6	5.5×10^{-11}	1.0	≈ 8	Recombination at interface[f]
n-ZnSe/p-GaAs	[7.23]	0.3	2×10^{-10}	1.4	?	Recombination?
n-Ge/p-GaAs	[7.21]	0.07	3.4×10^{-8}	1.68	?	Tunneling
n-Ge/p-GaAs	[7.21]	0.07	8.1×10^{-9}	1.17	?	Tunneling
p/n GaAs[g]	[7.120]	0	1.1×10^{-10}	2.0	14.8	Recombination
p/n GaAs[g]	[7.121]	0	3.8×10^{-12}	1.85[e]	12	Recombination/diffusion
			7.4×10^{-14}	1.54[e]		
	[7.121]	0	2.7×10^{-13}	1.75[e]	12	Recombination diffusion
			6.6×10^{-18}	1.15[e]		
n/p Si	[7.6]	0	5.5×10^{-6}	2.86[e]	11.6	Recombination/diffusion
			3.3×10^{-12}	1.0[e]		

[a] All data are at ≈ 300 K, η_s is the solar efficiency.
[b] $\varepsilon_{mf} \approx$ lattice mismatch.
[c] An equivalent diode quality factor A is given even though tunneling dominates in some cases. The tunneling exponent is then $\alpha = e/Ak(300 \text{ K})$.
[d] First named component is deposited on the second except for homojunctions.
[e] Indicates two branches of curve for single device.
[f] Calculated from data given for illuminated case.
[g] With AlGaAs window layer.

transport depend principally on the bulk properties of the components. In the section discussing MIS and SIS junctions we saw that there are ways to improve V_{oc} and solar efficiency by manipulation of the interface conditions in some cases, namely by the introduction of a thin insulating layer with the proper charge states.

The next section delves into the interface on a microscopic scale in order to elicit further clues as to how interface properties might be beneficially changed.

7.3 Interface Related Phenomena

In previous sections (7.1.3, 7.2.3) we discussed the effects of the interface on various HJ properties in terms of rather idealistic models, concentrating on the effects of the interfacial defects rather than the details of their origin. In this section, we focus on the physical nature of the interface.

It was previously stressed that the interface of a heterostructure is of paramount importance for the ultimate behavior of the device. In the region where two dissimilar materials come into intimate contact, we find a breakdown of periodicity, the most formidable defect that can be found in a crystalline structure. There is a drastic disruption of bulklike properties and new specific concepts come to play a very significant role: lattice misfit and misfit defects, interface states trapping carriers or acting as recombination centers, localized charges and dipoles, band-edge discontinuities, and jumps in the electrostatic potential and its derivative. We have a host of new interrelated phenomena with possibly predictable influence on the crystalline and energy configurations of the system and its transport properties.

Even an ideal interface confined to a crystalline plane will shape the transition region of the device and the overall behavior. Actually, the highly defective interface region extends into the transition region complicating the problem to a greater extent. Furthermore, the interface and its immediate neighborhood is characterized by a very distinctive electrochemistry, especially in the initial stage of formation. This governs alloying, intermediate compound formation, localized diffusion of defects and impurities, and precipitation in the defective regions. This occurs in layers where we find important gradients of chemical, electric, and mechanical potentials (if such macroscopic concepts can describe these very microscopic mechanisms), and presents situations that cannot be characterized by bulklike properties in an immediate way.

The characteristics of interface formation and behavior are truly microscopic in nature. Due to the lack of experimental techniques disclosing the microscopic aspects of the problem, the heterostructure interface has been for some time the object of pure speculation and in many cases oversimplification. With the application of the powerful surface and interface spectroscopic techniques in recent years, the microscopic processes associated with interface formation and behavior can be studied. Simultaneously, electron microscopy helps in the understanding of the defect configurations at the interface region. This information can be added to the results of a large variety of experimental techniques adapted or specific to surface research to build a complete description of the interface phenomena.

In the following pages we briefly review some new developments in interface physics and speculate on possible generalizations. Due to the limited extent of this work, we have confined ourselves to describing just a few new findings that we find basic, general, and certainly important in richness, and for possible extrapolations. We concentrate first on the metal-semiconductor (MS) interface since most of the available information deals with this system, due to its technological importance and the fact that is slightly simpler than the semiconductor-semiconductor (SS) interface. We then extend the description to intermediate compounds at the interface, and finally refer to the SS interface. To close, we refer to lattice mismatch and its effect on HJ behavior.

Before entering the subject we would like to point out that very basic mechanisms are described. As a consequence, the experimental techniques and material systems under consideration are highly ideal. We shall see that basic processes (for example, Fermi-level pinning) occur at a microscopic scale and at very early stages of interface formation. Therefore substrate preparation, cleanliness, and growing technique might well determine a priori the most important characteristics of the interface. This is especially important in extrapolating results from very ideal experiments to real life devices.

7.3.1 The Metal-Semiconductor Interface

a) Introductory Remarks

The basic considerations for experimental MS barrier heights were introduced in Sect. 7.2.4b. There we mentioned that if the interface did not play any role, the barrier height would be determined by the interchange of charge between the metal and the semiconductor as governed by the difference in their respective work functions. When the two-material system achieves equilibrium, the Fermi level is constant throughout, and its position is determined by bulklike properties, although some degree of surface dependence is unavoidable through the surface dependence of the work functions themselves. In this case the barrier height is given by (7.26). This expression expresses the basic Schottky model. The fact that barrier heights measured in metal-covalent semiconductor structures did not follow this model, also mentioned in Sect. 7.2.4b, was found very early in research on MS structures. The independence of barrier heights on the metal work function ϕ_M was indicative that the stabilization of the Fermi level was determined by other mechanisms. *Bardeen* [7.122] postulated that intrinsic surface states at the semiconductor surface were responsible for the Fermi-level pinning; further contact to the metal produced second-order effects. (A thorough introduction to semiconductor surfaces can be found in the book by *Many* et al. [7.59].) *Heine* [7.123] argued that, strictly speaking, localized semiconductor states cannot exist at an MS junction. But virtual resonance (metal-like) surface states can exist and account for the observed behavior. The study of the MS interaction has been the object of an immense amount of experimental and theoretical research and a substantial part of it (at least until a few years ago) related to intrinsic surface states. *Lindau* et al. [7.124] gives a brief but complete review.

The stabilization of the Fermi level at a given energy and consequently the dependency of the actual barrier height on the metal and semiconductor work

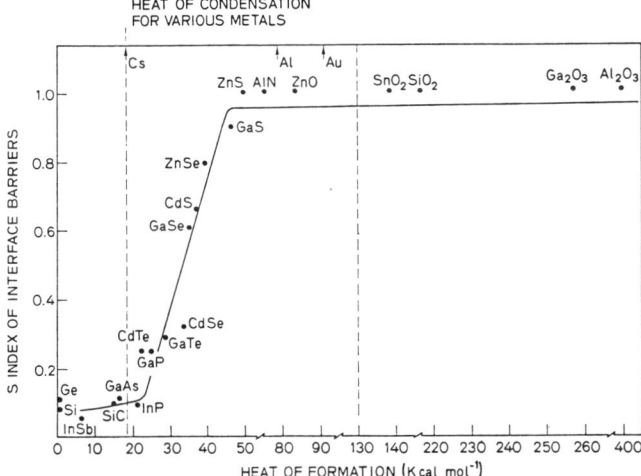

Fig. 7.18. The index of interface behavior S plotted against the chemical heat of formation. The values for S are those reported in [7.127]. The range for the heat of condensation for various metals is also shown. (After [7.124], courtesy of the American Institute of Physics)

functions have been customarily expressed by a linear equation of the form (see Sect. 7.2.4b):

$$\phi_b = S'(\phi_M - \chi_{sc}) + b', \tag{7.33}$$

where $S' = 1$ expresses the Schottky limit and $S' = 0$ complete stabilization with independence of the metal work function. Since the metal work function follows a linear relationship with its electronegativity X_M, the above equation can be written:

$$\phi_b = S(X_M - \chi_{sc}) + b \tag{7.34}$$

which has the advantage of using the more basic electronegativity instead of the metal work function which experimentally depends strongly on the surface conditions [7.125, 126].

In a paper which challenged the ingenuity of the researchers working in the field, *Kurtin* et al. [7.127] presented an extensive compilation of experimental MS barrier height data. The index of interface behavior S, defined by fitting the data with an expression like (7.34), was found to follow a trend, from $S \approx 0$ (complete pinning of the Fermi level for low-polarity semiconductors) to $S \approx 1$ (for high-polarity tetrahedral semiconductors and ionic compounds). When S was plotted as a function of the electronegativity difference ΔX for the semiconductor, a sharp transition was found at $\Delta X \approx 0.8$. We show in Fig. 7.18 a plot of the index of interface behavior as a function of the semiconductor heat of formation ΔH_F. This parameter is related to ΔX but provides a better basis for discussion in terms of interface chemistry.

This, among a set of empirical rules for Schottky barrier formation [7.128–132], constituted the base for discussion on MS interface behavior.

b) Ambiguities of the Parameter S

It is relevant to point out here that until recently some ambiguities of *Kurtin*'s curve were given little attention. *Schluter* [7.133] points out that $S = \partial \phi_b / \partial X_M$ has a limit for no pinning (the Schottky limit) that should be expressed as $S = S_0$, where S_0 is in principle different from one. S_0 corresponds to $S_0 = \partial \phi_M / \partial X_M$ which is a number between 2 and 3 for energies given in eV and electronegativities measured on the Pauling scale. Unfortunately in the expression for S_0, the quantity ϕ_M appears again with its dependence on surface conditions, a fact that we tried to avoid precisely by using the quantity X_M. The Schottky limit is then not represented by $S = 1$ but by $S = S_0 = 2-3$, and is dependent on the material system under consideration. Furthermore, such a limit has not been achieved even in compounds with high electronegativity difference. *Schluter* stresses that the scattering of the barrier height data demands a critical reexamination, making a linear fitting like (7.34) difficult and oversimplifying, and the abrupt transition of S rather ill defined.

We think that the nonlinearity of the barrier heights of different metals contacting a given semiconductor might be significant and that there may be a dependence of ϕ_b on the reactivity of the particular interfaces. The sharp transition in S must be closely reexamined. This sharpness is significant in the sense that it would provide a hint on the nature of the processes involved in MS interface formation. Most of the semiconductors shown in Fig. 7.18 are isoelectronic with the fourth column elements. Their essential properties, i.e., lattice parameter and band gap, vary smoothly from one to another as we compare them. However, there are other properties which show sharp transitions when the processes involved result from a delicate balance of energies, e.g., the self-compensation phenomenon. We comment on this point as the section proceeds.

c) The Origin of Surface States

Until recently the existence and role of intrinsic surface states in Schottky barrier formation has been the matter of controversy. Refined experiments in the last few years (see for instance [7.124] and references therein), have shown that at bare surfaces with a high degree of perfection there are no intrinsic states. This is the case for a large variety of the compounds and surfaces examined. Metal-induced phenomena are responsible for the pinning of the Fermi level at the interface [7.134–138]. These phenomena are truly microscopic and the MS interface is determined by very small coverage of metallic atoms, less than a monolayer. The reason for the controversy lasting so long can be easily understood. Ideal surfaces are unachievable. Observed surface

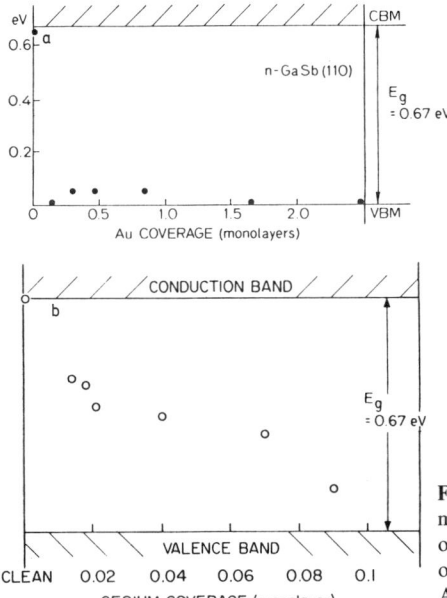

Fig. 7.19a and b. The Fermi level position for Au metal overlayers (**a**) and Cs metal overlayers (**b**) on the (100) n-GaSb surface shown as a function of coverage. (After [7.124], courtesy of the American Institute of Physics)

states are to be considered intrinsic if a high degree of perfection in surface preparation allows the assumption that all extrinsic causes have been eliminated. As pointed out in the thorough paper by *Spicer* et al. [7.134], there was general agreement at the III–V Interface Meeting of 1975 at Los Angeles in that the Fermi level at the (110) GaAs surface was intrinsically pinned near midgap. Several developments in subsequent years showed that this is not necessarily the case and pinning should be attributed to external causes rather than intrinsic properties. From the perspective of the present study, this result is doubly important. First, it discloses the basic properties of semiconductor interfaces. Secondly, it shows that unless close-to-ideal surfaces are prepared, Fermi-level pinning is unavoidable and arises from accidental processes and, paradoxically, at apparently the same position.

Lindau et al. [7.124] examined clean (110) surfaces of the III–V semiconductors GaAs, GaSb, and InP prepared by cleavage under UHV conditions ($p \approx 1 \times 10^{-10}$ Torr), on which overlayers of Au, Al, and In were evaporated. Photoemission measurements were then carried out using synchrotron radiation. The measured Fermi-level position for (110) GaSb as a function of Au coverage is shown in Fig. 7.19a. In less than 0.1 of a monolayer, a band bending over the entire band gap takes place, followed by a small variation between 0.1 and 1 monolayer, after which complete stabilization is achieved. The stabilization of the Fermi level starts taking place before a bulklike metal overlayer is deposited. Similar results are observed for Cs coverage, Fig. 7.19b, and for Au on (110) GaAs and (110) InP. By studying the photoemission spectrum of the

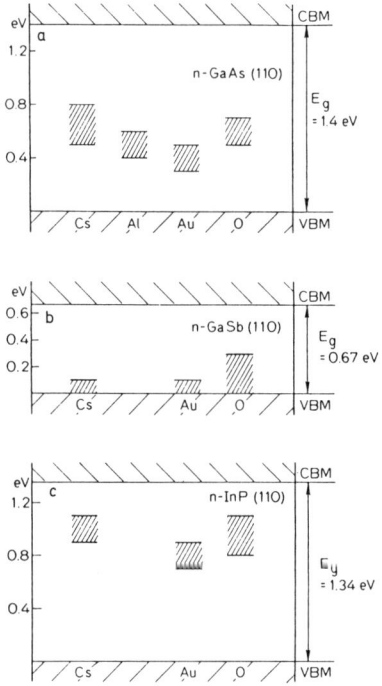
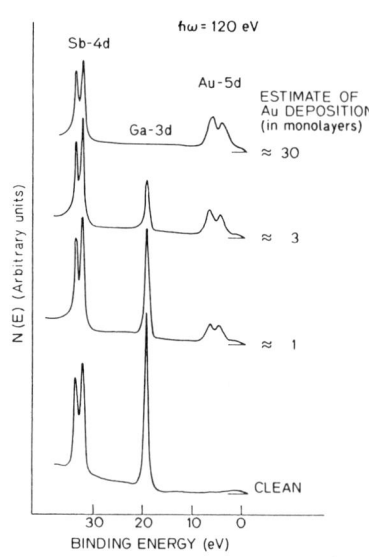

Fig. 7.20a–c. Fermi level positions for different adatoms: Cs, Al, Au, and O on the (100) surfaces of GaAs (**a**), GaSb (**b**), and InP (**c**). (After [7.124], courtesy of the American Institute of Physics)

Fig. 7.21. Photoemission spectra taken at a photon energy of 120 eV for (100) GaSb exposed to different amounts of Au. (After [7.124], courtesy of the American Institute of Physics)

system, it was determined that dispersed Au atoms rather than bulklike Au clusters were present. Another important fact is the apparent independence of the Fermi-level position on the nature of the overlayer. In Fig. 7.20 a schematic diagram of the Fermi-level position for the three III–V compounds for four different adatoms is shown. In all the cases the same early Fermi-level pinning is observed and occurs approximately at the same energy, even though the three adatoms are metals with very different electronegativities (Cs: 0.7, Al: 1.5, and Au: 2.4, Pauling scale), and one, O, is a nonmetal.

In Fig. 7.21 we show the energy distribution curves for (110) n-GaSb for various Au coverages. The photoemission curves were obtained at a radiation energy of 120 eV such that most of the signal comes from the last two molecular layers. We see a decrease of the Ga $3d$ signal, until complete disappearance, with increasing Au coverage; while the Sb $4d$ emission is constant, after an initial drop, even for coverages 15 times larger than the sampling depth. The results show that the surface layer of GaSb decomposes as a consequence of interface formation. Sb moves to the metal layer surface (as shown by Auger profiling), and a nonstoichiometric interface results. However, for InP the cation and anion signal decrease in a similar fashion, still giving sizeable emission for coverages 20 times the sampling depth. Decomposition is a fait accompli in these systems, and the decrease of both signals together in InP does not preclude the assumption of nonstoichiometry at the interface since such a slight difference in concentration is beyond the detection capability of the technique.

In these systems at least, we are dealing with strong interactions at the interface which modify the semiconductor surface chemically. This occurs very early and to a first approximation independently of the nature of the adatom. This fact would make Schottky barrier heights independent (within certain limits) of cleanliness and degree of perfection for instance, which is apparently the case [7.139–141]. *Lindau* et al. [7.124] hypothesize that the strong interaction produces defective interfaces and these defects are responsible for the Fermi-level pinning. The removal of constituent atoms towards the metal surface prevents healing the damage. The driving force for such a strong reaction is proposed to come from the heat of condensation of the metal. Depending on its comparison with the semiconductor heat of formation, the metal condensation will produce a high or low density of defects. The range of heat of condensation for various metals is shown in Fig. 7.18.

The fact that Ga deposited on *n*- and *p*-type GaAs [7.142, 143] shows no Fermi-level pinning can be considered to support a defect model for stabilization of the Fermi level. From this perspective it is interesting to note the similarity between the behavior of the index of interface behavior S and the extent of self-compensation both as functions of the semiconductor heat of formation. The extent of self-compensation might be defined as the ratio of the concentration of active dopant defects to the total concentration present. It is a simple empirical parameter and describes the "easiness" of moving the Fermi level position towards the band edges by extrinsic dopants in the bulk. As a general approximation, we can say that compounds lying in the $S=0$ region of Fig. 7.18 are easy to dope. Moving towards the ionic compounds there is, generally speaking, an increase in band gap and a decrease in cohesive energy. When an intrinsic defect is created in a crystal by breaking bonds, the energy of the system increases. The energy of the system is lowered by an exchange of electrons between the two sets of defects, however. When the band gap is comparatively large and the cohesive energy comparatively low, compensating defect creation becomes energetically favorable. Between no self-compensation for "covalent" materials and complete self-compensation for ionic compounds there is a "sharp" transition which lies in the same range as that for S. It is not unreasonable to think that a similar mechanism takes place at the interface. In this region it is easier to create defects for a variety of reasons. Less energy is required for defect creation and chemical interactions at the interface produce large amounts of local chemical energy. To illustrate the case we assume that the same kind of pinning defects are created in all compounds upon metal deposition. These defects are responsible for pinning the Fermi level in covalent semiconductors. The more ionic the compounds, the more easily interface self-compensation would occur and the less the position of the Fermi level would be determined by uncompensated states at the interface.

In the same way as the parameter S is a simple (and oversimplifying) concept to describe very complex processes taking place in MS barrier formation, self-compensation as described above is also a simple concept representing a complex set of defect electrochemical reactions. To verify the possible relation both should be thoroughly reexamined.

As was evident in the preceding pages, we think that to stress the role of defects at the interface is a most valuable approach. The driving force for defect formation is a much more complicated matter however. Chemical bonding cannot be disregarded in principle. We can expect that Au or Ag on GaAs, for example, won't react to a great extent at the interface, but more reactive metals like Al probably will, as seems to be the case [7.142]. If there is one general mechanism responsible for the Fermi-level pinning in the whole range of semiconductors, it should include the cases of both reactive and unreactive metals.

Brillson [7.144], adopting a different approach than *Lindau*, also reports the same early effects on the position of the Fermi level upon metallization. In his work, semiconductor compounds spanning a wider range in S, i.e., GaAs, CdSe, CdS, and ZnS, were examined. The interface behavior was characterized by Low Energy Electron Loss Spectroscopy (ELS), Surface Photovoltage Spectroscopy (SPS), and Ultraviolet Photoemission Spectroscopy (UPS). None of the surfaces examined showed intrinsic surface states which can account for Schottky barrier formation. Here a distinction between reactive and unreactive metals was made. The first produced detectable alloys at the interface but both sets gave rise to extrinsic surface dipoles responsible for the Fermi-level stabilization. The proposed interface reaction is of the type:

$$M + (1/x)CA \rightarrow (1/x)(M_xA + C), \tag{7.35}$$

where M, C, and A are metal, cation, and anion atoms respectively. The heat of formation for the reaction of (7.35) can be estimated from thermodynamic tables for the semiconductors and the most stable products M_xA. By comparison of both reactive and unreactive interfaces, a critical heat of formation H_R^c can be determined which constitutes the critical value for interface compound formation.

The extent of interface reactivity is reflected in ELS measurements. In Fig. 7.22 the ELS spectra are compared for Cu (reactive) and Ag (unreactive) on cleaved $(10\bar{1}0)$ CdS versus metal overlayer coverage. Cu on CdS exhibits a chemical shift of the MMM Cu Auger peak upon metal deposition indicating that a reaction had taken place. In this case, the ELS spectrum corresponding to cleaved CdS surface changes drastically with metal coverage from the very initial steps, showing a new set of features which do not correspond to a superposition of metal and semiconductor structures. The spectra change with

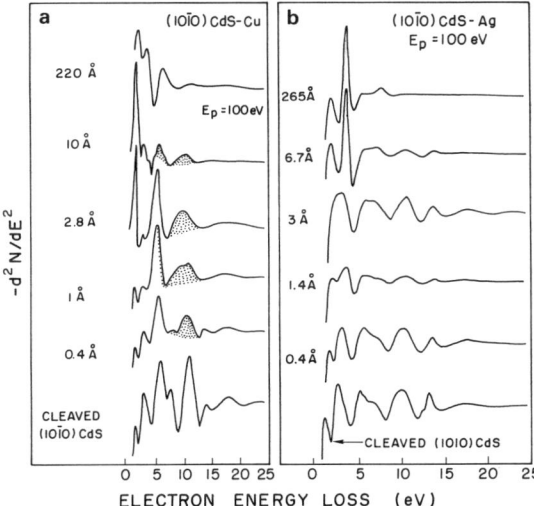

Fig. 7.22a and b. Electron-loss spectra of reactive and unreactive interfaces as a function of metal overlayer coverage: (a) Cu on cleaved (10$\bar{1}$0) CdS (features unrelated to bulk Cu or CdS are shaded) and (b) Ag on cleaved (10$\bar{1}$0) CdS. (After [7.144], courtesy of the American Institute of Physics)

coverage until the metallic structure dominates. In contrast, Ag on the same surface exhibits only a superposition of metal and semiconductor features. Al behaves as Cu in CdS, CdSe, and GaAs, while Au behaves as Ag in CdS and CdSe. From the set of ΔH_R values for $M_x A$ formation in these cases a critical interface heat of formation of about 0.5 eV per metal ion is evaluated. Most metals, though, are reactive following this criterion, and even more so for low-polarity semiconductors. Nonreactive systems were studied by SPS. It was found that at early stages of interface formation extrinsic levels appear within the forbidden energy gap. Each of these levels, on the four semiconductors, agrees with the ultimate Fermi level position to within ±0.1 eV. Fermi level positions were also studied by UPS and similar early stabilization was observed.

In principle, the distinction with respect to the reactivity of the interfaces complements and does not contradict a defect model for Fermi-level pinning. The defect model was proposed to explain results obtained for III–V compounds for which the barrier heights are almost independent of the nature of the adatoms. If we consider a wider range of polarity, where a larger range of barrier heights is found, a scheme which includes chemical bonding might account for: I) the lack of linearity of barrier heights versus metal reactivity, II) the fact that S saturates before the Schottky limit is achieved, and III) the anion rule [7.131].

Besides Fermi-level stabilization, the existence of an intermediate compound between the metal and the semiconductor is of primary importance for the electronic behavior of the device. The intermediate compound might be the result of a reaction between the metal and the semiconductor or a compound present accidentally at the semiconductor surface, e.g., an oxide present prior to metallization. In the first case, the low temperature bonding of metal atoms and semiconductor constituents has both theoretical and practical importance and several systems have been studied in relation to this problem. As an example, *Roth* and *Crowell* [7.145] studied transition-metal/Si compound formation using peak shape analysis on the Si $L_{2,3}VV$ Auger feature. Again, it is suggested that the heat of formation ΔH_R for silicide formation correlates with the extent of M–Si bonding found at the interface, as determined by Auger profiling. Particularly it is found that Pd reacts with Si forming predominantly Pd_2Si. The compound intermediates between the metal and Si so that the semiconductor interfaces the alloy rather than the metal. In contrast, in the case of Nb, even though some silicide is detected, mostly unreacted metal is in contact with the Si surface.

An example of the effects of compound formation at a semiconductor-semiconductor interface is that reported in [7.146] for CdTe/ZnSe HJs. The spectral response curves of J_{sc} for these junctions clearly show the presence a thin intervening layer of CdSe or ZnTe at the junction, depending on preparation. These devices were prepared at rather high temperatures (380–570 °C) by close-spaced vapor transport.

d) Experimental Evidence of Electrical Effects of Surface Quality

At this point it is interesting to mention research done by *Amith* et al. [7.141] on the behavior of Au Schottky barriers on (110) GaAs as a function of surface preparation. A set of samples with different degrees of cleanliness and order (as determined by Auger analysis and LEED) were prepared and the electronic characteristics, i.e., barrier height, JV, and CV characteristics, were measured. The barrier heights were found to be independent of surface preparation, consistent with our previous discussion. But a very different behavior was found regarding the capacitance measurements with even more drastic differences in the JV characteristics.

Among the variety of preparation conditions for *Amith*'s samples we discuss two as an example of how surface preparation has a strong influence on the electrical parameters without affecting the ultimate barrier height. These data also emphasize the hidden parameters that must be considered with regard to fabrication techniques. *Amith*'s sample surfaces 6 and 9 (Fig. 7.23) were cleaned and ordered by sputter etching and annealing before depositing Au contacts, without breaking vacuum. Samples 7 and 8 did not undergo the sputter "cleaning", and can be considered "dirty" and "disordered". The first set gave nonlinear plots of $1/C^2$ against V, high values of C, small depletion layer widths, and inordinately high (meaningless) values for V_{bi}. The second set behaved in a normal way. One possible reason is that sputter etching creates defects near the surface (Ar bombardment removes ions from the semiconductor surface, preferentially As). Annealing reorders the surface, but leaves the As vacancies in a region of about 15 Å from the surface. Typically, these As vacancies act as donors and produce an n^+ region in the neighborhood of the interface.

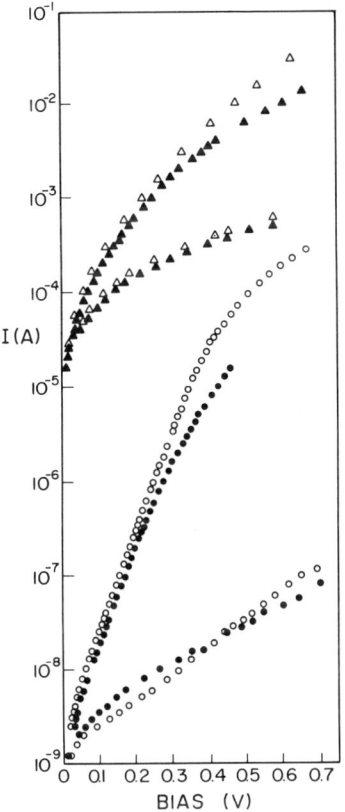

Fig. 7.23. Forward and reverse current vs voltage characteristics for representative samples of Au/GaAs (110) Schottky diodes. Samples 7 (●) and 8 (○) are considered to have "dirty" and "disordered" surfaces; samples 6 (▲) and 9 (▽) have "clean" and "ordered" surfaces. (Redrawn from [7.141], courtesy of the American Institute of Physics)

Capacitance measurements are very sensitive to the charge density in this region, and therefore the CV results must be reexamined, abandoning the constant doping assumption and including this extra doped layer. In addition, "dirtiness" and "disorder" do not seem to produce a significantly disrupted region such as to interfere with the capacitance results. We show in Fig. 7.23 the dark JV characteristics of the two sets. Again samples 7 and 8 give reasonable results. Their curves follow the thermionic emission model, the carrier concentrations extracted from them agree with the CV determinations, and the n factors are in the 1.2–1.3 range according to the model. The poor behavior of the sputter etched and annealed samples 6 and 9 is evident.

The effects of sputtering damage on the electrical characteristics of Mo/Si Schottky barrier diodes have been investigated by *Mullins* and *Brunnschweiler* [7.147]. Wafers of n-type Si, cut in the (111) plane, and having a resistivity of 0.2–0.5 Ω cm were cleaned following a standard chemical procedure. The diodes were produced by dc cathodic sputtering of Mo in an Ar atmosphere, at a pressure of about 0.1 Torr. The JV characteristics were found to depart gradually from ideality with increasing sputtering voltage for a given deposition time. Diodes produced with a cathode voltage of 1 kV for 60 min gave forward current characteristics close to ideal, with a barrier height of 0.74 eV. For the same deposition time, an increase in cathode voltage to 1.2 kV is enough to show noticeable departure from ideality, which increases dramatically with cathode voltage. For a given sputtering voltage the departure from ideality also increases with deposition time. Similarly, the CV characteristics of the devices showed increasing nonideality with both deposition parameters.

The experimental results were described assuming an exponential distribution of donorlike traps generated during the sputtering process. The density was assumed to be of the form $N_t(x) = N_{t0} \exp(-x/L)$, where N_{t0} is the trap density at the interface and L is a characteristic length. The experimental curves were fitted well by assuming a single trap level 0.43 eV below the conduction band and the values of N_{t0} and L can be estimated. At constant deposition time, N_{t0} was found to be independent of sputtering voltage, while L showed a linear dependence. Conversely, for a given sputtering voltage, N_{t0} showed a logarithmic relationship with deposition time and a constant characteristic length. The apparent independence of N_{t0} and cathode voltage suggested to the authors that the damage was produced mainly by electrons.

e) Oxide Interlayers

Among the most typical nonidealities found in semiconductor devices are native oxide layers in a thickness range which considerably affects the electronic behavior. Because of its technological importance, there is an extensive literature on the oxygen-semiconductor interaction. An example of the presence of native oxide as an intermediate compound is the research done by *Garner* et al. [7.148] on the $Au-Al_xGa_{1-x}As$ interface. The samples were prepared by chemical etching and then transferred to a vacuum system where Au was deposited as soon as a vacuum of 10^{-7} Torr was achieved. The results were compared with surfaces exposed to air for some weeks but not etched. Auger profiling in both cases showed the presence of an oxide layer of about 50 Å, the etched sample showing a more intense O Auger peak and even thicker oxide. This might be associated with a very reactive surface left after etching which grew an oxide layer during the transferring process, or merely the result of the etchant solution. The first hypothesis is very significant since this system tends to grow an oxide overlayer in times much shorter than any transferring process. The dependence of the oxide thickness on the surface index was also examined and a much thicker oxide was found on the (111) face than on the (110) face. By chemical shift data it was found that Al_2O_3 is the major oxide present.

The utilization of surface analytical techniques allows the investigation of not merely the presence and thickness of an intermediate oxide layer but, most important, the structural and chemical characteristics of the oxide. The defect structure, principally chemical composition and stoichiometry, determines the electronic interaction with the semiconductor while the crystalline structure determines the degree of perfection of the interface. *Revesz* [7.149] gives an interesting review of literature on the SiO$_2$/Si interface and suggests that the noncrystallinity of the thermal oxides grown on Si is responsible for the degree of perfection of the interface. This interface has attracted much attention due to its importance in modern electronics. The local atomic and chemical structure has been analyzed with interface techniques such as AES [7.150], ISS [7.151], and XPS [7.152]. In this last work, a transitional layer about one monolayer thick is found, consisting of three different and coexisting Si oxidation states, followed by a strained region with a different polymer structure than the rest of the amorphous SiO$_2$ layer.

The analysis of the oxide/semiconductor interface has also been applied to InP and GaAs by *Wilmsen* and *Kee* [7.53], using Auger and ESCA profiling. The concentration of the individual atoms and bonding types were followed through the oxide layer and across the interface region. In order to complete a characterization of the interface more information is needed but even from the initial sets of experiments of [7.153] important conclusions on the cation and anion bonding with oxygen could be drawn about these particular systems.

The set of techniques that have been applied during the last years to the rather thick oxides related to MOS structures can be also applied to the investigation of the thinner oxides which are extremely important in the MIS solar cell devices, as well as in practical, inexpensive heterojunction devices where thin intermediate native oxides might have an important influence on the overall behavior.

7.3.2 The Semiconductor-Semiconductor Interface

All the interface analysis techniques and many of the results mentioned in relationship to the M–S interfaces can be applied to the S–S interface, although in the latter case a higher degree of complexity is found. The initial stages of formation of a heterojunction are just beginning to be studied. One of the first studies of the system by photoelectron spectroscopy has been reported on Ge deposited epitaxially on (110) n-type and p-type GaAs surfaces [7.154]. The substrates were produced by cleavage in UHV, and Ge was evaporated at substrate temperatures $T_s = 350$–$525\,°C$. Synchrotron radiation was used as the photon source. The photon energy used determines a probing depth of about 3 atomic layers. We show in Fig. 7.24 the photoemission spectra for the case of heterojunction formation at $525\,°C$. The Ga $3d$ and As $3d$ signals decrease together rather uniformly with Ge coverage. The authors give several arguments rejecting the hypothesis of island formation, so equal bonding of Ge to Ga and As, within the limits of detection, is concluded. The coverage range is extended in Fig. 7.25. We notice two interesting facts. The Ga and As emissions persist for coverages much larger than the sampling depth; and the As emission remains strong while the Ga signal shows a continuous reduction, a situation

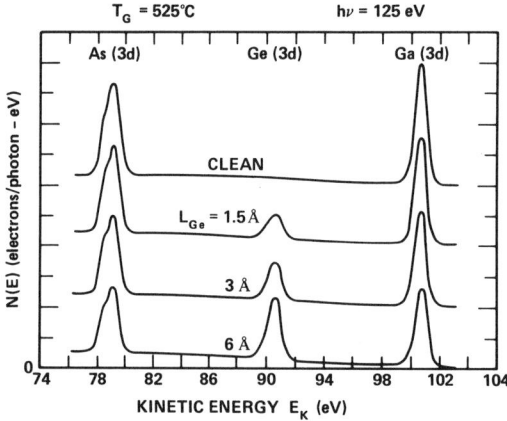

Fig. 7.24. Energy distributions normalized to constant core yield for $3d$ core electrons of As, Ge, and Ga photoemitted by 125 eV photons from (110) GaAs cleaved in situ and covered with equivalent thicknesses L_{Ge} of Ge up to 6 Å while held at a growth temperature T_G of 525 °C. The curves are offset vertically by one major division from each other for clarity. (After [7.154], courtesy of the American Institute of Physics)

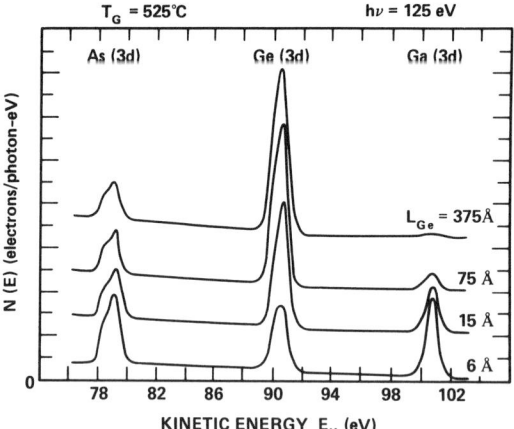

Fig. 7.25. Normalized energy distributions for $3d$ core electrons of As, Ge, and Ga photoemitted by 125 eV photons from GaAs (110) covered with Ge, $6\,\text{Å} \lesssim L_{Ge} \lesssim 375\,\text{Å}$, for $T_G = 525$ °C. The 6 Å data is the same as that shown in Fig. 7.24 for reference. The curves are offset by one major division from each other for clarity. The increasing emission of As relative to Ga and its persistence for overlayers much thicker than the photoelectron escape depth are indicative of interdiffusion of all the species. (After [7.154], courtesy of the American Institute of Physics)

that recalls Schottky barrier results on GaAs as previously quoted in this paper. Again, strong chemical disruption at the interface must be postulated and alloying must take place with preferential diffusion of As towards the Ge surface. By changing the photon energy, the probing depth changes, and a partial depth profiling can be made. This experiment leads to the conclusion that As is probably uniformly distributed in the Ge layer, rather than floating on the surface of the overlayer, as is apparently the case for metal overlayers. The photoemission spectra for systems grown at $T_s = 350$ °C show a different behavior: the Ga and As features keep the same proportion as in clean GaAs, but are very much reduced as compared to similar coverages at higher substrate temperatures, as in Fig. 7.26. A simple theory for the integrated area of the signals allowed *Bauer* to conclude that in the low T_s case, the composition changes from (110) planes containing all Ga and As atoms to complete Ge

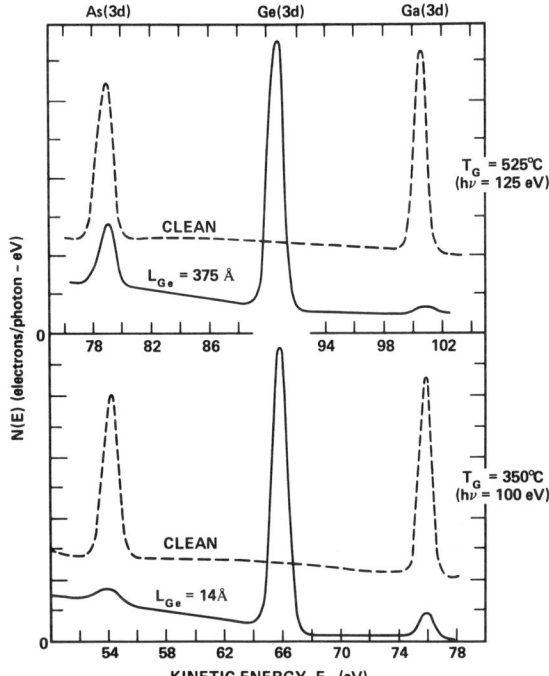

Fig. 7.26. Energy distributions normalized to constant core yield for 3d electrons of As, Ge, and Ga photoemitted from two (110) GaAs surfaces on which Ge was subsequently evaporated by MBE to a thickness corresponding to 375 Å when heated to 525 °C and to 14 Å when heated to 350 °C. The data of the upper panel is reproduced from Figs. 7.24, 25 and the curves are offset by one major division for clarity. Note the rapid attenuation of substrate emission in the soft X-ray region and the constant ratio of Ga to As emission when an abrupt interface is formed under the conditions in the lower panel. (After [7.154], courtesy of the American Institute of Physics)

layers over a distance comparable with the bonding distance, certainly a very abrupt transition; while for $T_s = 430$ °C and 525 °C, sizeable interdiffusion occurs. Therefore there is a transition for temperatures between 350 and 430 °C for measurable interdiffusion to take place. While a valence band discontinuity was not found for samples where interdiffusion was detected, the abrupt interface showed a sharp band discontinuity as measured by photoemission. The value obtained, $\Delta E_v = 0.7(+0.5, -0.3)$ eV, is consistent with previous values reported in the literature. The band structure was found to change over a distance of 5 Å.

Independently *Grant* et al. [7.155] studied the same system by the XPS technique on (111), (110), and (100) surfaces. Epitaxial layers (20 Å thick) of Ge were deposited by molecular beam epitaxy (MBE) on substrates held at $T = 425$ °C. It has been stressed that interface dipoles can produce energy band discontinuities that would depend on the crystallographic orientation of the interface plane. By measuring the difference in core level binding energies (by photoemission) for both sides of the heterojunctions with different surface indexes, *Grant* et al. found a systematic variation in ΔE_v. The maximum difference in ΔE_v of about 0.2 eV was found between the (111) and ($\bar{1}\bar{1}\bar{1}$) planes.

The concept of abruptness discussed above refers to a microscopic scale. When we define abrupt junctions as opposed to graded junctions, we usually refer to changes occurring over larger distances. The concept of the extent of

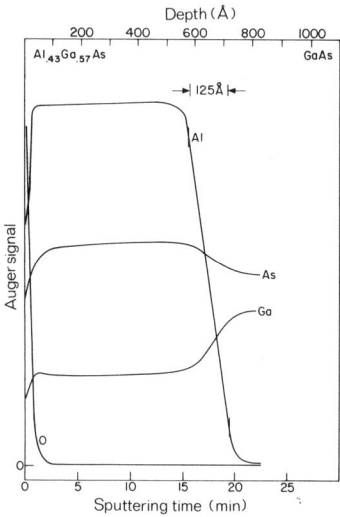

Fig. 7.27. Auger depth profile of an $Al_{0.43}Ga_{0.57}As$–GaAs heterojunction grown by LPE. $T_G \approx 800\,°C$. (After [7.148], courtesy of the American Institute of Physics)

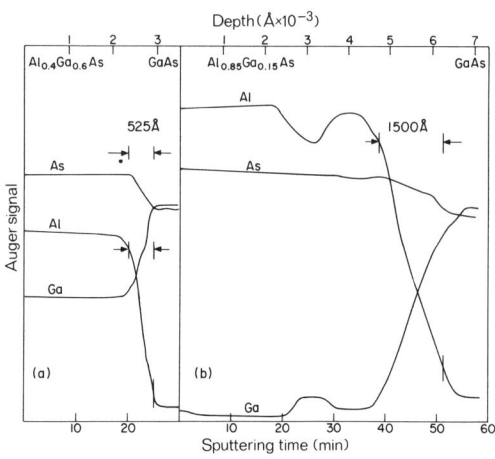

Fig. 7.28a and b. Auger depth profiles of samples of $Al_xGa_{1-x}As$–GaAs heterojunctions grown by two graded LPE growth techniques, labeled (a) and (b), corresponding to techniques G_1 and G_2 in [7.148]. Note the detection of an unintentional drop of Al concentration in (b). (After [7.148], courtesy of the American Institute of Physics)

abruptness is very important in solar cells, since the grading will determine the shape of the electrostatic potential across the transition region, the presence of band discontinuities, and the grading of intrinsic or extrinsic defects in the vicinity of the interface. All of these will govern the compromise between carrier injection and light generated carrier collection. The grading of a heterojunction can be designed and checked by different techniques, as for example by Auger profiling. Garner et al. [7.148] studied intentional grading in the $Al_xGa_{1-x}As/GaAs$ system for different values of x for samples grown by LPE at about 800 °C. The samples with no intentional grading (for different values of x and several thickness of layer) were found to have similar interface region widths, averaging 110 Å in thickness. A typical profile is shown in Fig. 7.27. The interface region widths were found to be slightly larger for samples grown at $T_s = 800\,°C$, (approx. 125 Å) than for those grown at $T_s = 750\,°C$ (100 Å). Intentional grading was produced by introducing additional amounts of Ga and Al into the melt. Graded junctions of a controlled width can be examined by Auger profiling and illustrative cases are shown in Fig. 7.28a, b. The first feature of note is the ability to control the interface transition region thickness. Secondly we notice in Fig. 7.28b a drop in the Al concentration in the region which was assumed to be of uniform composition. The photoresponse of the sample reflected the potential well associated with the dip on the Al concentration, an effect that would have been difficult to interpret without the chemical profiling technique.

7.3.3 Crystallographic Aspects of the Interface Region: Lattice Mismatch, Dislocations, and Electronic Behavior

Most of the semiconductors investigated for HJ devices belong to the group of tetrahedrally coordinated materials with closely related crystal structures which is advantageous for epitaxy. However, the lattice constants span a large range. When two different materials are interfaced, lattice coherence is generally lost to some degree due to the crystallographic mismatch [7.156]. The manner in which the misfit strain is accommodated determines the nature of strain fields and the concentration and morphology of metallurgical defects in the interface and its neighborhood. The problem has been investigated theoretically [7.156–158] by studying the strain interfacial energy between two dissimilar lattices. The final configuration will be the result of competing accommodation processes and depends on the materials, their surfaces, growth techniques, and experimental conditions in general.

Generally speaking, it is possible for the initial deposits of a material on a substrate of dissimilar crystallographic lattice to accommodate the misfit strain elastically. After a critical thickness, which depends inversely on the misfit strain [7.156, 157, 159], it is energetically favorable to accommodate the strain plastically, generating dislocations at the interface. For comparatively thicker films, practically all the strain is accommodated plastically [7.159], although some elastic strain might remain.

The importance of mismatch dislocations on HJ behavior was stressed early in the literature [7.35, 160]. A thorough discussion of the electronic behavior of dislocations can be found in the book by *Mataré* [7.50]. In principle, dangling bonds at the dislocation cores are to be expected. They give rise to energy levels within the forbidden band gap (for the range of energy gaps of interest in this paper). If the density is high enough the levels can show doping effects. In general, they are deep lying levels accounting for carrier trapping and fast recombination phenomena.

Considering just this simple picture, the presence of these levels at the interface suggests two immediate consequences. Acting as traps, the density of interface states produced by a lattice mismatch of just 1% ($\sim 10^{13}$ states cm^{-2}) is enough to produce appreciable band bending and to influence the energy band configuration at the interface and depletion regions. Even at lower densities, when acting as recombination centers, the levels provide a recombination pathway in a narrow region with strong influence on the transport of carriers across the junction.

a) Density of Dislocations

The original calculations on the number and configuration of mismatch dislocations were extended by *Holt* [7.161] for tetrahedrally coordinated structures in the three-dimensional heteroepitaxy case. A simple qualitative description is given in the paper by *Kressel* [7.162], who stresses in his analysis

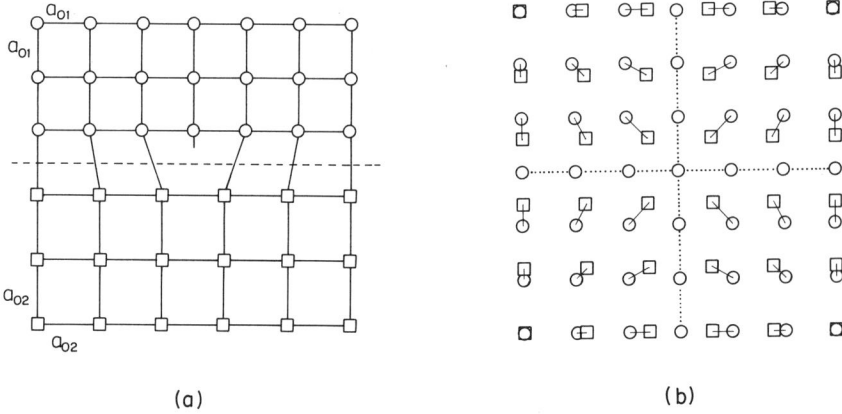

Fig. 7.29a and b. Schematic diagrams of the interfacing of two single cubic crystals with lattice parameters a_{01} and a_{02} at a (100) plane. Complete abruptness and plastic relaxation of the misfit strain is assumed. In (**a**) a view parallel to the interface plane illustrates the formation of a pure edge dislocation. A view perpendicular to the interface plane is shown in (**b**), illustrating the corresponding set of orthogonal dislocation lines. The circles and squares represent the interface layers of atoms in the corresponding materials

the role of interface defects in optoelectronic HJ devices. Figure 7.29 illustrates this simple analysis by showing the interfacing of two simple cubic structures with lattice parameters a_{01} and a_{02}. Assuming abruptness and complete plastic relaxation of the misfit strain, the mismatch gives rise to a set of orthogonal edge dislocations lying at the interface, with a linear density given by

$$\varrho_d = \frac{a_{01} - a_{02}}{a_{01} a_{02}} = \frac{\Delta a_0}{\bar{a}_0^2} \approx \frac{\Delta a_0}{a_0^2} \text{ cm}^{-1} \tag{7.36}$$

where $\bar{a}_0$ and a_0 are the geometric and arithmetic means of the lattice parameters respectively, the last expression being a good estimate for small differences. In the case of the sphalerite structure where the Burgers vector differs from the lattice parameter, the density of dislocations is expressed by

$$\varrho_d = \chi_{ijk} \left(\frac{\Delta a_0}{\bar{a}_0^2} \right) \approx \chi_{ijk} \left(\frac{\Delta a_0}{a_0^2} \right), \tag{7.37}$$

where χ_{ijk} is a geometrical factor depending on the crystallographic plane of interfacing. For (100) planes, with [110] type Burgers vector [7.161], the linear density is

$$\varrho_d = 2^{1/2} \left(\frac{\Delta a_0}{\bar{a}_0^2} \right) \approx 2^{1/2} \left(\frac{\Delta a_0}{a_0^2} \right) \text{ cm}^{-1}. \tag{7.38}$$

The surface density of dangling bonds can be estimated by

$$N_{ss} = \chi'_{ijk} \frac{a_{01}^2 - a_{02}^2}{a_{01}^2 a_{02}^2} = \chi'_{ijk} \frac{a_{01}^2 - a_{02}^2}{a_0^4} \approx 2\chi'_{ijk} \left(\frac{\Delta a_0}{a_0^3}\right) \text{cm}^{-2} \quad (7.39)$$

with χ'_{ijk} similarly defined. For (100) planes, the surface density is

$$N_{ss} \approx 8 \left(\frac{\Delta a_0}{a_0^3}\right) \text{cm}^{-2}. \quad (7.40)$$

Obviously matching of thermal expansion coefficients also plays an important role in the metallurgy of HJs, since perfect lattice match can only be achieved at one specific temperature.

If all the centers associated with dangling bonds are active as recombination centers, an estimate for the interface recombination velocity S_I is given by

$$S_I = v_{th} N_{ss} \sigma_c, \quad (7.41)$$

where v_{th} is the thermal velocity of the carriers and σ_c is the capture cross section of the centers.

The actual situation is by far more complicated as we see later. We just mention here that even in the simple model described above, all the misfit states are not expected to be electronically active. Rather, it is found that a large amount of compensation takes place and only a small fraction remain active and account for the observed phenomena as discussed by *Tansley* [7.30].

b) Dislocation Morphology

There is abundant literature on misfit defects covering a broad range of practical situations. A thorough discussion on epitaxy can be found in the book edited by *Matthews* [7.163]. In the case of epitaxy in heterojunctions, much effort has been invested in systems with small, controllable mismatch in a range close to perfect fit, and principally for the III–V compounds. These are close to ideal systems allowing for basic studies in interface-defect metallurgy. Technologically, they are of most practical importance due to their application to light-emitting diodes, solid-state HJ lasers, and high-efficiency solar cells. In these particular applications, misfit defects have a strong deleterious effect on the overall device performance, especially in the light-emitting case. In the following discussion, we concentrate mainly on results obtained for these systems to expose basic phenomena associated with lattice mismatch.

Olsen et al. [7.164] studied VPE $In_xGa_{1-x}P/GaAs$ HJs over a range of composition corresponding to an interval in misfit strain around perfect fit. They were able to study the region of misfit where the strain is accommodated elastically and to determine its effect on lattice parameter and band-gap shift, as well as the conditions for the onset of plastic deformation via generation and

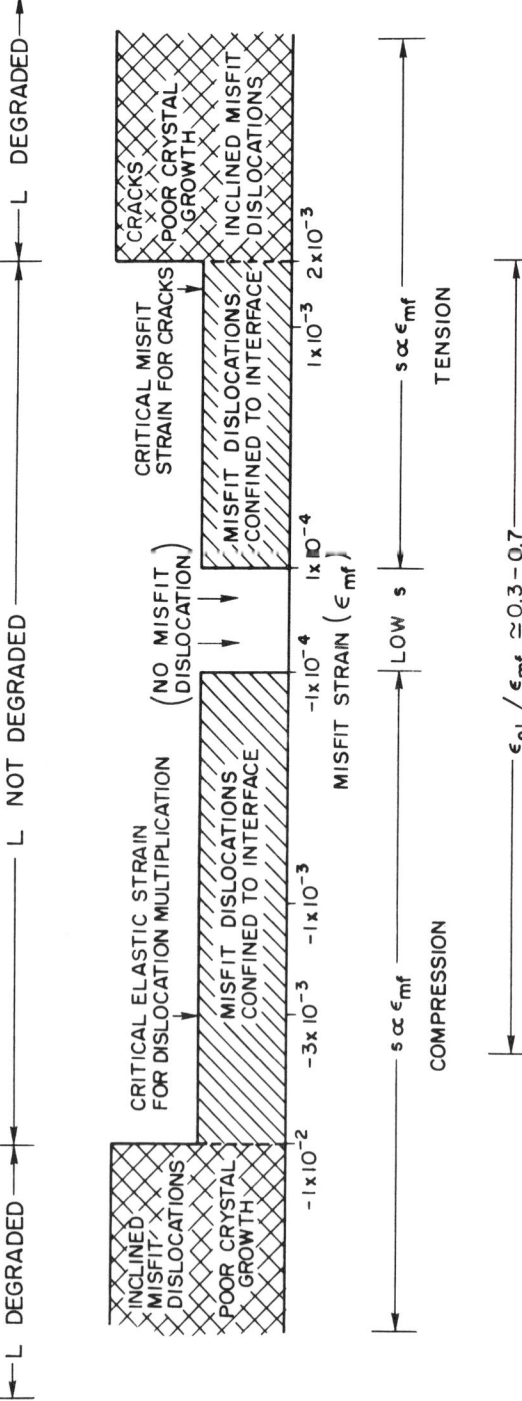

Fig. 7.30. Critical strains and stresses and their relationship to mechanical and electrical properties in III–V compound heterojunctions. (After [7.164], courtesy of the American Institute of Physics. L is the minority carrier diffusion length, ε_{el} is the elastic strain, and s is the interface recombination velocity

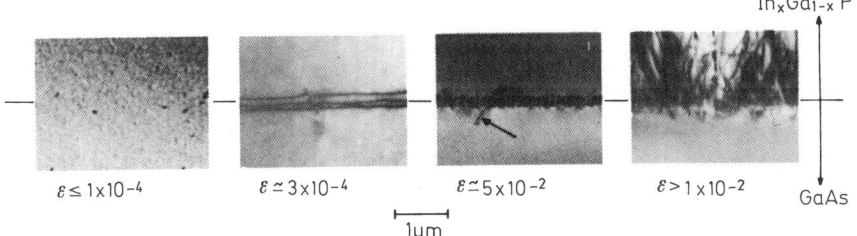

Fig. 7.31. Cross-sectional transmission electron microscope (XTEM) micrographs from VPE $In_xGa_{1-x}P/GaAs$ interface having increasing mismatch ε_{mf} as shown. The [100] growth axis is vertical. The structural effects of lattice mismatch can be observed. The arrow indicates a substrate dislocation which has been bent over at the interface due to lattice mismatch. (After [7.159], reprinted by permission of the Plenum Press)

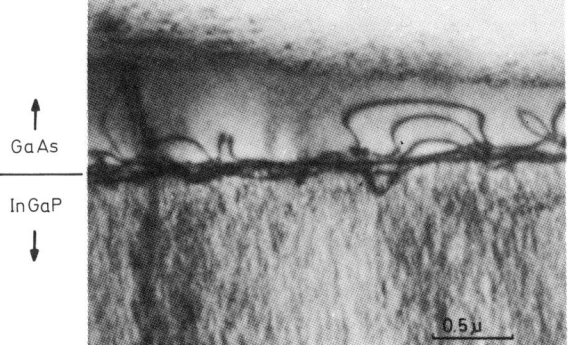

Fig. 7.32. Bright field {011} XTEM micrograph from a lattice mismatched $GaAs/In_xGa_{1-x}P$ interface of a VPE transmission photocathode. The [100] growth axis is vertical. A classic Frank-Read source in action can be seen. (After [7.159], courtesy of the American Institute of Physics)

multiplication of dislocations. Figure 7.30 illustrates their summary of critical strains and stresses for the different mechanisms of accommodation of mismatch strain and the relationship with mechanical and electrical properties in III–V compounds.

The effects of increasing mismatch on the VPE $In_xGa_{1-x}P/GaAs$ interface can be seen in Fig. 7.31 [7.159]. In general, there is a small interval around $\varepsilon_{mf} \sim 0$ for which the misfit strain ε_{mf} can be accommodated elastically without dislocation generation (see the first picture in the series shown in Fig. 7.31). The critical strain for dislocation generation is connected with the thickness of the overlayer as mentioned previously. The larger the misfit, the thinner the overlayer for which dislocations will begin to be observed. For large misfits, dislocation arrays can be seen even before a continuous film is grown [7.165, 166]. Dislocations can be generated by a variety of mechanisms [7.156, 159], including extension of substrate dislocations, accommodation of misorientation of coalescing nuclei, and plastic deformation of the film. Once they are introduced, multiplicative mechanisms can operate; an example is the Frank-Read source shown in Fig. 7.32. For small mismatch [7.159], ordered

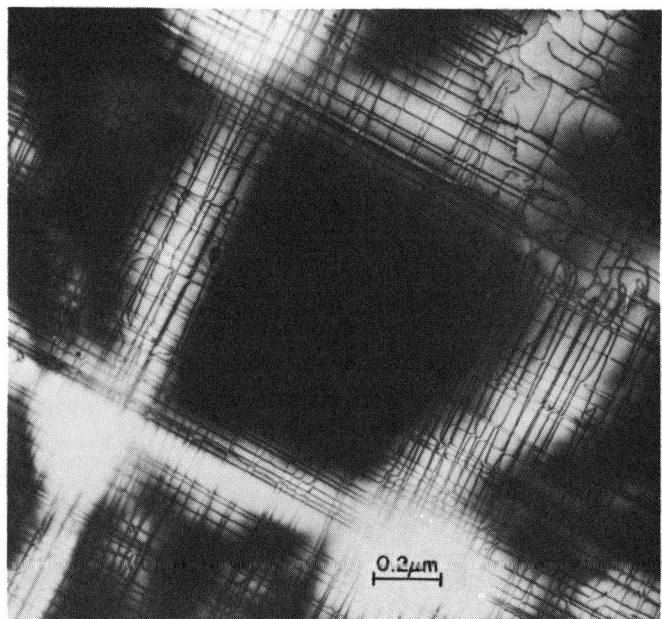

Fig. 7.33. TEM of the compositional graded region of a VPE (100) $In_xGa_{1-x}P/GaP$ sample. The dislocations lie along [011] directions. The square black regions are supposed to originate from surface undulations (cross-hatching) related to mismatch. (After [7.167], courtesy of North-Holland Publishing Company)

arrays of dislocations are observed at the interface. They are confined to a region close to the interface, but do not propagate along the growth axis. An orthogonal array of dislocations on a compositionally graded VPE $In_xGa_{1-x}P/GaP$ interface can be seen in Fig. 7.33.

Our original picture of pure edge dislocations located strictly at the interface is an oversimplification. The dislocation configurations are more complicated [7.167] because the phenomenon is three dimensional rather than planar. The dislocation arrays extend into the overlayer possibly changing the electrical properties of an appreciable portion of the depletion region. In addition, the Burgers vectors are not necessarily located in the interface plane (or parallel to it) and inclined dislocations are observed [7.156, 162, 167]. Above a critical strain, typically 1%, dense arrays of dislocations which propagate away from the interface along the growing direction are observed [7.159]. An example is (AlGa)As grown by LPE on GaP with a large misfit ($\varepsilon_{mf} \sim 3.7\%$) as shown in Fig. 7.34.

Cracking is another important phenomenon found in these systems. It has been observed that cracking in the overlayer tends to occur for films in tension (the epilayer has a smaller lattice parameter than the substrate) but less frequently in those grown in compression [7.159, 167, 168]. Cracking has been

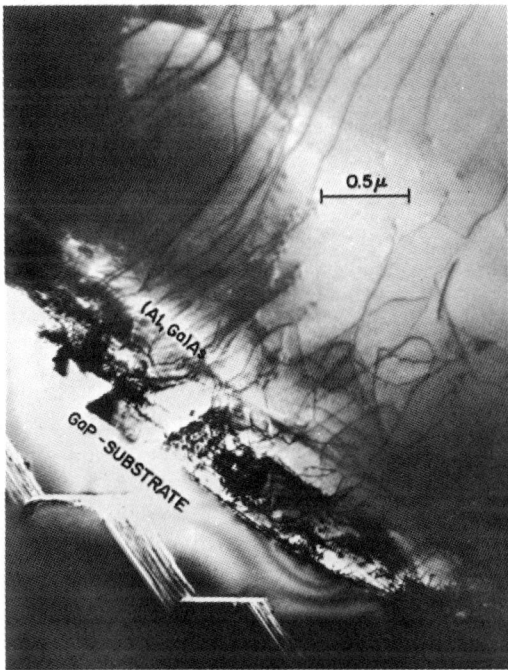

Fig. 7.34. {011} XTEM micrograph of (AlGa)As deposited directly on a (111) GaP wafer via LPE. The substrate epitaxial misfit is very large, $\varepsilon_{mf} \sim 3.7\%$. Note how the dislocation morphology is not arrayed in an orderly fashion and that many dislocations propagate along the growth axis. (After [7.167], courtesy of North-Holland Publishing Company)

Fig. 7.35. Bright field XTEM of the active GaAs region of a VPE double-heterojunction laser structure. The extensive decomposition at the top interface (but not the bottom) is to be noted. The top layer is in tension. Note the large number of vertically inclined dislocations in the top $In_xGa_{1-x}P$ layer (but not in the bottom) due to crack formation and subsequent propagation of dislocations down into the active GaAs region. (Courtesy of G.H. Olsen)

observed in VPE III–V heterostructures with a lattice mismatch of only 0.15% [7.167]. Cracking is caused by the pileups of mismatch dislocations producing microcracks [7.159] which then tend to be opened up into a fracture line by tension. The cracks are a source for further plastic deformation and the formation of dislocations in the substrate. This phenomenon is illustrated clearly in Fig. 7.35.

When the two materials forming an HJ separated by an interface populated with misfit dislocation are allowed to interdiffuse, the dislocations tend to distribute within the alloyed region in a fashion governed by their interaction with stress fields present [7.156].

c) Electronic Effects of Misfit Dislocations

The effects of mismatch defects on the optoelectronic behavior of HJs are manifold. Unfortunately, it is extremely difficult to correlate the defect structure with electronic parameters of the devices, and most often the models proposed are purely speculative. This is so mainly because this is a new field of research where most of the questions are still open. In addition, the electronic properties that are associated with misfit defects (for example, the interface recombination velocity) are not measured directly but are estimated from indirect measurements. The interpretation of the measurements and, consequently, the estimation depends on the theoretical model, and cannot be interpreted in a unique way.

We mention a few specific examples of the effects of misfit to illustrate the problem. The presence of inclined dislocations in the epilayer can reduce the carrier diffusion length, since they will impose an upper limit to $L_d \lesssim \varrho_d/2$ [7.162]. This is especially so in the dense dislocation arrays found for large misfit. As we mentioned previously, such dislocation arrays can be formed in the substrate as a consequence of the deposition of a mismatched epilayer, for example, when cracking occurs in tensioned deposits. In those cases a reduction of the diffusion length in the substrate material can also be observed [7.168].

A parameter strongly affected by misfit dislocations that has been investigated considerably is the interface recombination velocity S_I. We mention as an example here the results obtained by *Ettenberg* and *Olsen* [7.168] on the recombination properties of lattice mismatched $In_xGa_{1-x}P/GaAs$ heterojunctions. A series of $p\text{-}In_xGa_{1-x}P/p\text{-}GaAs/n\text{-}GaAs$ heterostructures were produced with variable thicknesses of $p\text{-}GaAs$. The short-circuit current I_{sc} and the injected carrier lifetime τ, which include as parameters L_d and S_I, were measured for the $p\text{-}GaAs$ layer as a function of its thickness d. By fitting the experimental data, values for L_d and S_I were obtained. Different (InGa)P compositions were deposited producing different values for the misfit strain, between 3×10^{-4} and 2.5×10^{-2}, both positive (epilayer in tension) and negative (compression). For the layers in compression (and very low misfit in tension), the recombination velocity determined at the heterojunction appeared to be linearly related to the lattice mismatch at the growth temperature, up to a mismatch of -2.5%. The parameter L_d appeared to be independent of mismatch for these samples and in agreement with the diffusion lengths previously reported for similarly grown $p\text{-}GaAs$. The values measured for the lattice mismatch and the estimated interface recombination velocity are plotted in Fig. 7.36 where the straight line represents the theoretical estimate given by the expression (7.41) as proposed by *Kressel* [7.162]. The fit is impressive but,

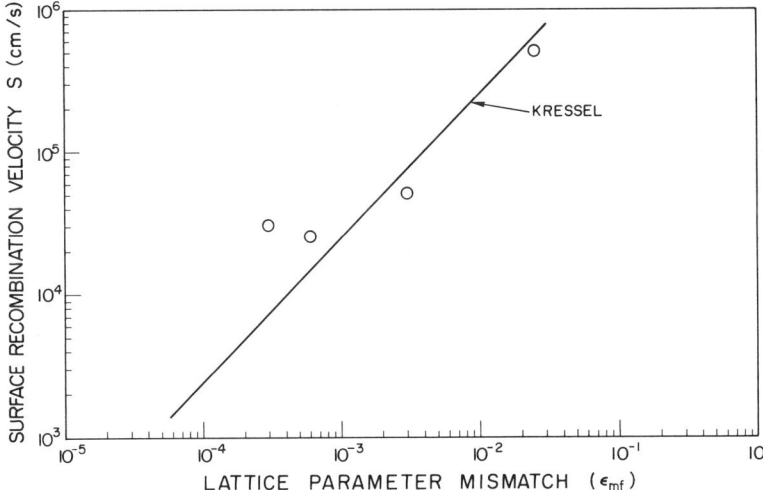

Fig. 7.36. Surface recombination velocity S_I as a function of lattice mismatch ε_{mf} (at growth temperature) for $In_xGa_{1-x}P/GaAs$ heterojunctions. The theoretical curve is obtained assuming: $N_{ss} = 8\varepsilon_{mf}/a_0^2$ for (100) planes, $\sigma_c = 10^{-15}$ cm^2, and $v_{th} = 10^7$ cm s^{-1}, giving $S_I \approx (2.5 \times 10^7)\varepsilon_{mf}$ cm s^{-1}. (After [7.168], courtesy of the American Institute of Physics)

as *Ettenberg* and *Olsen* point out, somehow fortuitous. The sample with a misfit of 3×10^{-4} did not show misfit dislocations and, in this case, the recombination should be attributed to other sources such as point defects concentrated at the interface due to the strain gradient or possibly effects associated with the elastic strain itself. Besides, compensation of dangling bonds has been reported quite generally making the model of fully active unsatisfied bonds dubious. Impurity accumulation and precipitation at regions containing mismatch dislocations has also been observed [7.169], a mechanism which makes compensation highly likely. In any case, a linear behavior of S_I on mismatch strain has been reported by several authors on III–V HJs for the same range of misfits [7.162, 170]. The absolute values of S_I are quite different from author to author and also depend, in each particular case, on the experimentally measured property from which S_I is estimated [7.171] and the way the data is interpreted [7.172].

Returning to the results reported by *Ettenberg* and *Olsen* [7.168], it is interesting to show the distinctive behavior of the samples in tension. For the sample with a mismatch of only $+0.5\%$, the measured data could not be fitted with a value of S_I predicted by the results shown in Fig. 7.36 and furthermore, not even with $S_I \to \infty$. A reasonably good fit could only be obtained by assuming an infinite recombination velocity and a shortened diffusion length in the GaAs. Cracking in the window layer and deterioration of the underlying GaAs layer was observed explaining the reduction in L_d and the "anomalously" high recombination velocity.

The fact that heterojunction devices show tunneling currents dominating in the low bias range may also be a consequence of mismatch. At higher voltages, more desirable forward-current mechanisms dominate. In the case of crystalline epitaxial HJs which are well matched, the range where tunneling mechanisms dominate can be limited to low enough voltages such as not to interfere with device performance. The less perfect the HJ, the larger the limiting voltage below which tunneling currents are observed. For large mismatch heterojunctions, the dark and light current characteristics are far from ideal and ad hoc models must be built correlating the particular characteristics of the interface region with the observed optoelectronic behavior.

It is customary to assume that a high recombination current at the interface of an HJ solar cell with large lattice misfit will produce a severe reduction in the short-circuit current. However, it is observed that even in systems with enormous mismatch, such as n-ZnO/p-CdTe (29% mismatch) [7.28], the short-circuit current is close to the theoretical prediction for perfect collection. The major drawback in these systems are the soft dark JV characteristics which can be associated with lattice mismatch. The presence of interface states produces soft diodes, often with unstable, light dependent forward characteristics, reducing the open-circuit voltage V_{oc} and the fill factor significantly. Much of the research in these cases is concentrated in improving the interface region in order to improve the "dark" transport mechanisms.

7.3.4 Summary

Interface phenomena are truly microscopic and it was demonstrated that highly ideal experimental designs and sophisticated experimental tools are needed to isolate and study the significant processes.

The nature of the Fermi-level stabilization in heterostructures appears to be clarified by the recognition that, in surfaces with a high degree of perfection, there are no intrinsic surface states which could account for the observed phenomena. Fermi-level pinning occurs however, and it is apparently governed by the following rules: I) the interface states responsible for the Fermi-level pinning are extrinsic; II) they are induced at the interface in the early stages of junction formation, when less than a monolayer has been deposited; and III) the dependence of the position of the Fermi-level at the interface on the nature of the overlayer is generally smaller than expected, and almost null for low-polarity semiconductors.

Experimental evidence led us to conclude that crystallographic and chemical disruption of the semiconductor interface occurs upon heterostructure formation. This disruption leaves a defective interface which has been suggested to be the main factor determining the energy band structure in the interface region. A connection with the concept of self-compensation could explain the behavior of the index of interface behavior S, yielding a model based strictly on defect formation. Also, it seems clear that alloying and even room-temperature compound formation cannot be disregarded.

In many cases, heterostructure formation is so subtle, however, that the main processes and the resulting structures might well be determined a priori by a set of experimental conditions rather than by the specific nature of the components.

The crystallography of the interface region is of major importance in HJs where phenomena induced by lattice mismatch can govern transport across the junction. We have reviewed briefly how misfit strain is released through the generation and multiplication of misfit dislocations. The descriptions given show the increasing complexity of dislocation morphology as lattice mismatch is increased. Although far from being understood, a strong connection between mismatched surface structure and electronic behavior can be seen. On the other hand, highly efficient HJ solar cells with huge lattice mismatch have been fabricated.

The presence of an "insulating" layer must also be considered, with important consequences with regard to the structural properties of the interface as well as to the electronic behavior of the device. Such an interfacial layer can relieve the mismatch strain by acting as an amorphous buffer between the two materials.

7.4 Conclusions

At the beginning of this chapter we showed the connection between the solar efficiency and the dark electrical transport properties of the heterojunction (HJ) solar cell {as represented, for example, by the equation $J = J_0[\exp(eV/AkT) - 1] - J_L$}. The maximum light-generated current J_L is fixed primarily by the semiconductor band gaps and, for good solar cells, the J_L values are currently in the range 85–95% of the theoretical maximum. Increasing the light-generated current from the absorber layer is a matter of material and processing control of the bulk semiconductor. However, the greatest potential for improving solar efficiency of HJs is by the control of J_0 and the diode quality factor A.

Despite the importance of such descriptive models as that above, the treatment of transport in HJs does not seem satisfying or unified at this time. Each junction type appears to be a more or less special case, often requiring its own model. The electron affinities of the components give only the roughest indication of the energy band structure of an HJ system. Furthermore, a knowledge of the energy band profiles is only a part of what is required to predict the transport properties of the HJ; the spatial and energy distributions of the centers involved in recombination, tunneling, and charge storage must also be known. Research on HJ, MS, and MIS structures shows that electrical transport is almost completely determined by the properties within the depletion layer and that, in most cases, the transport is even more specifically dominated by a much thinner region including the interface. This is true in all but the most carefully grown, low mismatch devices. It seems clear that a thorough understanding of the interface region is required to realize the promise of HJs for solar converters as well as for other HJ devices. This understanding must encompass both the metallurgical and the defect electronic properties as well as the transport behavior of the depletion region. The flavor of this sort of unified approach is illustrated by the paper by *Lindholm* et al. [7.173] on Si *pn* junction solar cells.

The experimental work to which we referred in order to support our conclusions about interface behavior was necessarily confined to ideal cases because of the complexity of the interface region and the diversity of the processes involved there. The basic "driving forces" acting ultimately to produce the junction structure are summarized here: I) the difference in work functions across the junction (which plays a basic, although many times minor role); II) microscopic pinning of the Fermi level at the interface (as influenced by chemical reactivity of the HJ components and the resulting charged interface states and dipoles); III) the defect structure in the vicinity of the interface (including the effects of point, line, and surface defects, and the preferential segregation of dopants and impurities there); and IV) the presence of intervening "insulating" layers (which may contain trapped charge, and can act as buffering layers to accomodate lattice mismatch).

We feel that there are underlying principles, as yet undisclosed (as, for example, suggested by the "fortuitous" parallelism between the equal solar efficiency curves and the relation between $\log J_0$ and $1/A$ for various diodes seen in Fig. 7.10). It remains, of course, to bridge the gap between these theories and results for ideal systems and the application to practical devices which can be mass produced. That educated manipulation of the interfacial properties can bring about substantial improvements in solar converters is aptly shown by the recent high solar efficiencies attained in MIS cells, for example [7.111]. All of the considerations above demonstrate the need for coordinated research, bringing together surface, solid-state, and device physics.

Acknowledgements. The authors wish to express their special appreciation to G.H. Olsen for fruitful discussion and for providing many of the illustrations for the latter part of the review. We wish to thank warmly I. Lindau, C.M. Garner, and R.S. Bauer for helpful discussions and illustrations and R.H. Bube for critical reading of the manuscript. We appreciate the careful review of the manuscript by F. Ponce and T. Chynoweth.

The authors appreciate the support of the Materials Science Program, Division of Basic Energy Sciences, of the Department of Energy for support for a portion of the research reported herein.

References

7.1   Paul Rappaport: RCA Rev. **20**, 373 (1959)
7.2   M. Wolf: Proc. IRE **48**, 1246 (1960)
7.3   H. Hovel: "Solar Cells", in *Semiconductors and Semimetals, Vol. 11*, ed. by R.K. Willardson, A.C. Beer (Academic Press, New York 1975)
7.4   P.T. Landsberg: Solid-State Electron. **18**, 1043 (1975)

7.5 S.M. Sze: *Physics of Semiconductor Devices* (Wiley-Interscience, New York 1969)
7.6 M. Wolf, H. Rauschenbach: Advanced Energy Conver. **3**, 455 (1963). Reprinted in *Solar Cells*, ed. by C. Backus (IEEE, New York 1976) p. 146
7.7 C.R. Fang, J.R. Hauser: In *13th Photovolt. Spec. Conf. Proc.* (IEEE, New York 1978) p. 1306 and p. 1318
7.8 V.L. Dalal, A.R. Moore: J. Appl. Phys. **48**, 1244 (1977)
7.9 J.G. Fossum, E.L. Burgess: In *12th Photovolt. Spec. Conf. Proc.* (IEEE, New York 1976) p. 737
7.10 K. Seeger: *Semiconductor Physics* (Springer, Berlin, Heidelberg, New York 1973)
7.11 R. Stratton: J. Appl. Phys. **40**, 4582 (1969)
7.12 G. Persky, D.J. Bartelink: Phys. Rev. B**1**, 1640 (1970)
7.13 T.W. Sigmon, J.F. Gibbons: Appl. Phys. Lett. **15**, 320 (1969)
7.14 E.A. Davies: J. Phys. Chem. Solids **25**, 201 (1964)
7.15 *Proc. Int. Conf. Hot Electrons in Semiconductors*, Denton, Texas, 1977. Published in Solid-State Electron. **21** (1978)
7.16 C.T. Sah, R.N. Noyce, W. Shockley: Proc. IRE **45**, 1228 (1957)
7.17 K.W. Mitchell, A.L. Fahrenbruch, R.H. Bube: J. Appl. Phys. **48**, 4365 (1977)
7.18 A.L. Fahrenbruch, R.H. Bube: J. Appl. Phys. **45**, 1264 (1974)
7.19 A. Rothwarf: "Theoretical Prospects of the CdS/Cu_2S Solar Cell," Technical Report NSF/RANN/AER 72-03478 A04/TR 76/1 (Institute of Energy Conversion, Newark, Delaware 1976)
7.20 K.W. Mitchell, A.L. Fahrenbruch, R.H. Bube: Solid-State Electron. **20**, 559 (1977)
7.21 A.R. Riben, D.L. Feucht: Int. J. Electron. **20**, 583 (1966)
7.22 J.F. Womac, R.H. Rediker: J. Appl. Phys. **43**, 4129 (1972)
7.23 J. Aranovich: Private communication
7.24 H.M. Manasevit, L.L. Hess, P.D. Dapkus, R.P. Ruth, J.J. Yeng, A.G. Campbell, R.E. Johanson, L.A. Moudy, R.H. Rube, L.B. Fabick, A.L. Fahrenbruch, M.-J. Tsai: In *13th Photovolt. Spec. Conf.* (IEEE, New York 1978) p. 165
7.25 S. Wagner, J.L. Shay, K.J. Bachmann, E. Buehler: J. Appl. Phys. **47**, 614 (1976)
7.26 J.L. Shay, S. Wagner, M. Bettini, K.J. Bachmann, E. Buehler: IEEE Trans. Electron Devices **24**, 483 (1977)
7.27 A. Yoshikawa, Y. Sakai: Solid-State Electron. **20**, 133 (1977)
7.28 A.L. Fahrenbruch, J. Aranovich, F. Courreges, T. Chynoweth, R.H. Bube: In *13th Photovolt. Spec. Conf. Proc.* (IEEE, New York 1978)
7.29 M. Wolf, H. Raushenbach: Adv. Energy Convers. **3**, 455 (1963)
7.30 T.L. Tansley: "Heterojunction Properties," in *Semiconductors and Semimetals*, Vol. 7, Part A, ed. by R.K. Willardson, A.C. Beer (Academic Press, New York 1971) p. 294
7.31 L.J. Van Ruyven: "Phenomena at Heterojunctions", in *Annual Review of Materials Science*, Vol. 2, ed. by R.A. Huggins (Annual Reviews, Palo Alto, CA 1972) pp. 501–527
7.32 A.G. Milnes, D.L. Feucht: *Heterojunctions and Metal-Semiconductor Junctions* (Academic Press, New York 1972)
7.33 R.L. Anderson: IBM J. Res. Dev. **4**, 283 (1960)
7.34 U. Dolega: Z. Naturforsch. **18**a, 653 (1963). [In German]
7.35 W.G. Oldham, A.G. Milnes: Solid-State Electron. **7**, 153 (1964)
7.36 A.R. Riben, D.L. Feucht: Solid-State Electron. **9**, 1055 (1966)
7.37 S.C. Choo: Solid-State Electron. **11**, 1069 (1968)
7.38 H.J. Hovel: In *10th Photovolt. Spec. Conf. Proc.* (IEEE, New York 1973) p. 34
7.39 P.J. Anderson: Electron. Lett. **13**, 496 (1977)
7.40 H. Kroemer: In *CRC Critical Reviews in Solid-State Sciences* (Chemical Rubber Co., New York 1975) pp. 555–564
7.41 T.S. TeVelde: Solid-State Electron. **16**, 1305 (1973)
7.42 P.F. Lindquist, R.H. Bube: J. Appl. Phys. **43**, 2839 (1972)
7.43 A. Rothwarf, L.C. Burton, H.C. Hadley, G.M. Stork: In *11th Photovolt. Spec. Conf. Proc.* (IEEE, New York 1975) p. 476
7.44 K.W. Böer: To be published

7.45 W.G. Oldham, A.G. Milnes: Solid-State Electron. **6**, 121 (1963)
7.46 D. Cheung, S.Y. Chiang, G.L. Pearson: Solid-State Electron. **18**, 263 (1975)
7.47 E.D. Hinkley, R.H. Rediker: Solid-State Electron. **10**, 671 (1967)
7.48 P.C. Newman: Electron. Lett. **1**, 265 (1965)
7.49 O. Weinreich, H.F. Mataré, B. Reed: *Solid State Physics in Electronics and Telecommunications* (Academic Press, New York 1960) Vol. 1, Part 1, 97
7.50 H.F. Mataré: *Defect Electronics in Semiconductors* (Wiley-Interscience, New York 1971)
7.51 C. Van Opdorp, J. Vrakking: Solid-State Electron. **10**, 955 (1967)
7.52 L.J. Van Ruyven, J.M.P. Papenhuijzen, A.C.J. Verhoeven: L.J. Van Ruyven: Phys. Solid-State Electron. **8**, 631 (1964); Status Solidi **3**, K109 (1964)
7.53 J.P. Donnelly, A.G. Milnes: Solid-State Electron. **9**, 174 (1966)
7.54 G.L. Miller, D.V. Lang, L.C. Kimerling: "Capacitance Transient Spectroscopy", in *Annual Reviews of Materials Science*, Vol. 7, ed. by R.A. Huggins (Annual Reviews, Palo Alto, CA 1977) p. 377
7.55 C.T. Sah: IEEE Trans. Electron Devices **24**, 410 (1977)
7.56 J.P. Donnelly, A.G. Milnes: IEEE Trans. Electron Devices **14**, 63 (1967)
7.57 J.M. Borrego, R.J. Gutmann, S. Ashok: Solid-State Electron. **20**, 125 (1977)
7.58 M. Schulz, N.M. Johnson: Appl. Phys. Lett. **31**, 622 (1977); and Solid State Commun. (to be published)
7.59 A. Many, Y. Goldstein, N.B. Grover: *Semiconductor Surfaces* (North-Holland, Amsterdam 1971)
7.60 M. Ettenberg, G.H. Olsen: J. Appl. Phys. **48**, 4275 (1977)
7.61 M. Ettenberg, H. Kressel: J. Appl. Phys. **47**, 1538 (1976)
7.62 G.A. Acket et al.: "GaAs and Related Compounds", Inst. Phys. Conf. Ser. No. **24**, 181 (1975)
7.63 R.H. Dean, C.J. Nuese: IEEE Trans. Electron Devices **18**, 151 (1971)
7.64 M. Schulz, N.M. Johnson: Appl. Phys. Lett. **31**, 622 (1977)
7.65 J. Stannard: To be published
7.66 A.R. Riben: Thesis, Carnegie Institute of Technology (1965)
7.67 A.G. Chynoweth, W.L. Feldmann, R.A. Logan: Phys. Rev. **121**, 684 (1961)
7.68 R.H. Rediker, S. Stopek, J.H.R. Ward: Solid-State Electron. **7**, 621 (1964)
7.69 S.J.T. Owen, T.L. Tansley: J. Vac. Sci. Technol. **13**, 954 (1976)
7.70 J.P. Donnelly, A.G. Milnes: Proc. IEE (London) **113**, 1468 (1966)
7.71 J. Tuac: Rev. Mod. Phys. **29**, 308 (1957)
7.72 M. Wolf: Proc. IRE **48**, 1246 (1960)
7.73 P. Emtage: J. Appl. Phys. **33**, 1950 (1962)
7.74 Y. Marfaing, J. Chevallier: IEEE Trans. Electron Devices **18**, 465 (1971)
7.75 A. Bouazzi, Y. Marfaing, J. Mimila-Arroyo: Proc. CEC Photovoltaic Solar Energy Conf. (Reidel, Dordrecht 1978) p. 628
7.76 R.L. Moon, L.W. James, H.A. Van der Plas, Y.G. Chai, G.A. Antypas: In *13th Photovolt. Spec. Conf. Proc.* (IEEE, New York 1978) p. 859
7.77 K.W. Böer, C.E. Birchenall, I. Greenfield, H.C. Hadley, T.L. Lu, L. Partain, J.E. Phillips, J. Schultz, W.F. Tseng: In *10th Photovolt. Spec. Conf. Proc.* (IEEE, New York 1973) p. 77
7.78 A.E. Van Aerschodt, C. Capart, K.H. David, M. Fabricotti, K.H. Heffels, J.J. Loferski, K.K. Reinhartz: IEEE Trans. Electron Devices **18**, 471 (1971)
7.79 C. Wagner: Phys. Z. **32**, 641 (1931)
7.80 W. Schottky, E. Spenke: Wiss. Veröff. Siemens-Werke **18**, 225 (1939)
7.81 H.A. Bethe: MIT Radiation Lab. Rep. 43–12 (1942)
7.82 S.M. Sze: *Physics of Semiconductor Devices* (Wiley and Sons, New York 1969) Chap. 8
7.83 C.R. Crowell, S.M. Sze: Solid-State Electron. **9**, 1035 (1966)
7.84 C.R. Crowell, V.L. Rideout: Solid-State Electron. **12**, 89 (1969)
7.85 V.L. Rideout, C.R. Crowell: Solid-State Electron. **13**, 993 (1970)
7.86 M.J. Turner, E.H. Rhoderick: Solid-State Electron. **11**, 291 (1968)
7.87 C.R. Crowell, V.L. Rideout: Appl. Phys. Lett. **14**, 85 (1969)
7.88 A.Y.C. Yu, E.H. Snow: Solid-State Electron. **12**, 155 (1969)
7.89 M.A. Green, J. Shewchun: Solid-State Electron. **16**, 1141 (1973)

7.90   P. Viktorovitch, G. Kamarinos: *Inst. Phys. Conf. Ser. No.* **22**, 36 (London 1974)
7.91   J.E. Parrot: *Inst. Phys. Conf. Ser. No.* **22**, 20 (London 1974)
7.92   G.H. Parker, C.A. Mead: Appl. Phys. Lett. **14**, 21 (1969)
7.93   B. Pellegrini: Solid-State Electron. **17**, 21 (1974)
7.94   J.M. Wilkinson: *Inst. Phys. Conf. Ser. No.* **22**, 27 (London 1974)
7.95   F.A. Padovani, R. Stratton: Solid-State Electron. **9**, 695 (1066)
7.96   H.C. Card, E.H. Rhoderick: Solid-State Electron. **16**, 365 (1073)
7.97   H.C. Card, E.H. Rhoderick: J. Phys. D**4**, 1589 (1971)
7.98   E.H. Rhoderick: *Inst. Phys. Conf. Ser. No.* **22**, 3 (London 1974)
7.99   E. Spenke: *Electronic Semiconductors* (McGraw-Hill, New York 1958)
7.100 R.J. Stirn, Y.C.M. Yeh: Appl. Phys. Lett. **27**, 95 (1975)
7.101 A.H.M. Kipperman, M.H. Omar: Appl. Phys. Lett. **28**, 620 (1976)
7.102 *12th IEEE Photovolt. Spec. Conf.*, Sect. 12: "MIS and Schottky Barrier Cells" (Baton Rouge 1976) pp. 847–928
7.103 S.J. Fonash: J. Appl. Phys. **46**, 1286 (1975)
7.104 S.J. Fonash: In *11th Photovolt. Spec. Conf.* (IEEE New York 1975) p. 376
7.105 S.J. Fonash: J. Appl. Phys. **47**, 3597 (1976)
7.106 R. Childs, J. Fortuna, J. Geneczko, S.J. Fonash: 12th IEEE Photovolt. Spec. Conf. (Baton Rouge 1976) p. 862
7.107 S.J. Fonash, T.E. Sullivan, R. Childs, J. Ruths: 13th IEEE Photovolt. Spec. Conf. (Washington, D.C. 1978) p. 645
7.108 R. Childs, J.M. Ruths, T.E. Sullivan, S.J. Fonash: J. Vac. Sci. Technol. **15**, 1397 (1978)
7.109 J. Shewchun, M.A. Green, F.D. King: Solid-State Electron. **17**, 563 (1974)
7.110 M.A. Green, F.D. King, J. Shewchun: Solid-State Electron. **17**, 551 (1974)
7.111 J.A. St. Pierre, R. Singh, J. Shewchun, J.J. Lofersky: 12th IEEE Photovolt. Spec. Conf. (Baton Rouge 1976) p. 847
7.112 J.P. Ponpon, P. Siffert: J. Appl. Phys. (to be published) (Sept. 1978)
7.113 J.P. Ponpon, P. Siffert: Rev. Phys. Appl. **12**, 427 (1977)
7.114 P. de Visschere, H. Pauwels: In *Photovolt. Solar Energy Conf.* (Luxembourg 1977) p. 330
7.115 J. Shewchun, J. Dubow, A. Myszkowski, R. Singh: J. Appl. Phys. **49**, 855 (1978)
7.116 G.B. Stringfellow: J. Vac. Sci. Technol. **13**, 908 (1976)
7.117 C.T. Sah: IRE Trans. Electron Devices **9**, 94 (1962)
7.118 C.H. Henry, R.A. Logan, F.R. Merritt: J. Appl. Phys. **49**, 3530 (1978)
7.119 F.F. Morehead: J. Appl. Phys. **37**, 3487 (1966)
7.120 Lawrence Chin-Chiu Shen: Thesis, Stanford University (1976)
7.121 H.J. Hovel, J.M. Woodall: In *10th Photovolt. Spec. Conf. Proc.* (IEEE, New York 1973) p. 25
7.122 J. Bardeen: Phys. Rev. **71**, 717 (1947)
7.123 V. Heine: Phys. Rev. **138**, A 1689 (1965)
7.124 I. Lindau, P.W. Chye, C.M. Garner, P. Pianetta, C.Y. Su, W.E. Spicer: J. Vac. Sci. Technol. **15**, 1332 (1978)
7.125 M. Aven, C.A. Mead: Appl. Phys. Letters **7**, 8 (1965)
7.126 R.K. Swank, M. Aven, J.Z. Devine: J. Appl. Phys. **40**, 89 (1969)
7.127 S. Kurtin, T.C. McGill, C.A. Mead: Phys. Rev. Lett. **22**, 1433 (1969)
7.128 C.A. Mead, W.G. Spitzer: Phys. Rev. **134**, A 713 (1964)
7.129 C.A. Mead: Solid-State Electron. **9**, 1023 (1966)
7.130 T.C. McGill: J. Vac. Sci. Technol. **11**, 935 (1974)
7.131 J.O. McCaldin, T.C. McGill, C.A. Mead: Phys. Rev. Lett. **36**, 56 (1976)
7.132 J.O. McCaldin, T.C. McGill, C.A. Mead: J. Vac. Sci. Technol. **13**, 802 (1976)
7.133 M. Schluter: J. Vac. Sci. Technol. **15**, 1374 (1978)
7.134 W.E. Spicer, I. Lindau, P.E. Gregory, C.M. Garner, P. Pianetta, P.W. Chye: J. Vac. Sci. Technol. **13**, 780 (1976)
7.135 A. Huijser, J. van Laar: Surf. Sci. **52**, 202 (1975)
7.136 J.E. Rowe, S.B. Christman, G. Margaritondo: Phys. Rev. Lett. **35**, 1471 (1975)
7.137 G. Margaritondo, S.B. Christman, J.E. Rowe: Phys. Rev. B**14**, 5396 (1976)
7.138 J.E. Rowe, G. Margaritondo, S.B. Christman: Phys. Rev. B**15**, 2195 (1977)

7.139 R.R. Varma, A. McKinley, R.H. Williams, I.G. Higgenbotham: J. Phys. D **10**, L 171 (1977)
7.140 R.H. Williams, R.R. Varma, A. McKinley: J. Phys. C **10**, 4545 (1977)
7.141 A. Amith, J.L. Yeh, P. Mark: J. Vac. Sci. Technol. **15**, 1344 (1978)
7.142 R.Z. Bachrach: J. Vac. Sci. Technol. **15**, 1340 (1978)
7.143 J.M. Woodall, C. Lanza, J.L. Freeouf: J. Vac. Sci. Technol. **15**, 1436 (1978)
7.144 L.J. Brillson: J. Vac. Sci. Technol. **15**, 1378 (1978)
7.145 J.A. Roth, C.R. Crowell: J. Vac. Sci. Technol. **15**, 1317 (1978)
7.146 F. Buch, A. Fahrenbruch, R.H. Bube: J. Appl. Phys. **48**, 1596 (1977)
7.147 F.H. Mullins, A. Brunnschweiler: Solid-State Electron. **19**, 47 (1976)
7.148 C.M. Garner, Y.D. Shen, J.S. Kim, G.L. Pearson, W.E. Spicer: J. Vac. Sci. Technol. **14**, 985 (1977)
7.149 A.G. Revesz: J. Non-Cryst. Solids **11**, 309 (1973)
7.150 J.S. Johannessen, W.E. Spicer, Y.E. Strausser: Appl. Phys. Lett. **27**, 452 (1975)
7.151 W.L. Harrington, R.E. Honig, A.M. Goodman, R. Williams: Appl. Phys. Lett. **27**, 644 (1975)
7.152 F.J. Grunthaner, J. Maserjian: J. Vac. Sci. Technol. **15**, 1518 (1978)
7.153 C.W. Wilmsen, R.W. Kee: J. Vac. Sci. Technol. **15**, 1513 (1978)
7.154 R.S. Bauer, J.C. McMenarnin: J. Vac. Sci. Technol. **15**, 1444 (1978)
7.155 R.W. Grant, J.R. Waldrop, E.A. Kraut: J. Vac. Sci. Technol. **15**, 1451 (1978)
7.156 J.W. Matthews: "Coherent Interfaces and Misfit Dislocations", Chap. 8 in [Ref. 7.163]
7.157 J.H. van der Merwe, C.A.B. Ball: "Energy of Interfaces Between Crystals", Chap. 6 in [Ref. 7.163]
7.158 N.H. Fletcher, K.W. Lodge: "Energy of Interfaces Between Crystals: An Ab-Initio Approach", Chap. 7 in [Ref. 7.163]
7.159 G.H. Olsen, M. Ettenberg: "Growth Effects in the Heteroepitaxy of III–V Compounds", in *Crystal Growth: Theory and Techniques, Vol. II* (Plenum, New York 1978)
7.160 W.G. Oldham: Ph.D. Thesis, Carnegie Instit. of Technol., Pittsburgh, Penn. (1963)
7.161 D.B. Holt: J. Phys. Chem. Solids **27**, 1053 (1966)
7.162 H. Kressel: J. Electron. Mater. **4**, 1081 (1975)
7.163 J.W. Matthews (ed.): *Epitaxial Growth* (Academic Press, New York 1975)
7.164 G.H. Olsen, C.J. Nuese, R.T. Smith: J. Vac. Sci. Technol. **15**, 1410 (1978)
7.165 M.J. Stowell: "Defects in Epitaxial Deposits", Chap. 5 in [Ref. 7.163]
7.166 M.S. Abrahams, C.J. Buiocchi, J.F. Corboy, Jr., G.W. Cullen: Appl. Phys. Lett. **28**, 275 (1976)
7.167 G.H. Olsen: J. Cryst. Growth **31**, 223 (1975)
7.168 M. Ettenberg, G.H. Olsen: J. Appl. Phys. **48**, 4275 (1977)
7.169 H. Kasano, S. Hosoki: J. Appl. Phys. **46**, 394 (1975)
7.170 C.J. Nuese: J. Electron. Materials **6**, 253 (1977)
7.171 J.S. Kim: Ph.D. Dissertation, SEL-78-014, Stanford University (1978) p. 66
7.172 R.J. Nelson: J. Vac. Sci. Technol. **15**, 1475 (1978)
7.173 F.A. Lindholm, C.T. Sah, M.P. Godlewski, H.W. Brandhorst: *12th IEEE Photovolt. Spec. Conf. Proc.* (Baton Rouge, LA 1976) p. 1

Several Papers on Solar Energy Physics Published in *Applied Physics*

W. A. Anderson, J. K. Kim: Reliability studies on MIS solar cells. Appl. Phys. **17**, 401 (1978)

E. Bucher: Solar cell materials and their basic parameters. Appl. Phys. **17**, 1 (1978)

G. H. Derrick, R. C. McPhedran, D. Maystre, M. Nevière: Crossed gratings: A theory and its applications. Appl. Phys. **18**, 39 (1979)

P. de Visschere, H. J. Pauwels: The influence of tunneling effects on the efficiency of heterojunction solar cells. Appl. Phys. **15**, 413 (1978)

V. M. Fridkin, B. N. Popov, K. A. Verhovskaya: The photovoltaic and photorefractive effects in KDP-type ferroelectrics. Appl. Phys. **16**, 313 (1978)

A. Goetzberger: Fluorescent solar energy collectors: Operating conditions with diffuse light. Appl. Phys. **16**, 399 (1978)

A. Goetzberger, W. Greubel: Solar energy conversion with fluorescent collectors. Appl. Phys. **14**, 123 (1977)

A. Goetzberger, O. Schirmer: Second stage concentration with tapers for fluorescent solar collectors. Appl. Phys. **19** (May, 1979)

H.-J. Hoffmann: On the density of localized levels in amorphous silicon. Appl. Phys. **18**, 429–431 (1979)

H. F. Mataré, G. A. Wolff: Concentration enhancement of current density and diffusion length in III–V ternary compound solar cells. Appl. Phys. **17**, 335 (1978)

R. C. McPhedran, D. Maystre: On the theory and solar application of inductive grids. Appl. Phys. **14**, 1 (1977)

K. Toda, M. Oikawa, K. Hisano: Photovoltaic effect in thin film of $Pb(Zr, Ti)O_3$ evaporated on stainless steel. Appl. Phys. **17**, 207 (1978)

M. Tomkiewicz, H. Fay: Photoelectrolysis of water with semiconductors. Appl. Phys. **18**, 1 (1979)

F. Demichelis, E. Minetti-Mezzetti: Concentrator for solar air heater. Appl. Phys. **19**, (June/July, 1979)

F. Demichelis, G. Russo: Cavity-type surfaces for solar collectors. Appl. Phys. **18**, 307 (1979)

Additional References with Titles

Chapter 3

The references below refer to the text on p. 88

R. Brako: Optical properties of composite media. J. Phys. C. **11**, 3345 (1978)
D.R. McKenzie, R.C. McPhedran: Exact modeling of cubic lattice permittivity and conductivity. Nature **265**, 128 (1977)
R.C. McPhedran, D.R. McKenzie: The conductivity of lattices of spheres. I. The simple cubic lattice. Proc. R. Soc. London A**362**, 45 (1978)
D.R. McKenzie, R.C. McPhedran, G.H. Derrick: The conductivity of lattices of spheres. II. The body centered and face centered cubic lattice. Proc. R. Soc. London A**362**, 211 (1978)
R. Ruppin: Validity range of the Maxwell-Garnett theory. Phys. Status Solidi b **87**, 619 (1978)

The references below refer to the text on p. 110

J.C.C. Fan: Selective-black absorbers using sputtered cermet films. Thin Solid Films **54**, 139 (1978)
R. Marchini, R. Gandy: Selective absorption properties of lead sulfide-aluminum coatings as a function of lead sulphide thickness. J. Appl. Phys. **49**, 390 (1978)
C.M. Horwitz: Solar-selective globular metal films. J. Opt. Soc. Am. **68**, 1032 (1978)
G.L. Harding, D.R. McKenzie, B. Window: The D.C. sputtering of solar-selective surfaces onto tubes. J. Vac. Sci. Tech. **13**, 1073 (1976)
L.C. Botten, I.T. Ritchie: Improvements in the design of solar selective thin film absorbers. Opt. Commun. **22**, 421 (1977)
D.R. McKenzie: Gold black and gold cermet absorbing surfaces. Gold Bull. **11**, 49 (1978)
D.R. McKenzie: Effect of substrate on graphite and other selective surfaces. Appl. Opt. **17**, 1884 (1978)
D.R. McKenzie: Gold, silver, chromium, and copper cermet selective surfaces for evacuated solar collectors. Appl. Phys. Lett. **34**, 25 (1979)
D.M. Trotter, H.G. Craighead, A.J. Sievers: Design of selective surfaces for solar energy collection. Sol. Energy Mater. **1**, (1979)
H.G. Craighead, R. Bartynski, R.A. Buhrman, L. Wojcik, A. J. Sievers: Metal/insulator composite selective absorbers. Sol. Energy Mater. **1** (1979)
B. Window, I.T. Ritchie, K. Cathro: A study of selective electroplated chromium black. Appl. Opt. **17**, 2637 (1978)
A. Ignatiev, P. O'Neill, G. Zajac: The surface microstructure optical properties relationship in solar absorbers: black chrome. Sol. Energy Mater. **1** (1979)
C.M. Lampert: Microstructure of a black chrome solar selective absorber. Sol. Energy Mater. **1**, (1979)
J.I. Gittleman, E.K. Sichel, Y. Arie: Composite semiconductors: selective absorbers of solar energy. Sol. Energy Mater. **1** (1979)

Chapter 6

B.K. Gupta, O.P. Agnihotri: Structural investigations of spray-deposited CdS films doped with Cu, In, and Ga. Phil. Mag. B **37** (5), 631–633 (1978)

S.R. Das, V.D. Vankar, Prem Nath, K.L. Chopra: The preparation of Cu_2S films for solar cells. Thin Solid Films **51**, 257–264 (1978)

K.L. Chopra: High efficiency spray-deposited CdS/Cu_2S thin film solar cells. Private communication

Subject Index

Abruptness 309, 310
Absorber, simple blackbody 10
Absorber temperature 10
Absorptance
  actual 11
  hemispherical 9
  of merit 11
  solar 9
Absorption 93, 267
  angle of maximum 82
  coefficient 148, 149, 186, 187
  cross section 86
  edge, fundamental 26
  free carrier 27
  infrared 75
  intrinsic 15, 42
  length 257
  profile 26
Absorptivity 66, 69, 72, 75, 103
  diffuse 75
  maximum total 57
  solar 107–109
  spectral 69, 73
  specular 75
Acceptance angle 13, 39
Activation energy barriers 131
Adsorbed ionic charge 117
Agglomeration 47, 48
Air mass 0, 1 (AM 0, AM 1) 173, 173n
Aluminum 22, 42
  oxide 42, 48
Amorphous semiconductors 23 *see also* Silicon, amorphous
Anderson model for heterojunction 273–277, 292
Annealing 29, 201, 203
Antimony oxide 42
Auger
  analysis 302, 304, 306, 307, 310
  recombination 192, 209

Backwall CdS–Cu_xS cells 219
Band bending 148

Band-edge
  discontinuities 275, 295, 309, 310
  position 118, 120, 140
Band-gap transition 28
Bandpass reflection filters 7
Band-profile discontinuity 3
Barium titanate 160
Barrier height 285–289, 303
Blackbodies 69, 79
Blackbody
  centroid frequency 74
  distribution 106
  spectrum 71, 84
Black chrome 13, 43–46, 108
Black mirrors 7, 44
Black nickel 45
Blacks
  gold 33, 108
  metal 107
  selective 1, 34, 35, 42–44
Bohm and Pines result 64
Boltzmann's constant 12
Born and Oppenheimer separation of nuclear and electronic motion 59
Boundary condition 264, 265
Brillouin zone boundaries 20
Bruggeman effective medium theory 88

Capacity-voltage characteristics 241, 275, 279, 280, 304, 305
Carbon 194
Carrier
  concentration, minority 264
  cross sections 184, 199
  lifetime 2, 173, 189, 210, 260, 263
  mobility 260
  transport equation 258
CdS 25, 128, 143, 146, 150, 152, 268, 281, 302
  bulk 220
  films 220, 224
CdSe 143, 151, 152
CdTe 162, 265, 268, 281
Centroid frequency 70, 74

Subject Index

Cermets 108, 109
Charge transfer 18, 126
Chemical barriers 131
Chemical spray deposition 215
Chemical vapor deposition (CVD) 27, 31, 32, 39, 47–51, 213
Chromium oxide 36, 44, 48
Clausius-Mosotti relation 87
Coating 2, 36, 37, 39, 40, 60, 108
  composite 58, 93, 97, 109
  electroplated 43, 45
  interference 36, 38
  selective 35, 58
Cobalt oxide 42, 44
Collection functions 267, 269
Composite
  coating 58, 93, 97, 109
  films, graded 109
  layers 20
  materials 1, 2, 23
  media 87–89, 91, 92, 99, 100, 104, 108
  structures 34, 91
Concentration ratio 8, 10
Conductivity
  ac 102
  dc 101, 103, 104
  electron 103
  lattice 103
Converter
  flat-plate 10
  single-material 17, 19
Copper 22, 203, 302
  chloride 216
  dispersion-hardened 97, 98, 101–103
  oxides 42–44, 142
  sulfides 216, 217, 228, 229, 233, 235, 236, 276
Coulomb interaction 62–65
Current-voltage characteristics 126, 239, 261, 262, 265, 269, 273–275, 284, 304, 305

Dangling bonds 29, 30, 313, 319
Debye theory 85
Decomposition 134, 135, 137, 139, 145
  anodic 136, 141, 147
  cathodic 136
  reaction 134, 137
Depletion layer 263–266, 270, 275, 284
Dielectric function 23, 28, 67, 89, 99, 100, 102
Diffusion
  coefficient 197
  equations 259
  length 3, 173, 186, 187, 257, 266
  temperature 206
Diode parameters 261, 270–272, 283

Dipole-dipole coupling 108
Dipping process 216
Direct-gap material 26, 263, 264
Dislocations 23, 205, 311–322
Disorder 20, 23
Dispersoid 98, 99
Doping concentration 192, 209
Drude
  metal 88, 101, 104
  model 72, 98, 100–103
  relaxation time 23
Drude-Zener theory 19

Ebanol 43
Efficiency 267
  charge separation 128
  collection 175
  conversion 6, 9, 109, 164, 179, 258
  maximum 107
  quantum 257, 261, 270
  solar 261, 263, 269, 293, 294, 321
  storage 167, 169
Electrical properties 225, 229, 233
Electrode, two-layer n-type semiconductor 163
Electrodeposition 34, 217
Electrolysis 133, 169
Electromagnetic response 67
Electromagnetic waves 65, 69, 87
  TE 80
  TM 80
Electron affinities 273, 275, 276, 285
Electronegativity 28, 297
Electron-electron interaction 61, 64
Electron gas 69
Electron-hole pair 105, 106
Electron-phonon collisions 23, 73
Electron transfer 18, 122, 124–126
Ellipsoids 88, 89
Elysium 16
Emission, infrared 109
Emissivity 57, 65, 69, 73, 78, 80, 82, 84, 97, 100–104, 106, 108, 109
  hemispherical 79, 81, 82, 109
  normal 78, 81, 109
  spectral 69, 72, 74
  thermal 69, 77, 107
Emittance 9, 12, 24
Epitaxy 313
  molecular beam 309
Exchange current density 124, 125, 167
Extinction coefficient 91, 94, 95

Fe_2O_3 160
Fe_3O_4 43

Fermi
  energy   135
  statistics   64
  temperature   58
Fermi level   123, 135, 140
  pinning   275, 277, 278, 286, 289, 293, 296, 299, 301–303, 320
  stabilization of   299, 301–303, 320
Figure of merit   9, 84
Fill factor   258, 269, 320
  curve   179
Flux
  amplification   12
  densities   13
Franck-Condon effect   23
Frank-Read source   315
Frontwall CdS–Cu_xS cells   219

GaAs   128, 151, 152, 261, 264, 318
GaP   152, 162
Ge/GaAs   277
Gettering   195, 206
Glow discharge, rf   30
Gold   22, 182, 185, 203, 299, 300
  blacks   33, 108
  particles   92
  smoke   33, 92
Graded-index profiles   39, 51
Graded junctions   109, 282
Grain boundaries   23, 98

Hagen-Rubens result   73, 74
Helmholtz double layer   118
Heterojunctions   3, 213, 218, 257–326
Heterostructures, double   280
HfC_x films   18
Holstein effect   77, 78
Homojunctions   260, 265, 271, 272
Hybridization   18
Hydrogen   197

Impedence, surface   71–73
Indium oxide   25, 40
Injection level   182, 199
Insulators   65, 66
Integrated circuitry   2
Interband
  critical points   20
  transitions   15, 19, 24, 26, 82
Interface   2, 257, 295
  metal-insulator-semiconductor (MIS)   283, 291
  metal-semiconductors (MS)   295–307
  semiconductor-semiconductor (SS)   307–310

Interface
  behavior   297, 298, 311–320
  states   277, 279, 280, 286, 295
Interfacial
  defects   3
  dipoles   282
  insulating layer   287
Interference   93
  coatings   36, 38
  effects   35, 44
  filter   15
  stack   57
Interstitials   23
Iron   201
Irrotational
  motion   68
  vector fields   62, 63

Kink sites   137
Kirchhoff's law   69, 72
Kramers-Kronig relations   16
$KTaO_3$ 160

Lattice
  mismatch   293, 311, 313
  resonance   102
  vibrations   27, 65, 98, 104
Layer crystals   144, 145, 153
Line-focussing systems   13
Longitudinal motion   68
Loss
  by thermalization of carriers   265
  coolant   13
  energy   160
  free energy   167
  interfacial recombination   267
  photogenerated current   280
  reflection   6
  reradiation   6, 11
Low energy electron loss spectroscopy   302

Manganese oxide   42
Maxwell-Garnett theory   2, 34, 58, 87, 88, 97, 98, 108–110
Maxwell's equations   63, 67, 69, 71, 72
Metal
  composite   88
  mesh   58
  particles   58, 86
Metal-metal interaction   19
Metals   19, 24, 66, 72, 75, 97
  free-electron-like   66, 94
Mie theory   85, 86
Mirror
  dark   25, 26, 41, 42
  heat   25, 39

334 Subject Index

Misfit strain   311, 315
Modes
  optic   66, 102–104
  sphere   96
  surface   68, 69
Molybdenum   17, 22, 50, 51
MoS_2 143–145
Mott-Schottky relation   118
Mott-Zener result   73
Multilayer
  coatings   57
  films   23
  solar cells   169
  stacks   47, 51

Nb   96
Nb_2O_5 160
Nickel   22, 95
  blacks   44
  oxide   42
NiS–ZnS coating   44

One-electron approximation   61, 62
Optical
  constants   91
  engineering   16
  processes, temperature dependence of   21, 23, 24
  properties   84, 98
  transmission: photoconduction   226
Oscillator strength   26
Overvoltages   131, 166
Oxide interlayers   306
Oxygen   194, 204

Particles, small   86, 87, 93
PbS
  films   109
  photocell   106
Phase diagram of Cu–S system   228
Phonons   23, 78
Photochemical cells   2
Photoconductive decay method   180
Photocorrosion   142
Photocurrent-voltage curves   127, 128, 151, 155
Photodecomposition   133, 136, 137, 143, 146, 156
Photoelectrochemical cells   149, 154, 158
Photoelectrolysis   2, 167, 168
Photoelectrolytic cell   132, 168
Photogenerated current   2, 258, 263, 266, 321
Photosynthesis   115
Photothermal conversion   5, 6, 11, 109
Photovoltage   128, 129
  surface   185
  surface, spectroscopy   280, 302

Photovoltaic cells   2, 3, 238, 257, 258
Planck's law   69
Plasma
  edge   16, 18, 19, 40
  frequency   23, 24, 64, 87, 91, 100
  oscillations   62, 65
Plasmon
  effects, surface   2
  frequency   91
Plasmons   64, 69
Platinum   22
Polarizability of valence electrons   27
Polarization   63, 65, 68, 88, 89
Polycrystalline materials   3
Porous materials   15
Power characteristics   150, 164

Quality factor   285, 289, 294
Quantum yield   127
Quasi-Fermi energies   131
  level   132, 260, 261, 263, 265, 266, 271, 285
Quasi-neutral region   258, 261, 263, 264, 273
Quenching   202

Radiation   2, 80
Ratio a/e 9
Receivers   12, 13
Recombination   261, 265–267, 279–281, 284
  Auger   192, 209
Recombination
  centers   182, 263, 272
  probalities   184
  rate   260, 263
  surface   3
  velocity   183, 267, 313, 318
  generation mechanism   266, 271
Redox
  battery   153, 159, 169
  energy levels   123
  reactions   146
  systems   143, 150
Reflectance   25
Reflection
  multiple   32
  specular   76
Reflectivity   72, 109, 187
Reflector, metal   15
Reflectometer   48
Refractive index   27, 42, 51, 85
Refractory metals   50–52   see also Transition metals
Regenerative cells   147–149, 152
Relaxation
  frequency   24, 74
  region   73, 74, 76
  time   77, 92, 93, 98, 260

Subject Index

Resistivity, dc 74, 78, 82, 97, 98
Resonance 69, 87, 90, 95
   frequency 93, 94
   particle 94, 100
Response function 87
Reststrahl 98
Retardation effects 69
Rhenium
   trioxide 21
   whiskers 33
Richardson constant 284
RuO_2 167

Sah, Noyce, and Shockley (SNS) theory 271, 272
Scattering 16, 18, 20, 86
   reflective 15, 32
   resonant 15, 32–34
   surface 72, 76–78
Schottky
   barrier 165, 298, 301, 304, 308
   diode 279, 284, 290, 292, 305
   limit 298, 303
Selective surfaces 84, 105, 109 *see also* Blacks, Coating
Se^{2-}/Se_2^{2-} redox couple 151
Shockley diode 266, 273
Shockley–Read model 184, 263
Silane 29, 30, 47
Silicon 47, 181, 189, 194, 266, 278
   absorber 29, 32, 40, 47
   amorphous 29–32, 51, 52
   crystalline 2, 30, 51
Si–MgO 109
Si_3N_4 48
SiO_2 48
Silver 22
Sintered layers 215
Skin depth 71, 76
Skin effect
   anomalous 72, 76, 77
   classical 76, 77
Smokes 33, 92, 109
SnO_2 25, 40, 142, 160
Sodium tungsten bronzes 21
Solar
   absorbers 25, 29, 32, 40, 47
   cells 105, 174, 257, 262, 287
   collector 91, 93
   energy 106
   energy conversion 1
   flux intensity 12
   power 107
   selectivity 96
   spectrum 59, 91, 93, 94, 96, 105, 106

Solenoidal
   motion 68
   vector fields 62, 63
Space-charge layer 118
Spectrally selective absorbers 6, 12, 24
Spectral response 243
Spectral selectivity 1, 6, 14, 15, 32, 57–114
Sphere
   aggregated 108
   homogeneous 84
   mode 90, 91
   particles 91
   resonances 91
Spray method 218
Sputtering 29, 30, 214
$SrTiO_3$ 160, 162
S^{2-}/S_2^{2-} 151
Stability 109
Stabilization 49, 50, 145
Stagnation temperature 107
Stainless steel 42
Stefan–Boltzmann law 8
Storage
   cell 155
   energy, direct 153
Structural effects 28
Sulfides 161
Sum rules 20
Superposition 258, 262
Surface
   charge, point of zero 122
   morphology 43, 44
   properties 276, 277
   quality 304
   roughness 32, 69
   states 120, 146, 298
   texture 32, 34, 42, 58
Surf-ride 76
Sweep-out 182, 192
Swirls 206

Tandem
   absorber-reflector 24, 26, 28, 36, 39, 47, 50
   action of free and bound electrons 14
   stacks 39, 43, 48
Te^{2-}/Te_2^{2-} 151
Textural
   effects 28, 35, 45
   profile of surface 32
Thermal
   emission spectrum 8
   energy 106
   expansion mismatch 47
   reradiation 106

Thermionic emission   284
Thin films   2, 3, 49, 213
Tin oxide   25, 40, 142, 160
Titanates   142
Titanium oxide   42, 142, 146, 159, 162
Transformer   106, 107
Transition-metal
  carbides, refractory   18
  coatings   58
  compounds, refractory   19
  nitrides, refractory   18
  oxides   20, 21, 43
Transition metals   2, 24, 91, 94, 95
Transport
  control   290
  equation   259
  parameters   294
Transverse motion   68
Trap   183
Tungsten   22, 50, 96
Tunneling   3, 257, 281, 284, 289, 320
Two-thirds rule   286

Ultraviolet photoemission spectroscopy   302

Vacancies   23, 197, 202
Vacuum evaporation   29
Vibrational energy   60

Water   156–159, 169
W–MgO   109
Wide-gap material   28, 273
Window
  effect   3, 273
  layer   259, 262
Windows, laser   23
WO_2 92
WO_3 142, 160
Work function   275, 285, 296, 297
WS_2 143

X point   24

$YFeO_3$ 160

Zinc oxide   118, 142
Zirconium oxide   42, 160

Applied Physics

A monthly journal

Board of Editors
S. Amelinckx, Mol. **V.P. Chebotayev,** Novosibirsk
R. Gomer, Chicago, IL., **H. Ibach,** Jülich
V.S. Letokhov, Moskau, **H.K.V. Lotsch,** Heidelberg
H.J. Queisser, Stuttgart, **F.P. Schäfer,** Göttingen
A. Seeger, Stuttgart, **K. Shimoda,** Tokyo
T. Tamir, Brooklyn, NY, **W.T. Welford,** London
H.P.J. Wijn, Eindhoven

Coverage
application-oriented experimental and theoretical physics:

Solid-State Physics *Quantum Electronics*
Surface Sciences *Laser Spectroscopy*
Solar Energy Physics *Photophysical Chemistry*
Microwave Acoustics *Optical Physics*
Electrophysics *Integrated Optics*

Special Features
rapid publication (3-4 months)
no page charge for **concise** reports
prepublication of titles and abstracts
microfiche edition available as well

Languages
mostly English

Articles
original reports, and short communications review and/or tutorial papers

Manuscripts
to Springer-Verlag (Attn. H. Lotsch), P.O. Box 105 280
D-6900 Heidelberg 1, F.R. Germany

Place North-American orders with:
Springer-Verlag New York Inc., 175 Fifth Avenue,
New York, N.Y. 100 10, USA

Springer-Verlag
Berlin
Heidelberg
New York

Springer Series in
Solid-State Sciences
Editors: M. Cardona, P. Fulde, H. J. Queisser

Volume 1
C. P. Slichter
Principles of Magnetic Resonance
2nd, revised and expanded edition 1978
115 figures, X, 397 pages
ISBN 3-540-08476-2

Contents: Elements of Resonance. – Basic Theory. – Magnetic Dipolar Broadening of Rigid Lattices. – Magnetic Interactions of Nuclei with Electrons. – Spin-Lattice Relaxation and Motional Narrowing of Resonance Lines. – Spin Temperature in Magnetism and in Magnetic Resonance. – Double Resonance. – Advanced Concepts in Pulsed Magnetic Resonance. – Electric Quadrupole Effects. – Electron Spin Resonance. – Summary. – Problems. – Appendices.

Volume 2
O. Madelung
Introduction to Solid-State Theory
Translated from the German by B. C. Taylor
1978. 144 figures. XI, 486 pages
ISBN 3-540-08516-5

Contents: Fundamentals. – The One-Electron Approximation. – Elementary Excitations. – Electron-Phonon Interaction: Transport Phenomena. – Electron-Electron Interaction by Exchange of Virtual Phonons: Superconductivity. – Interaction with Photons: Optics. – Phonon-Phonon Interaction: Thermal Properties. – Local Description of Solid-State Properties. – Localized States. – Disorder. – Appendix: The Occupation Number Representation.

Volume 3
Z. G. Pinsker
Dynamical Scattering of X-Rays in Crystals
1978. 124 figures, 12 tables. XII, 511 pages
ISBN 3-540-08564-5

Contents: Wave Equation and Its Solution for Transparent Infinite Crystal. – Transmission of X-Rays Through a Transparent Crystal Plate. Laue Reflection. – X-Ray Scattering in Absorbing Crystal. Laue Reflection. – Poynting's Vectors and the Propagation of X-Ray Wave Energy. – Dynamical Theory in Incident-Spherical-Wave Approximation. – Bragg Reflection of X-Ray I. Basic Definitions. Coefficients of Absorption: Diffraction in Finite Crystal. – Bragg Reflection of X-Rays. II. Reflection and Transmission Coefficients and Their Integrated Values. – X-Ray Spectrometers. Used in Dynamical Scattering Investigation. Some Results of Experimental Verification of the Theory. – X-Ray Interferometry. Moire Patterns in X-Ray Diffraction. – Generalized Dynamical Theory of X-Ray Scattering in Perfect and Deformed Crystals. – Dynamical Scattering in the Case of Three Strong Waves and More. – Appendices.

Volume 4
Inelastic Electron Tunneling Spectroscopy
Proceedings of the International Conference and Symposium on Electron Tunneling, University of Missouri-Columbia, USA, May 25–27, 1977
Editor: T. Wolfram
1978. 126 figures, 7 tables. VIII, 242 pages
ISBN 3-540-08691-9

Contents: Review of Inelastic Electron Tunneling. – Applications of Inelastic Electron Tunneling. – Theoretical Aspects of Electron Tunneling. – Discussions and Comments. – Molecular Adsorption on Non-Metallic Surfaces. – New Applications of IETS. – Elastic Tunneling.

Volume 5
F. Rosenberger
Fundamentals of Crystal Growth
Macroscopic Equilibrium and Transport Concepts
ISBN 3-540-09023-1

Contents: Thermodynamics. – Phase Transitions and Phase Diagrams. – Crystal Growth and Phase Diagrams. – Mass Transport and Heat Transfer. – Segregation.

Volume 6
R. P. Huebener
Magnetic Flux Structures in Superconductors
1979. 99 figures, 5 tables. Approx. 270 pages
ISBN 3-540-09213-7

Volume 7
E. N. Economou
Green's Functions in Quantum Physics
1979. 49 figures, 2 tables. Approx. 270 pages
ISBN 3-540-09154-8

Contents: Green's Function in Mathematical Physics. Green's Functions in One-Body Quantum Problems. Green's Functions in Many-Body Systems. – Appendices.

Volume 8
Solitons and Condensed Matter Physics
Proceedings of the Symposium on Nonlinear (Soliton) Structure and Dynamics in Condensed Matter. Oxford, England, June 27–29, 1978
Editors: A. R. Bishop, T. Schneider
1978. 120 figures, 4 tables. XI, 341 pages
ISBN 3-540-09138-6

Contents: Introduction. – Mathematical Aspects. Statistical Mechanics and Solid-State Physics. – Summary.

Springer-Verlag
Berlin Heidelberg New York